I0890886

Cases on STEAM Education in Practice

Judith Bazler
Monmouth University, USA

Meta Van Sickle
College of Charleston, USA

A volume in the Advances in Educational Technologies and Instructional Design (AETID) Book Series

IGI Global
DISSEMINATOR OF KNOWLEDGE
www.igi-global.com

Published in the United States of America by
IGI Global
Information Science Reference (an imprint of IGI Global)
701 E. Chocolate Avenue
Hershey PA 17033
Tel: 717-533-8845
Fax: 717-533-8661
E-mail: cust@igi-global.com
Web site: http://www.igi-global.com

Copyright © 2017 by IGI Global. All rights reserved. No part of this publication may be reproduced, stored or distributed in any form or by any means, electronic or mechanical, including photocopying, without written permission from the publisher.
Product or company names used in this set are for identification purposes only. Inclusion of the names of the products or companies does not indicate a claim of ownership by IGI Global of the trademark or registered trademark.

Library of Congress Cataloging-in-Publication Data

Names: Bazler, Judith, editor.
Title: Cases on STEAM education in practice / Judith Bazler and Meta Van Sickle, Editors.
Description: Hershey, PA : Information Science Reference, [2017]
Identifiers: LCCN 2016059609| ISBN 9781522523345 (hardcover) | ISBN 9781522523352 (ebook)
Subjects: LCSH: Science--Study and teaching--Methodology. | Interdisciplinary approach in education. | Science and the arts.
Classification: LCC LB1585 .C33 2017 | DDC 372.35/044--dc23 LC record available at https://lccn.loc.gov/2016059609

This book is published in the IGI Global book series Advances in Educational Technologies and Instructional Design (AETID) (ISSN: 2326-8905; eISSN: 2326-8913)

British Cataloguing in Publication Data
A Cataloguing in Publication record for this book is available from the British Library.

All work contributed to this book is new, previously-unpublished material. The views expressed in this book are those of the authors, but not necessarily of the publisher.

Advances in Educational Technologies and Instructional Design (AETID) Book Series

ISSN:2326-8905
EISSN:2326-8913

Mission

Education has undergone, and continues to undergo, immense changes in the way it is enacted and distributed to both child and adult learners. From distance education, Massive-Open-Online-Courses (MOOCs), and electronic tablets in the classroom, technology is now an integral part of the educational experience and is also affecting the way educators communicate information to students.

The **Advances in Educational Technologies & Instructional Design (AETID) Book Series** is a resource where researchers, students, administrators, and educators alike can find the most updated research and theories regarding technology's integration within education and its effect on teaching as a practice.

Coverage

- Game-Based Learning
- Classroom Response Systems
- Web 2.0 and Education
- Collaboration Tools
- Higher Education Technologies
- E-Learning
- Online Media in Classrooms
- Instructional Design
- Virtual School Environments
- Hybrid Learning

IGI Global is currently accepting manuscripts for publication within this series. To submit a proposal for a volume in this series, please contact our Acquisition Editors at Acquisitions@igi-global.com or visit: http://www.igi-global.com/publish/.

The Advances in Educational Technologies and Instructional Design (AETID) Book Series (ISSN 2326-8905) is published by IGI Global, 701 E. Chocolate Avenue, Hershey, PA 17033-1240, USA, www.igi-global.com. This series is composed of titles available for purchase individually; each title is edited to be contextually exclusive from any other title within the series. For pricing and ordering information please visit http://www.igi-global.com/book-series/advances-educational-technologies-instructional-design/73678. Postmaster: Send all address changes to above address. Copyright © 2017 IGI Global. All rights, including translation in other languages reserved by the publisher. No part of this series may be reproduced or used in any form or by any means – graphics, electronic, or mechanical, including photocopying, recording, taping, or information and retrieval systems – without written permission from the publisher, except for non commercial, educational use, including classroom teaching purposes. The views expressed in this series are those of the authors, but not necessarily of IGI Global.

Titles in this Series

For a list of additional titles in this series, please visit: www.igi-global.com

Integrating an Awareness of Selfhood and Society into Virtual Learning
Andrew Stricker (The Air University, USA) Cynthia Calongne (Colorado Technical University, USA) Barbara Truman (University of Central Florida, USA) and Fil Arenas (The Air University, USA)
Information Science Reference • copyright 2017 • 383pp • H/C (ISBN: 9781522521822) • US $200.00 (our price)

Intercultural Responsiveness in the Second Language Learning Classroom
Kathryn Jones (Lamar University, USA) and Jason R. Mixon (Lamar University, USA)
Information Science Reference • copyright 2017 • 327pp • H/C (ISBN: 9781522520696) • US $175.00 (our price)

Handbook of Research on Transformative Digital Content and Learning Technologies
Jared Keengwe (University of North Dakota, USA) and Prince Hycy Bull (North Carolina Central University, USA)
Information Science Reference • copyright 2017 • 453pp • H/C (ISBN: 9781522520009) • US $230.00 (our price)

Digital Tools for Seamless Learning
Süleyman Nihat Şad (Inonu University, Turkey) and Martin Ebner (Graz University of Technology, Austria)
Information Science Reference • copyright 2017 • 398pp • H/C (ISBN: 9781522516927) • US $200.00 (our price)

Multiculturalism and Technology-Enhanced Language Learning
Dara Tafazoli (University of Córdoba, Spain) and Margarida Romero (Université Laval, Canada)
Information Science Reference • copyright 2017 • 339pp • H/C (ISBN: 9781522518822) • US $190.00 (our price)

Handbook of Research on Writing and Composing in the Age of MOOCs
Elizabeth A. Monske (Northern Michigan University, USA) and Kristine L. Blair (Youngstown State University, USA)
Information Science Reference • copyright 2017 • 457pp • H/C (ISBN: 9781522517184) • US $270.00 (our price)

IGI GLOBAL
DISSEMINATOR OF KNOWLEDGE
www.igi-global.com

701 E. Chocolate Ave., Hershey, PA 17033
Order online at www.igi-global.com or call 717-533-8845 x100
To place a standing order for titles released in this series,
contact: cust@igi-global.com
Mon-Fri 8:00 am - 5:00 pm (est) or fax 24 hours a day 717-533-8661

Table of Contents

Detailed Table of Contents

Out of a conversation between two long-time colleagues—each a science educator and practicing artist, emerged the question, "What does it mean to STEAMify a lesson, and why would a teacher actually choose to do such a thing, other than, say, for-grant-writing-purposes? Their science selves really liked the idea of a STEAM system, acted upon by forces, both from the outside and from within, and with energy flowing and cycling, all the while transforming grey matter in ways that sustained the teaching/learning process. When it came to their art, however, their dialogue followed pathways grooved by long years of practice and hard work in their respective fields. One author is a seasoned vocalist, trained in the nuances of both individual and group vocal performance as well as the attendant dimensions of music, its composition and phraseology. The other is a painter, poet, and novelist, shaping words, color, and line to tell stories and communicate ideas. What was significant to each was that their artistic habits of mind had shaped their axiology, transforming their ways teaching.

across educational horizons. It reads as ethnography and has been performed in community, in association with the most caring of souls, with the goal of achieving a more inclusive/empowering, aesthetic science education and a deep appreciation of the importance of the creative arts in the learning process. It presents two research-based iterations in STEAM education in practice: 1) the creation of arts-integrated middle school ocean science curricula and 2) the development of a pedagogical tool called the "Know"tation as a way for teachers and students to make learning visible and integrate the languages of science throughout the process of inquiry. The cases described in this chapter apply many features of the STEAM model developed in Chapter 1 of this book.

This chapter explores the use of creative movement to extend meaning to inquiry-based science investigations. This process embraces the addition of A to STEM to realize the impact of STEAM. The chapter builds on the import of scientific and physical literacy, interdisciplinary learning, and the power of kinesthetic engagement. Students become active collaborative agents within a dynamic model using creative movement to bring meaning to the science of simple machines. We utilize a Working Words into Movement strategy to help students use their past experiences and motor memory to explore, interpret, and engage with as they seek understanding of simple machines. A Midwest urban elementary school provides the context for a unit plan culminating in a dance performance. The foundational ideas presented within this unit can be enacted within any classroom by creative movement (physical education or dance) specialists, science specialists, or classroom generalists. It follows with a presentation of science content on simple machines exploring the disciplinary core idea of force and motion.

Connects the use of creative dramatics with engaging students previously disengaged with STEM classes. Elucidates a variety of reasons how creative dramatics allows STEM teachers and their students to succeed. Each of the assertions is backed by citations of research studies, classroom practice, and details of the theoretical underpinnings. Also included are detailed descriptions of three effective classroom methodologies using theater in STEM classes, along with specific examples of each that include student and teacher interactions.

Section 3
Theme Based

Incorporating STEAM into classroom practice can be both challenging and daunting. "Imagioneering" a new Mission: Space showcases a way to utilize STEAM into the environmental earth science classroom. Using the familiarity many students have with Disney, critical analysis is applied to the design of a famous park ride. Ride-layout is critiqued and improved upon via student collaboration. Students then use their ability to analyze design to build a structure that fits a particular purpose. This project is geared towards the high school level and students redesign the exterior and line-queue for Mission: Space.

The purpose of the chapter is to provide an exemplar of an inquiry-based unit on pollination for designing and implementing constructivist instructional practices while simultaneously providing outstanding teacher preparation. The unit on pollination was developed by preservice teachers through a partnership between the Monmouth Conservation Foundation and the Monmouth University School of Education. Through collective action, these institutions were able to enhance student learning on a vital part of the science curriculum, provide a rich clinical experience for pre-service teachers, and to familiarize teachers with a more constructivist approach to pre-school science instruction.

Teachers can find art in their science lessons if they know what to look for. The art in many science lessons is in the form of visual representations, which can be seen to exist on a continuum ranging from very concrete and iconic to very abstract and symbolic. There should be a progression through a lesson that guides students in identifying, interpreting, and analyzing different visual representations, and ultimately help them know how to create their own. The teacher should be very deliberate in his/her selection, generation, and/or use of visual representations. Five critical things the teacher should be mindful of include: (1) how the learner's mind processes visual information, (2) what visual literacy is, (3) what is meant by cognitive loading, (4) different types of visual representations, and (5) augmenting student learning by using a variety of visual encryptions.

The next generation science standards promote the teaching of engineering skills including the designing, testing and building of models. Tower building can provide a real-world experience that not only provides the students with physics and mathematics through motion and stability but the explanation of the use of models and the engineering practice of design, redesign and testing of these models. Tia Pilskow used the project of building a tower with her middle school students in order to provide a cooperative team long-term project. She focused first on the design using background information on existing towers. She required each team to design their tower first using graph paper and scale. This process stressed the need for Art and Mathematics in the STEAM project. The science, technology and engineering also played a major part in the design. The case included in this article expands her process by including a cost analysis attempting to promote real world engineering.

This is a case in which students build paper airplanes from templates provided by the instructor as well as those which they design themselves. They extend their ideas on flight by using the principles developed in the paper airplane race to power simulated rockets made out of balloons. They consider such variables as materials, mass and design to see which combination of material design and mass are most effective in constructing an airplane or a rocket that flies the fastest and the furthest in a competition. Contestants are rated on team consideration of variables needing to be controlled in order to have a fair assessment of the designs. When the designs are agreed upon and constructed, a race is conducted. The ideas developed in the paper airplane competition are then used to design a rocket carrying a paper airplane capable of flying across the classroom in the fastest time with the most direct route. This is a simulation of the space shuttle flights. The parameters of the races are developed by the participants.

Preface

The first goal of this book is to assist methods professors in the STEAM methods classroom. The second goal is to assist teachers who are ready for STEAM methods in their classroom. The cases in the book provide techniques and practices for STEAM in the classroom.

STEAM is a current "hot" topic in science education. This book assists teachers and educators with cases that are examples of STEAM teaching. This collection of chapters includes a theory underlying the STEAM teaching model.

Science, technology, engineering and mathematics education (STEM) has been implemented for many years. The most current standards for science education, Next Generation Science Standards (NGSS), includes engineering and environmental sciences. Science education has implicitly included arts education (STEAM) in many forms including, models, diagrams, graphic novels, acting out phenomenon and so on. We view STEAM as a way to include more students in the process(es) of learning and understanding science. Research also supports STEAM because approximately 75% of the nerve pathways in the human brain are dedicated to processing sensory information (reception, decoding and storage). This information is particularly true about visual information. If we want students to learn STEM content more deeply and be able to use the information it behooves us to use methods that will increase the probability of understanding the content.

The research on STEAM education is in its infancy. In education, research is generally based on either a testable model or an ability to identify and describe salient features of practice. This book focuses on case studies of practices and a resultant model that can be further tested and refined/refuted. Each of the chapter authors and book editors are prominent members of a variety of professional organizations that focus on teaching in the STEAM areas.

The book is subdivided into four sections. Section 1 is the introduction to STEAM and its underlying theory. The introductory chapter is the first attempt by the editors to answer the questions, "Why this book?" and "What is STEAM?" The second chapter is a conversation between two educators using the theme, unanswered questions, for the development of a STEAM teaching model. Section 2 provides the readers with cases that include descriptions of art practices. The case chapters include information about graphic novels, trade books, theatre, movement, fine art and visualization. Section 3 focuses on theme based chapters. Lessons include topics from tower building, air planes, bee pollination, catapults, insulators, engineering, ramps and sliders, marine science, to imagineering. Section 4 uses case information from the book chapters to explain differentiating lessons for special needs students. Finally the conclusion sums up and highlights special features from the cases.

We would like to thank all of the case authors for not only writing such interesting chapters but also for the time they spent peer-reviewing their colleagues' chapters. The peer-review process strengthened every chapter in this book.

We hope that this book motivates you to move forward utilizing STEAM practicing.

Judith A. Bazler
Monmouth University, USA

Meta Van Sickle
College of Charleston, USA

Acknowledgment

The editors would like to acknowledge the hard work of editing by Kayla Gomez, Monmouth University Graduate Student and Alexis Patsalo, Terese Peterka, Ceaser Nzuki, College of Charleston Graduate Students. We also thank all of our authors and peer reviewers for their hard work and persistence in the chapter writing processes. Both Judy and Meta commend Diane Cudahy for her support and thoughtful ideas throughout the process of creating this book.

In addition, I (Dr. Judith A. Bazler) would like to thank Kirk Melnikoff, Lara Vetter, and Finley Melnikoff for their constant support.

Introduction

INTRODUCTION

STEAM is a teaching model that is emerging in the United States through initiatives which focus on a perceived global need for individuals skilled in utilizing both divergent and convergent thinking. The reasoning behind these initiatives may be related to U.S. economic strategy, namely the need to continue the high rate of patents in the U.S. People in the U.S. invent the next generation of product and the remainder of the world buys the patent(s) and produces the good(s). However, the rate at which other countries are producing patents is increasing while the rate at which the U.S. is creating patents is decreasing.

The previous teaching model, Science, Technology, Engineering and Mathematics (STEM), was originally named Science, Mathematics, Engineering and Technology (SMET). Dr. Peter Faletra, a director from the Office of Science division of Workforce Development for Teachers and Scientists, suggested the name change from SMET to STEM. The acronym arose in common use shortly after an interagency meeting on science education held at the US National Science Foundation chaired by the then NSF director Rita Colwell. The acronym SMET was not considered memorable and thus was changed to STEM. SMET was the first approach to integrating content areas thought necessary for increasing knowledge base through critical and creative thinking. One of the first NSF projects to use the acronym was STEMTEC, the Science, Technology, Engineering and Math Teacher Education Collaborative at the University of Massachusetts Amherst, which was founded in 1997.

Science, technology, engineering and math (STEM), as a teaching model, is a currently recognized application in cross-content-area disciplines to create knowledge as a whole (Committee on Prospering in the Global Economy of the 21st Century, 2005). STEM-focused teaching is described as a way to increase academic rigor in schools and introduce students to skills and knowledge that are believed to be of growing importance to tomorrow's workforce. The focus on STEM education extends beyond federal agencies such as NSF and the U. S. Department of Education to include non-profits, businesses and community partners (Brady, 2014). The goal of STEM initiatives in classrooms is broader than creating more scientists, engineers, or mathematicians (Vasquez, Sneider, & Comer, 2013), with the intent being to develop capable students who can use information from the four disciplines and function in a highly technological world. Consequently, STEM education and research centers can be found in countries worldwide (Delaney, 2014).

Early rhetoric about STEM education was to increase diversity in the associated fields (Tsui 2007; Reyes, 2012), but STEM careers remain significantly homogenous and women and minorities hold only 28% and 10%, respectively, of STEM jobs (National Science Board, 2014). The plan to diversify the STEM workforce so that it reflects multiple perspectives, knowledge, and experiences, requires we alter the way we think about and teach STEM. This effort "begins in elementary school and extends through the professoriate and the rest of the workforce" (Educational Testing Service (ETS), 2015, p. 3).

Making the change from STEM to STEAM focusses on the development of critical thinking that challenges a personal dialogue between the eyes, mind and hands of a student. Historically, the term STEM (science, technology, engineering and mathematics) began to be used in education when immigration discussions about the lack of career ready personnel from the U.S. for high-tech jobs became the norm (Jolly, 2014). Skills required of today's graduates for many jobs in the U.S. include more in-depth knowledge of science and mathematics, along with the ability to integrate and apply that knowledge in fields such as technology and engineering. To integrate the knowledge across disciplines, critical thinking, creativity, communication, entrepreneurship, and collaboration are all skills needed. The power of STEM was deemed to be higher learning outcomes through the integration of the disciplines versus separating the disciplines.

STEAM fully integrates the arts to give students opportunities to employ critical thinking, creativity and communication in new ways by pulling elements of each and using them in a cross-cutting manner. For example, the general concept of design is a feature of all the disciplines. How design is used in each varies to a degree, but equally sets criteria for all the work produced. If one takes the design principle of "space" then the manner in which space is used in all the disciplines becomes the criteria for the lesson. An example across the disciplines in the design of a product could be, math: area, science: scale, engineering: dimensions or "fit", technology: graphic positioning, and art: perspective. Each discipline is using space within the lesson and each use is important to the outcome.

In order to do STEAM, a rethinking of the one-dimensional school format, including class times, assessment, and curriculum order and separation needs to be considered. Ruth Catchen (Jolly, 2014), a STEAM educator, says that "engaging students' strengths using art activities increases motivation and the probability of STEM success" (Jolly, 2014). She also suggests that art enhances STEM by providing direct applications through communication and expression. In his Scientific American guest blog, Steven Ross Pomeroy (2012) discusses how art and science align. He points out that "noble laureates in the sciences are seventeen times likelier than the average scientist to be a painter, twelve times as likely to be a poet, and four times as likely to be a musician (Pomeroy, 2012)." A TED speaker suggested that science and art are manifestations of the same thing (Pomeroy, 2012). Dr. Jerome Kagan (Pomeroy, 2012), Harvard University, suggests that the arts contribute to learning by combining motor skills, perceptual representation and language. STEM advocates stress that STEM already involves art, including design, communication, and social studies (Pomeroy, 2012). STEM curriculum does not exclude arts, rather they are used implicitly for the engineering challenges. The STEM focus is on rigorous science and math and some believe that including the arts would lessen the content. STEAM advocates suggest that the arts are a necessary component for the foundation. Thus the authors in this book propose that the purpose of STEAM is to unite art with the other STEM content in real situations.

A series of legislation and acts since the mid 2000's reveal a social/political pattern of involvement in K-12 curriculum with a focus on STEAM. For example, Jim Langevin introduced a House Resolution 319 that "expresses the sense of the House of Representatives that adding Art and design into federal programs that target STEM (science, technology, engineering and mathematics) fields encourages innovation and economic growth in the United States." NSF also funded three workshops called from STEM to STEAM (Maeda, 2012). The National Endowment of Arts hosted an event in order to promote STEAM (Maeda, 2012). In 2006, the United States National Academies recommended increasing America's talent pool by improving K-12 science and mathematics education and strengthening the skills of teachers through additional training in science, mathematics and technology, thereby enlarging the pipeline of students prepared to enter college and graduate with STEM degrees (Maeda, 2012).

WHY ADD THE "A" TO STEM

Each approach to STEAM has proponents from arts are the lead, to science is the lead. Among these approaches to teaching in a STEAM model are approaches using the method of one discipline as being the lead and the others following. As is often the case in education, the temptation to promote the content one knows best/most is prevalent. Such promotion is especially true if one content area has been reduced or eliminated from the curriculum as has historically been the case with the removal of the arts. STEAM, as an integrated approach to teaching, has many merits in the form of discipline specific content in particular (Maeda, 2012; Pomeroy, 2012; Jolly, 2014) and thus the visual metaphor we will use is that one is teaching using capital and non-capital letters to illustrate which is considered the leading content.

Thus, steAm would be the teaching model where art for art's sake is the leader. Mathematics teachers approach to such a lesson is steaM, technology teachers focus on sTeam, engineering teachers model stEam, and finally, science teachers would use Steam. Thus, we have a visual metaphor for each of the disciplines being the lead for a lesson. In contrast, the cases in this book are designed with the intent that each content area is appropriately used within a lesson and none are considered the lead discipline, but rather, are one of five disciplines incorporated, each with its unique features and importance for the success of the lesson. Collaboration across the disciplines is the model.

While the editors of this book propone that individualism is a highly regarded social value it is rare to see many approaches in a classroom that a teacher uses to strive to attain a personalized approach for each child to learn that *begin* with the lesson plan. Most individualized plans begin the process by seeking the child's strengths and then working to use the models/forms of teaching that will supposedly best aid the learning. In this book, the intent is to talk about STEAM teaching as a way to reach the largest number of students in a class through their strength first and allow them to develop all the other information through this strength. There is an inherent beauty in allowing each child to enter the study through their love/strength and yet ultimately attain all the goals or standards of a lesson. It is also a way to personalize the lesson through content strengths, rather than, through a specific pedagogy. Such an approach to personalizing the lesson has not been previously described.

Sizer, (1996) notes that students learn differently because they are different and they grow more distinctive as they mature. From early work in education learning theory, Vygotsky (1928) noted that because students cannot begin at the same place, they cannot end at the same place, no matter how intentional or well-designed their school may be. Thus the general pattern is that high scoring students often have a reason to learn more academic material, students scoring in the middle may see no obvious reason for focusing their mental energy on class work, and students scoring at the bottom often have a long history of school failure. It seems that success generates success and that learning is often based on the individual.

In identifying and describing aspects of personalized learning, Clarke and Fraser (2003) noted in a study that.

The daily routine of high school makes it impossible for teachers to know their students well or to recognize their individual strengths and needs. Such student anonymity is prevalent in most American high schools today and is antithetical to a positive learning experience for students and teachers alike. Through personalizing both teaching and learning, students can move beyond anonymity to become recognized and invested partners in achieving success in their classrooms.

Personalized learning is not the same as individualized learning. Personalized learning requires the student to choose the active direction of the lesson; while individualized learning requires that the school tailor the curriculum to scaled assessments based on prescribed interests and abilities. The difference between individualization and personalization lies in who has the control over the lesson, the learner or the teacher. However, in collaborative learning the control over the lesson is not imperative because the unique jobs of each individual adds to the project. In the STEAM model we are proposing that both individualized/personalized learning and the ability to work with others is paramount to the success of the project. A key question to ask to determine if one is using personalized learning v. individualized learning is, "How much does the student direct the process of his or her own learning?" The success that the product provides for each individual relies on individuals receiving feedback and becoming engaged in the process while viewing the process through a flow (Csikszentmihalyi, 1997).

In addition, NASBE (2002) notes that to personalize learning:

Learning is student-centered. Every student feels a personal connection to the school and is known well by at least one adult. Students participate in determining how they will meet learning standards and their personal goals.

While they were describing a high school reform model it is still apparent that personal connections between a learner and a teacher is a requirement. If this personal connection does not occur, then the learning decreases, if it exists at all. Further, when studying Noddings (1992) it is noted that *both* the teacher and the learner need to *care for* each other, ideas, living and nonliving things. When learning opportunities include the need to care for each other, ideas, living and nonliving things then the learning is personalized and most effective.

Yet in the face of overwhelming evidence that all learning is personal and largely idiosyncratic, we continue to design impersonal curricula that is distilled into discrete subjects and then ask groups of students to learn everything in the same way and at the same time. However, students continue to express greater interest in ideas and questions than in disciplines. Personal aspects of learning effect what is learned and how it is organized for use in individual minds. "All new learning depends on prior learning," David Ausubel wrote in 1968. "Ascertain such and teach accordingly."

Studies from the 1990's (Lee, et al., 1995; Newmann, et al., 1992; Stigler & Hiebert, 1999) have identified, "the importance of active participation in learning as a major determinant of performance on tests and in grades." In addition, Scales and Leffert (1999) identify forty different developmental assets among adolescents that determine ways students engage in learning. These assets, include such factors as: "family support, parent involvement in school, positive peer influence, time at home, bonding to school, personal power, sense of purpose, and others, is firmly connected to background characteristics and conveys advantages or disadvantages to individual students in a school setting."

Early stage research in STEAM education (Stigler, Park, Kim, & Kim, 2012) indicates that it increases motivation, engagement, and effective disciplinary learning in STEM areas. It is important to note that students strengthen their learning within the disciplines, and more importantly, between disciplines, through opportunities to explore and make connections between art, music, mathematics, science, and more. These same early studies suggest STEAM-based curricula produce a higher percentage of students who remain interested in pursuing careers that require knowledge and skills from the fields of math and science (Masata, 2014). It is believed that students, who might be uninterested in science and math, are drawn to these disciplines through a transdisciplinary approach to learning, that encourages the integration of various modes of inquiry (i.e., creative process, technological problem solving) that students may find more interesting and accessible (Vanasupa et al., 2007). STEAM curriculum is centrally rooted in real problems and incorporates all of the disciplines in the process of solving the problem(s). STEAM programs are designed to build upon multiple ways of thinking and demonstrating knowledge, thereby attracting students who might not see themselves as interested in one of the STEM disciplines. STEAM-based curricula, remain rooted in curriculum standards *and* engage the learner through the range of disciplines in a manner that promotes understanding while solving a relevant problem. To illustrate the difference between a STEM and a STEAM based lesson, consider the following:

- **STEM Teaching:** The objectives of the unit are to infer an earthquake's epicenter from seismographic data, explain how earthquakes result from forces inside earth, and to identify and illustrate geologic features of (choose your state, region, or fracking location) and other regions of the world. Our class will use imagery and topographic maps to analyze data. From this data, you will draw a map of the (state, region or fracking location) marking the epicenters of the earthquakes and identify and describe fault lines or lack of fault lines.
- **STEAM Teaching:** Last week there was an earthquake of a 2.0 magnitude in (state, region or fracking location)! While most earthquakes are minor and we cannot even feel the earth shifting, major earthquakes cause serious damage and have the potential to harm and kill people. Your task is to determine whether or not a major earthquake (6.0 or higher) could ever take place in (state, region or fracking location), predict the epicenter, or other sources of energy to cause the earthquake and discuss the potential impacts. You and your team will gather evidence to support your answer. Your task is to

communicate this information to the local city council to help them decide if the public should be informed. Visualization tools or videos expressing your perspectives are encouraged.

WHERE DOES STEAM FIT INTO THE CURRICULUM?

The editors left the answer to this question up to the authors of the chapters and specifically to the readers. We believe that teachers are competent to know when to use STEAM lessons and can use the cases in this book as examples to help them begin. We organized the chapters in the following manner so that each reader can decide where to start their reading:

Section 1: Introduction and a model-based ethics and aesthetics of STEAM as the underlying pedagogy.
Section 2: Includes chapters that provide specific arts techniques for use within the lesson.
Section 3: Includes chapters that begin with a theme for study and use all five disciplines to conduct the study.
Section 4: Is two chapters about inclusion practices for the special needs learner and taking stock or concluding thoughts based on a cross-analysis of the case chapter.

Meta Van Sickle
College of Charleston, USA

Judith Bazler
Monmouth University, USA

REFERENCES

Ausubel, D. (1968). *Educational psychology: A cognitive view*. New York, NY: Holt, Rinehart and Winston.

Boyer, E. (1983). *High school: A report on secondary education in America*. New York, NY: Joanna Colter Books.

Brady, J. (2014, September 5). STEM is incredibly valuable, but if we want the best innovators we must teach the arts. *The Washington Post.* Retrieved from https://www.washingtonpost.com/news/innovations/wp/2014/09/05/stem-is-incredibly-valuable-but-if-we-want-the-best-innovators-we-must-teach-the-arts/?utm_term=.7dfe19789cff

Clarke, J. (1999). Growing high school reform: Planting the seeds of systemic change. *NASSP Bulletin, 83*(April), 1–9. doi:10.1177/019263659908360601

Clarke, J. (2001). Teaching each and all: Harnessing dynamic tension in school reform. *NASSP Bulletin, 85*(626), 75–79. doi:10.1177/019263650108562608

Clarke, J. (2003). Personalized learning and personalized teaching. In J. DiMartino, J. Clarke, & D. Wolk (Eds.), *Personalized learning: Preparing high school students to create their futures*. Lantham, MD: Scarecrow Press.

Clarke, J., & Agne, R. (1997). *Interdisciplinary high school teaching: Strategies for integrated learning*. Boston: Allyn and Bacon.

Clarke, J., Aiken, J., & Sullivan, M. (1999, September). Interactive leadership in high school innovation. *NASSP Bulletin, 83*(608), 98–111. doi:10.1177/019263659908360811

Clarke, J., Bossange, J., Erb, C., Gibson, D., Nelligan, B., Spencer, C., & Sullivan, M. (2000). *Dynamics of change in high school teaching: A study of innovation in five Vermont professional development schools*. Providence, RI: Northeast and Islands Regional Educational Laboratory at Brown University.

Clarke, J., & Fraser, E. (2003). Making learning personal: Educational practices that work. In J. DiMartino, J. Clarke, & D. Wolk (Eds.), *Personalized learning: Preparing high school students to create their futures*. Lantham, MD: Scarecrow Press.

Clarke, J., Sanborn, S., Aiken, J., Cornell, N., Goodman, J., & Hess, K. (1998). *Real questions/ real answers: Focusing teacher leadership on school development*. Alexandria, VA: ASCD Publications.

Clarke, J. H. (1990). *Patterns of thinking: Integrating learning skills in content teaching*. Boston, MA: Allyn & Bacon.

Clinchy, E. (1996, December). Reforming American education from the bottom to the top: Escaping academic captivity. *Phi Delta Kappan*, 268.

Cohen, D. (1995). What is the system in systemic reform? *Educational Researcher, 24*(9), 11-17.

Committee on Prospering in the Global Economy of the 21st Century. (2005). *An agenda for American science and technology.* National Academy of Sciences, National Academy of Engineering, and Institute of Medicine.

Cotton, K. (2001). *New small learning communities: Findings from recent literature.* Seattle, WA: Northwest Regional Educational Laboratory.

Csikszentmihalyi, M. (1997). *Finding flow: The psychology of engagement with everyday life.* New York, NY: Basic Books.

Cushman, K., Steinberg, A., & Riordan, R. (2001). *Rigor and relevance: Essential ideas about connecting school and work.* Boston, MA: Jobs for the Future Inc.

Darling-Hammond, L., Ancess, J., & Ort, S. (2002). Reinventing high school: Outcomes of the Coalition Campus Schools Project. *American Educational Research Journal, 39*(3), 639–673. doi:10.3102/00028312039003639

DiMartino, J. (1997). *Study of Dropouts at West Warwick High School.* Providence, RI: Northeast and Islands Regional Educational Laboratory at Brown University.

DiMartino, J., & Clarke, J. (2000). *Personal learning plans and high school renewal.* Paper presented at NASSP Annual Conference, San Antonio, TX.

DiMartino, J., Clarke, J., & Wolk, D. (Eds.). (2003). *Personalized learning: Preparing high school students to create their futures.* New York, NY: Scarecrow Press.

Fosnot, C. T. (1989). *Enquiring teachers, enquiring learners: A constructivist approach for teaching.* New York, NY: Teachers College, Columbia University.

Friedrichs, A. (2000). *Continuous learning dialogue: An ethnography of personal learning plans* (Dissertation). Burlington, VT: University of Vermont.

Fullan, M. (1993). *Change forces.* London, UK: Falmer Press.

Fullan, M., Bennett, B., & Rolheiser-Bennett, C. (1990). Linking classroom and school improvement. *Educational Leadership, 47*(8), 12.

Gardner, H. (1990). *Frames of mind, the theory of multiple intelligences.* New York, NY: Harper Collins.

Gibson, D., & Clarke, J. (2000). *Growing toward systemic change: Developing personal learning plans at Montpelier High School.* Providence, RI: Northeast and Islands Regional Educational Laboratory at Brown University.

Goldberger, S., & Kazis, R. (2000). *Revitalizing high schools: What the school-to-career movement can contribute*. Boston, MA: American Youth Policy Forum, Institute for Educational Leadership, National Association of Secondary School Principals, Jobs for the Future.

Jolly, A. (2014). STEM vs. STEAM: Do The Arts Belong? *Education Week*.

Keefe, W., & Jenkins, J. (2002). A special section on personalized instruction. *Phi Delta Kappan*, *83*(6), 440–448. doi:10.1177/003172170208300609

Kyle, R. (1993). *Transforming our schools*. Lessons from the Jefferson County Public Schools.

Lambert, L. (1998). *Building leadership capacity in schools*. Alexandria, VA: ASCD Publications.

Lee, V., Bryk, A., & Smith, J. (1993). The organization of effective secondary schools. Review of Research in Education, 19, 171-267. doi:10.2307/1167343

Lee, V., & Smith, J. (1994a). High school restructuring and student achievement: A new study finds strong links. *Issues in Restructuring Schools, 7*(1-5), 16.

Lee, V., & Smith, J. (1994b). Effects of high school restructuring and size on gains in achievement and engagement for early secondary students. *Issues in Restructuring Schools, 7*(1-5), 16.

Lee, V., & Smith, J. (2001). *Restructuring high schools for equity and excellence*. New York, NY: Teachers College Press.

Lee, V., Smith, J., & Croninger, R. (1995). *Another look at high school restructuring. Issues in Restructuring Schools (Report #9)*. Madison, WI: Center on Organization and Restructuring of Schools.

Legters, N., Balfanz, R., & McPartland, J. (2002). *Solutions for failing high schools: Converging visions and promising models. Paper Commissioned by the Office of Vocational and Adult Education*. U.S. Department of Education.

Levine, E. (2002). *One kid at a time: Big lessons from a small school*. New York, NY: Teachers College Press.

Littke, D., & Allen, F. (1999). Whole school personalization: One student at a time. *Educational Leadership*, *57*(1), 24–29.

Maeda, J. (2012). *STEM to STEAM: Art in K-12 Is Key to Building A Strong Economy*. Edutopia.

Maine Commission on Secondary Education. (1998). *Promising Future A call to improved learning for Maine's secondary students*. Augusta, ME: Author.

McCarthy, B. (1990, October). Using the 4MAT system to bring learning styles to schools. *Educational Leadership*, *48*(2), 31–37.

McLaughlin, M. (1990). The Rand change agent study revisited: Macro perspectives and micro realities. *Educational Researcher*, *19*(9), 11–16. doi:10.3102/0013189X019009011

McLaughlin, M., & Talbert, J. (1990, November). Constructing a personalized school environment. *Phi Delta Kappan*, 230–235.

Meier, D. (1995). *The power of their ideas*. Boston, MA: Beacon Press.

Meier, D. (1996, September). The big benefits of smallness. *Educational Leadership*, *54*(1).

Morris, D. (1997). Adrift in a sea of innovations: A response to Alexander, Murphy and Woods. *Educational Researcher*, *26*(4), 22–26.

Murnane, R., & Levy, F. (1995). *Teaching the new basic skills principles for educating children to thrive in a changing economy*. Cambridge, MA: The Free Press.

National Association of Secondary School Principals. (1996). *Breaking Ranks: Changing an American institution*. Reston, VA: NASSP.

National Association of State Boards of Education. (2002). *Most likely to succeed Policymaking in support of a restructured high school*. Alexandria, VA: NASBE.

National Commission on Excellence in Education. (1983). *A nation at risk: A report to the nation*. Washington, DC: U.S. Government Printing Office.

National Education Commission on Time and Learning. (1994). *Prisoners of time: Report of the National Commission on Time and Learning*. Washington, DC: U.S. Government Printing Office.

Newmann, F., Wehlage, G., & Lamborn, S. (1992). *The significance and sources of student engagement. Student engagement and achievement in American secondary schools* (F. M. Newmann, Ed.). New York, NY: Teachers College Press.

Noddings, N. (1992). *The challenge to care in schools: An alternative approach to education*. New York, NY: Teachers College Press.

Piaget, J. (1964). *Judgment and reasoning in the child*. Patterson, NJ: Littlefield Adams and Company.

Pomeroy, S. R. (2012). From STEM to STEAM: Science and Art go Hand-to-Hand. *Scientific American Blog*.

Prairie, C. (2002). *Draft of a personalization rubric. (Personal Correspondence)*. Bloomington, IN: Indiana University.

Raywid, M., & Oshiyama, L. (2000). Musings in the wake of Columbine: What can schools do? *Phi Delta Kappan*, *81*(6), 444–449.

Reyes, M.-E. (2012). Increasing diversity in STEM by attracting community college women of color. In B. Bogue & E. Cady (Eds.), *Apply research to practice (APR) resources*. Retrieved from https://www.engr.psu.edu/awe/ARPAbstracts/CommunityCollege/ARP_WomenofColorCommunity%20College_Abstract%20Final.pdf

Savage, R. N., Chen, K. C., & Vanasupa, L. (2007). Integrating project-based learning throughout the undergraduate engineering curriculum. *Journal of STEM Education: Innovation and Research*, *8*, 1–13.

Scales, P., & Leffert, N. (1999). *Developmental assets: A synthesis of the scientific research on adolescent development*. Minneapolis, MN: Search Institute.

Senge, P. (1990). *The fifth discipline: The art and practice of the learning organization*. New York, NY: Currency Doubleday.

Sizer, T. (1984). *Horace's compromise: The dilemma of the American high school*. Boston, MA: Houghton Mifflin.

Sizer, T. (1992). *Horace's school: Redesigning the American high school*. Boston, MA: Houghton Mifflin.

Sizer, T. (1996). *Horace's hope: What works for the American high school*. Boston, MA: Houghton Mifflin.

Sizer, T. (1999). No two are quite alike. *Educational Leadership*, *57*(1), 12–17.

Steinberg, A. (2001). *Forty-three valedictorians: Graduates of the Met talk about their learning*. Boston, MA: Jobs for the Future.

Stigler, J., & Hiebert, J. (1999). *The teaching gap: Best ideas from the world's teachers for improving education in the classroom*. New York, NY: Simon and Schuster.

Tsui, L. (2007). Effective strategies to increase diversity in STEM fields: A review of research literature. *The Journal of Negro Education*, *76*(4), 555–581.

United States Department of Education. (1999a). *New American high schools: In their own words*. Washington, DC: The Kamber Group.

United States Department of Education. (1999b). *Aiming high, strategies to promote high standards in high schools*. Berkeley, CA: MPR Associates.

United States Department of Education. (1999c). *New American high schools at the leading edge of reform*. Berkeley, CA: MPR Associates.

United States Department of Education. (1999d). *Key reform strategies: An overview of research findings*. Berkeley, CA: MPR Associates.

United States General Accounting Office. (2002, February). *School dropouts Education could play a stronger role in identifying and disseminating promising prevention strategies* (GA0-02-240). Author.

Vasquez, J. A., Comer, M., & Sneider, C. (2013). STEM lesson essential, grades 3-8: Integrating science, technology, engineering, and mathematics. Portsmouth, NH: Heinemann.

Vermont Department of Education. (2002). *High schools on the move*. Montpelier, VT: DOE.

Vygotsky, L. (1986). *Thought and language* (A. Kozulin, Trans.). Cambridge, MA: MIT University Press. (Original work published 1926)

Wang, M., Haertel, G., & Walberg, H. (1993). Toward a knowledge base for school learning. *Review of Educational Research*, *63*(3), 249–294. doi:10.3102/00346543063003249

Windschitl, M. (2002). Framing constructivism in practice as the negotiation of dilemmas. An analysis of the conceptual, pedagogical, cultural and political challenges facing teachers. *Review of Educational Research*, *72*(2), 131–175. doi:10.3102/00346543072002131

Woodrow Wilson Foundation National Commission on the Senior Year. (2001). Raising our sights: No high school senior left behind. Princeton, NJ: Author.

Section 1
Introduction to STEAM

Chapter 1
Musing on Unanswered Questions

Meta Van Sickle
College of Charleston, USA

Merrie Koester
University of South Carolina, USA

EXECUTIVE SUMMARY

Out of a conversation between two long-time colleagues—each a science educator and practicing artist, emerged the question, "What does it mean to STEAMify a lesson, and why would a teacher actually choose to do such a thing, other than, say, for-grant-writing-purposes? Their science selves really liked the idea of a STEAM system, acted upon by forces, both from the outside and from within, and with energy flowing and cycling, all the while transforming grey matter in ways that sustained the teaching/learning process. When it came to their art, however, their dialogue followed pathways grooved by long years of practice and hard work in their respective fields. One author is a seasoned vocalist, trained in the nuances of both individual and group vocal performance as well as the attendant dimensions of music, its composition and phraseology. The other is a painter, poet, and novelist, shaping words, color, and line to tell stories and communicate ideas. What was significant to each was that their artistic habits of mind had shaped their axiology, transforming their ways teaching.

DOI: 10.4018/978-1-5225-2334-5.ch001

Copyright ©2017, IGI Global. Copying or distributing in print or electronic forms without written permission of IGI Global is prohibited.

INTRODUCTION

Why should just one more in an interminable line of educational acronyms or slogans actually matter? Is STEAM really *different*, or is it another fad? That the idea of STEAM has even emerged in the first place implies some underlying and global lack of creativity and imagination in both science and STEM education; and yet, we have personally observed many such teachers employing highly artistic teaching styles, flexibly improvising to adapt to changing situations and student needs and driving home the message that in the end (as Eisner has said), education is the act of inventing yourself through hard work, practice, failing, and trying again. There is no kit with instructions for the making of the self. Such an effort begins with and is sustained by the individual. We are calling for a kind of STEAM education that develops the habits of mind associated with making art, which includes the practice of critique and care throughout the learning/making process. Such STEAM educators would recognize that what educators Hetland, Winner, Veenema, and Sheridan (2013) describe as "studio thinking" are highly congruent with the Science and Engineering Practices (NGSS Lead States, 2013). Figure 1.

We advocate for a kind of STEAM that builds in time for critique to help students learn to observe, interpret, explain, and evaluate (Hetland, Winner, Veenema, & Sheridan, 2013). Through the imagery of musicianship— be that improvisation, composition, or the interpretation of its performance—we will introduce a model of teaching science as aesthetic inquiry that we think can move students away from alienation and towards affiliation with science and STEM education. Our aesthetic model foregrounds knowing as communal and value-laden, rather than as a static, detached, objective process of naming and identifying. Aesthetic inquirers recognize that the creative process (always present at the heart of science) spirals outward through cacophony and chaos, and that neat, harmonious compositions may not

Figure 1. Comparison of art studio habits of mind and NGSS science and engineering practices

Art Studio Habits of Mind

- OBSERVING
- ENVISIONING
- INNOVATING THROUGH EXPLORATION AND DESIGN
- REFLECTIVE SELF-EVALUATION
- ENGAGING AND PERSISTING THROUGH MISTAKES
- CRITIQUING

Source: Hetland, Winner, Veenema, and Sheridan (2013)

Science and engineering practices (NGSS Standards)

- Ask questions and define problems.
- Develop and use models.
- Design solutions.
- Obtain, evaluate, and communicate information.

Source: NGSS Lead States (2013)

always be the result, especially as new and yet unanswered questions emerge. Finally, we have conceptualized this form of STEAM education as a kind of communal *making*—one that unites the head, heart, and hands as ideas and learning artifacts are constructed. We fully recognize that our approach may evoke responses ranging from dissonance to resonance and that there is no one best way to *do* STEAM education.

THE RESONANCE/DISSONANCE DIALECTIC

When people are deeply connected in a positive way with one another, they are often described as being *tuned into* one another, *on the same wavelength*, or in complete *harmony*. Their emotional and intellectual *frequencies* are in *resonance*, and they can often anticipate the other's next *phrase*. They listen to one another attentively and appreciatively, until and unless they are interrupted by a noise, which suddenly knocks them out of synch and forces them to attend to the distraction. They may even become irritable. How dare an intruder ruffle the smooth *texture* of their discourse and introduce *tension* into their relationship! Do they really have to *listen* to such noise? There are many educators who may criticize STEAM education as not-science or not-STEM. We recognize that we may not even approach equilibrium with these individuals. For them, even the idea of STEAM introduces too much cognitive *dissonance* and threatens their sense of should and ought. But we wonder, too, if the arts were a significant part of their own education, and if so, how were such experiences organized for them?

Liora Bresler (2005) is a musicologist who has devoted her career and scholarship to exploring what musicianship can teach educational researchers. She has reminded us of the "complexity of translating anything, but particularly a fluid form into a fixed one" and that the power of performances (be it an educational seminar or a musical performance) reside in the degree to which they establish connection and interaction (p. 169). How is the room arranged? Does it allow for movement? The expression of feelings? The development of empathy and trust? When conceptualized as *music* creation, what are the "forms, rhythms, textures, and timbres" of the performance (p. 170)?

Inspired by such writing, we asked, "What if science and STEM educators acquired the capacity to compose and perform lessons as would a musician? How might performances be enacted so that independent learning continued in the time and space outside the classroom, too? What science standards might be taught through this kind of teaching/learning performance? Our musings were piling up, our notes strewn everywhere. We could feel ourselves toppling into entropy—until

Meta suddenly blurted out, "Why do we science teachers so often avoid the ethics and aesthetics (axiology) of teaching?"

A DIALOGUE COMMENCES

Merrie: Meta, I've often heard you talk about these things. What exactly do you mean?

Meta: That's just it. In science, we're zeroed in on the exactly. The philosopher in me is troubled by this approach. Philosophers view a topic from many different perspectives. They have conversations about how it is that we come to know anything and whether that knowing matters or not. When you teach, unless you marry the content with life—with what matters—deep learning cannot occur. But teaching in this way can be a very messy and complex process, because it has to do with feelings. Musicians work at shaping feelings into forms that touch the heart. To achieve the effects they want, they also know they must practice and persist through many revisions of their work as they experiment with different possibilities. Science educators have a lot to learn from musicians!

Merrie: Wow! That's a compelling premise. What if we were to show through a musical performance (that is, provide authentic evidence) that the claim you just made about employing musicianship to teach science can be supported? Better yet, what if we could figure out a way for our readers to interact with that performance and, at the same time, employ science and engineering practices? Do you know of a piece that could do those things?

Meta: I do! Have you ever heard of the composer, Charles Ives?

Merrie: No, but I'm sure going to tell me!

Meta (laughing): Do you have your laptop? Pull up the PBS link for their Music Mavericks program called "The Unmixed Question": http://musicmavericks.publicradio.org/features/feature_unmixed_question.html [This file could take a long time to download, so readers, you may want to go get your coffee. Please do not skip over this part of this chapter!]

Merrie: Got it! It says, "Okay, let's play!" I love this already. [She clicks on PLAY and what looks like an audio mixing board pops up on screen]. WHAT???? It's a virtual mixing board! It says here in the written description that "the piece was created using binaural 3D positioning software to move the musical parts around in an imaginary hall." That sounds like music GPS!

Meta: That's essentially what it is. They use two microphones to create a 3D stereo sound that makes you feel like you're in the performance hall. Charles Ives accomplished a similar effect by positioning members of the orchestra in different physical spaces. He was a musical genius. He'd put the horn section offstage sometimes and the winds in places you'd never normally find them.

Merrie: So, he was engaging in science and engineering practices, intentionally changing one variable at a time to document the experimental effect(s).

Meta: Of course! This is what musicians have to do every day. A good choral director is constantly going after different effects, depending on the feelings and sounds he or she wants to evoke. For example, the arrangement of the singers would be different if the conductor wants the audience to feel surrounded by sound vs. the room being filled with sound.

Merrie: So, here, in this simulation model, it looks like we have three different kinds of voices—a trumpet, the winds (clarinets, flutes, saxophones, etc.) and the strings (violin, viola, cello, bass, etc.). There's also this "miscellaneous" voice we can vary.

Meta: Right, but you get to control what you hear. Try it!

Merrie: [listening and playing with the different control knobs]: Here is where I would ask our readers to enter into this experience and ask, "How many different independent variables can be manipulated in this experience? How many experimental questions might be generated and then explored through science and engineering practices? The observational data might take the form of narrative writing and the use of descriptive adjectives or symbolic emoticons. They might make different kinds of expressive lines or body movements to communicate what's happening with them. Students might be asked to record their feelings when they hear each voice by itself or describe what happens when that voice is combined with others. Learners might work together to create different kinds of "noticing" charts to visually represent their observations. I can tell you that I personally wanted to silence those jabbering winds. They were so annoying! My emoticon would be scowling.

Meta: (laughing) Just think of the conversations this experience might start –among both teachers and students. For example, how many times do we teachers have to persist in the face of annoying distractions? Which ones do we pay attention to? What if those winds were students who were desperately trying to tell you that what you were doing wasn't working for them? Perhaps what you thought was perfectly suitable music was for them an ugly noise. These students would not feel cared for at all.

THE ROLE OF CARE

Merrie: OMG. You're right! This activity has powerful implications for the study I am doing with dyslexic science learners. Several have already confessed that they will go to great lengths to call attention away from the fact that they can't read. They say it's especially embarrassing when a science teacher asks them to read directions out loud or be a scribe in a group activity; so they'll act like the class clown, pretend to be sick, or even start a fight. When I asked what kind of science learning they did enjoy, they all said the kind where they get to solve the problem on their own. They absolutely hate doing worksheets and claim that printed directions make them cringe! (Koester, in press).

Meta: Indeed, arts-integrated lessons tend not to marginalize the struggling reader. In this simulation, all students will likely enjoy playfully experimenting with sound energy and then personally arriving at the best combination of voices for their sensibilities. They'll also acquire an appreciation for how musical conductors use dynamics to achieve entirely different effects. Teachers can start dialogues about how sound engineers must work closely with the conductor and how both STEM and the Arts knowledge, skills, and practices are required to make a musical performance work. A care for ideas will emerge (Noddings, 1992).

Merrie: This dialogue is getting more exciting by the minute! As it turns out, most of the dyslexic individuals in my study are also deeply involved in some kind of art, whether it's music, painting, theatre, or martial arts. They like the idea of making new worlds in which they play a starring role. One profoundly dyslexic art professor in my study shared that science had always made her feel stupid, but that her art made her feel powerful. I recently went to a community forum facilitated by an organization called Understood.org. I learned to my amazement that 1 in 5 students struggle with learning and attention issues, often making it difficult for them to read (http://www.adcouncil.org/Our-Campaigns/Education/Learning-Attention-Issues). This means that if we teach science through mostly talking, telling, and reading, we may be marginalizing fully 20% of our students! No wonder there has been a GREAT TURN OFF to science (Trefil, 2008).

Meta: That's another reason why arts-informed science education has such great potential. The arts offer other ways in, pathways to understanding and the freedom that comes with being included in the teaching/learning process. However, when you introduce the arts, you also enter into the world of feelings, aesthetics, and creating caring, empathic relationships—topics which can make a lot of science educators uncomfortable.

Musing on Unanswered Questions

Merrie: What you have just said reminds me of a statement by Sartre (1976):

Authenticity consists in having a true and lucid consciousness of the situation, in assuming responsibilities and risks that it involves.

We are most definitely asking science teachers to take risks and to teach toward outcomes that may not be measurable with a number or test score. For many (if not most) science teachers, lecturing is an acceptably authentic way of being a science teacher; whereas being asked to teach in an "artistic" way may provoke considerable concern, especially for the teacher who claims to have no artistic "talent." Furthermore, they may balk at the extra time an arts-integrated approach requires. In order for them to risk experimenting with an aesthetic paradigm, science teachers will want examples which show that it works. They'll need to see that their science standards can be taught; otherwise, we won't get much traction. Lederman & Lederman (2015) call on science education researchers to clearly define their theoretical frameworks. For us, this means we must answer two questions: 1) What problems might be solved by experimenting with this kind of approach, and 2) Why is our approach feasible?

Meta: Scientists like their numbers and their stats, but in a science teaching performance, things often happen that can't be measured. The simulation we just experienced was an example of a deeply aesthetic, open-ended, personalized form of inquiry—so unlike the prescriptive science education lab with its attendant worksheets and linear progression toward right answers. The simulation also taught the participant a lot about nature of science (NOS), through which discoveries emerge far less methodically (and more naturalistically) than is often presented in the science education canon (NGSS Lead States, 2013).

Merrie: This one STEAM lesson—through which we explore science and engineering standards through music-informed inquiry—could serve as a springboard into dialogue about energy transfer, properties of sound, form and function, cause and effect, and patterns—all cross-cutting principles in the NGSS standards (2013). If all these standards can be "covered," then the time spent on an aesthetic approach will have been well invested indeed! By "noticing what there is to be noticed"—as we engage in participatory arts exploration of content—Green (2001) has reminded us that "we create new connections in experience, allow new patterns to be perceived and open new vistas of understanding" (p. 6). The teacher does not necessarily need any artistic talent, but collaboration with talented students or a creative arts specialist can produce some phenomenal results.

Meta: I completely agree! An aesthetic approach is not only feasible but desirable in the science classroom. In Bresler's (2005) model of music-informed inquiry, a musicianship approach opens spaces for "improvisation, empathy, and embodiment" (p. 169). The Ives spatial music simulation is not just about playing with the controls on the sound mixer. Otherwise, it's just another fun technology toy, app, or game. You have to have dialogue about what is happening! For example, what does manipulating the technology signify for the one at the controls? For the student, those jabbering winds could signify anything from a nagging mother to a little brother who won't be quiet. You just said yourself that they reminded you of the student who ceaselessly interrupted you, possibly to call attention away from the fact that he or she could not read.

Merrie: These matters of signification are important points of dialogue and reflection. In the midst of a learning experience, the "image" that takes shape in my "mind's eye" can be completely different than what's in yours. Finson and Pederson (2011) provide a comprehensive argument for why and how, for example, visual data are so important in science education. They are a way for teachers to see student thinking and the types of connections they have made—or not.

Meta: Yes! In our simulation, spatial music becomes both a visual and aural metaphor for the classroom, which is a highly complex and dynamic setting where many voices are competing for attention. In the "Unanswered Question" performance, there are serene moments throughout, played by the strings. The trumpet sounds at interspersed moments. Did you hear the clarion call of the question? Just as you likely picked up on many sound patterns in the performance of this spatial music, caring, perceptive teachers also pick up on patterns of discourse and behavior of their students.

Merrie: We keep returning to the role of care in the teaching/learning process. Care is a significant component of an aesthetic approach to education (Siegesmund, 2010). In your own writing on the caring science teacher (Van Sickle & Spector, 1996), you state that caring teachers notice how their students are responding to what's happening in the classroom.

Meta: That's right. You were talking earlier about your research on the dyslexic learner. Caring teachers realize that if they teach only through talking and telling, they may lose, or marginalize some 20% of their students. The caring teacher finds ways to move all students forward, finding ways to challenge them, and offers constructive critique to validate them. They identify each child's strengths and what matters to him or her and then use that information as on-ramps into the learning. Noddings (1992) talked a lot about this.

Merrie: And so, each learner can shape a different type of experience for him or herself…

Meta: Yes, but only if we don't over-value instant success and right answers—something we do all too often in science classes.

Merrie: So often, we hardly wait at all before we just tell them the right answers, too. Why should they bother to do the work of making their own ideas? No one enjoys being told what to think all the time (Duckworth, 2006).

Meta: Yes! Emergent, creative thinking is characteristic of the nature of science and art. If you look back at the chart you made that compares artistic habits of mind with science and engineering practices, you'll see that artists and scientists all make a lot of observations before they develop their questions. When you read Nobel laureate literature, they talk about "playing" with ideas, just as artists play with materials, notes, dance moves, voice modulation, etc. They all do thought experiments and mess around with notions. They all get worked up, angry, defensive, happy, or despondent when they fail; but we rarely hear about the feelings, the fun, or the fear of failure when we read science papers.

Merrie: I think this is a really important point you're making. It's as if we're dehumanizing the nature of science in our classrooms by making it seem less messy and more predictable than it is in authentic practice. By contrast, art teachers preach persistence in the face of failure, which they conceptualize as an integral and essential part of the learning process.

Meta: In our choral group, we learn to enjoy the process, make a ton of mistakes, make improvements, and persist through hours of rigorous practice. And we do that so that the performance moment can be all that it can be. That's pure joy, and it's at the heart of what it means to learn in an aesthetic space.

Merrie: You're right! My athletic coaches' favorite expression was always, "No pain. No gain." And, I've been blessed to have deeply caring coaches. I always wanted to work harder to improve my performance because they made me feel like I mattered. Coleman's (2003) idea of human capital is very relevant here because it's acquired when one learns a new skill or gains new knowledge. However, Adjibolosoo's (1995) construct of aesthetic capital is even more to our point. Adjibolosoo explained that "aesthetic or artistic capital brightens or enriches people's' lives" (p. 181). According to Uhrmacher (2010), aesthetic capital may cause a student to feel or act differently and to undergo a radically different way of being in the world (in Koester, 2015).

Meta: It's a teacher or coach's ethical responsibility to set high expectations and keep you moving forward and upward, all the while making it clear that it's your responsibility to get there. You can't produce an aesthetic transformation—the pleasure and even the feeling of power derived from a beautiful experience—without also having an ethic of care. When you practice an ethic of care, it's comparable to the Hippocratic oath of doing no harm. An ethic of care always asks the question of what happens in a relationship. When there is an ethic of care, a system can thrive. When students, the primary actors in an educational system, don't feel cared for, they stop interacting, and they stop learning (Newmann, 1981). They become alienated from the adults and the ideas. They may even retaliate against the equipment and each other. They'll either act out or shut down…

Merrie: …like so many students who struggle to understand the language of science.

Meta: Yes! When they feel included, the winds will settle down and stop interrupting so much. There is a movement toward affiliation with each other and the content, and there will be resonance in the classroom. Tenets of care will have been met.

Merrie: Can you explain more what you mean by tenets of care?

Meta: I think the best explanations can be found in Nel Noddings's work (2003). She talks about the difference between caring about and caring for. I can care about many things, and take no action. If I'm going to care for something or someone, however, I actually have to do something. I can care for others, in this case, our students. I can care for other living things. I can care for objects, too. This latter tenet is very important in your classroom because you need your students to take care of the equipment you're using…

Merrie: …what about care for the self?

Meta: Caring for self is truly important, because the mind and body have to be healthy in order to grow and develop. If you don't care for your mind, you are, by default, limiting yourself and your possibilities of becoming. You certainly can't pass tests and college entrance exams if your mind isn't prepared.

Merrie: This is amazing and beautiful to me, Meta. In my workshops with teachers, I often bring in an inflatable brain. You saw me pull one out at a conference recently.

Meta: I do remember! You had us imagine a huge, brain-shaped hot air balloon, which could only lift off if the teaching practices (in the basket below the balloon) did not bring the system down…

Merrie: …Yes, and the question was, "What's in your science teaching basket?" (See Figure 2).

Meta: An ethic of care and aesthetic transformation cannot be separated from one another in practice…

Merrie: …so one's the balloon, and the other's the air in the balloon. Without both, there's no semiotic lift off, that is to say, no deep meaning-making and no upward and outward movement in the system.

Meta: Now, you're talking!

Merrie: It seems to me that you're also talking about flow. I've felt this so many times when I'm painting or writing.

Figure 2. What's in your teaching basket?

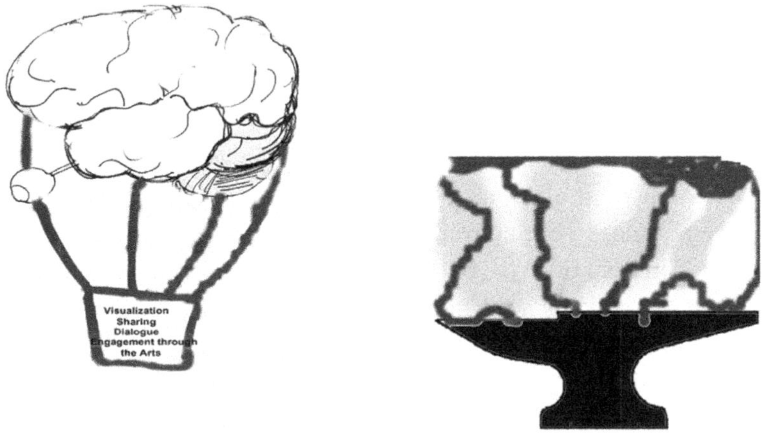

Meta: Yes! Csíkszentmihályi Mihályi, an educational psychologist, writes about flow (1997). In flow, emotions are aligned with the challenge at hand. When flow happens in a classroom, the best thing you can do is get out the students' way and let it happen. It results in deep, joyful learning. It's the place where ethics and aesthetics merge into a beautiful teaching/learning performance.

Merrie: It seems that we're talking about a process that moves in spirals. Making art happens in the same way. You spiral in and out of flow, but you keep on moving, attending to what's happening in the moment, until a problem is solved.

Meta: And this takes time, work, and persistence. It takes grit. Angela Duckworth gave a TED talk on grit that I share with all my students. Here's the URL: https://www.ted.com/talks/angela_lee_duckworth_grit_the_power_of_passion_and_perseverance#t-34479

Merrie: Wait. Let's go to it now. [We hope our readers will do the same.]

Meta: The most fascinating thing Angela Duckworth and her colleagues uncovered in their research was that grit has very little to do with IQ or talent. Instead, it begins with passion and moves forward through perseverance. The "gritty" individual does not quit when bored or tired, or when they fail (2007).

Merrie: Yes! And hard work can require a tremendous expenditure of energy. It's as if the teacher (acting as a conductor) is a heat source that initiates and sustains a convective process suffused with aesthetic energy (Koester, 2015).

Meta: Agreed. I think the concept of energy transfer and flow is a topic about which the arts have a lot to teach us all. In the arts, energy is transferred from the performers to an audience, who are transformed, transported, and moved. Knowing that they will be the source of this energy, performers are told to rest before they go onstage…

Merrie: This is why caring teaching can be so exhausting.

Meta: Yes, and I would add that even more energy is required when the teacher has the courage to say to the student(s) working on a project or learning a new skill: "You're not there yet" (Dweck, 2006). Persistence and hard work are key mindsets for learning, and there can be harm done if praise is given too often for minimal improvement. On the other hand, the worst thing a teacher/coach/conductor can do is allow a student to just give up. Caring is complicated.

Merrie: Several times in our conversation, you have moved us forward by saying "Yes, and" instead of "Yes, but." I recently heard Erica Halverson give a talk at AERA. A practicing actor herself, she explained that these "Yes, and" affirmatory communications are the mainstay of excellent improvisational theatre. In her work, she advocates for the use of improvisation techniques as a "new framework for the design of learning environments," all the while recognizing that doing so may make many educators somewhat uncomfortable (2016). In my own participatory action research, I often encounter teachers who say they are worried that they draw like nine-year-olds. But then, they plunge right in. Significantly, they find that making themselves vulnerable is perceived by their students as a profound act of care.

Meta: Once again, you've described an ethic of care—the honest relationship to the other.

VISUALIZING THE PROCESS OF STEAMIFICATION

Merrie: This book is about STEAM teaching. This chapter is about how doing STEAM might look and feel when that happens. We've brought our readers into a music simulation that allows them to experiment with different features of a spatial music performance, and each has likely experienced the simulation in his or her own way as a factor of which variables they chose to manipulate and the sounds they hypothesized they might be able to achieve. As sound engineers, they used technology to shape different audio outcomes. And, if they wanted to re-produce a particular sound effect, they needed to carefully document their experimental procedures. Depending on their technological and mathematical skills, they might have been able to deduce multiple relationships between variables.

Meta: Indeed. I think we've made an excellent case for why taking the time to employ an arts-integrated approach to STEM teaching/learning is worth the investment. Music is a powerful medium for teaching principles of energy transfer and experimental processes, but I think it has even broader applications. I personally see music composition as a metaphor for the art of teaching. [Picking up a pen and starting to draw]. (See Figure 3).

If you look at this drawing, everywhere you see two dots (a colon), I drew a new spiral. Each colon signifies the beginning of an individual student's experience—wherever he or she may be at the moment. The process is organic, temporal, and spatial.

Merrie: …which means that both time and space have to be factored in by the teacher for growth and development to occur. It may seem like not much is happening (that is, there may be a lot of noise and even confusion to the casual observer), while, in fact, many possible pathways for learning are opening up…

Meta: …and, in a musical performance, there are many different voices, which can be instruments, humans, and even the building itself. There are times for individual voices to be heard and other times when they must blend in harmony, or be altogether silent.

Merrie: Liora Bresler (2005) talked about this when she described music dynamics, or the perception of loud and soft. She explained that context is essential. A loud sound seems louder if it arrives on the heels of silence or a soft tone. Likewise, those interruptions by the winds in the simulation contrasted so vigorously with the strings and the horn that I, at least, had to lower their volume even to hear the other voices.

Meta: That's why music is such a great metaphor for the teaching/learning process. Its dimensions of form, melody, dynamics, and texture, as well as the sheer amount of work that must go into its performance, can be applied in the classroom. But none of this can take place in the absence of an ethic of care…

Figure 3. Thinking through drawing

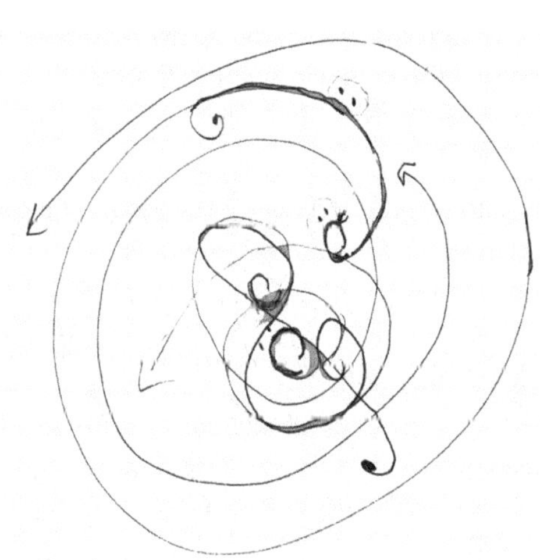

Merrie: …which is at the heart of all aesthetic inquiry (Siegesmund, 2010). And so, we've spiraled back to our beginning premise: Aesthetic inquirers recognize that the creative process (always present at the heart of science) spirals outward through cacophony and chaos, and that neat, harmonious compositions may not always be the result, especially as new and yet unanswered questions emerge. This process is not a linear, inorganic, or prescriptive one. Any number of different outcomes are possible…

Meta: …which should then be critiqued in a way that is empowering and moves the process forward and deepens the learning.

Merrie: There is nothing at all static about the STEAMification process we're describing.

STEAM as Critical Making

Meta: Yes. We want our model to exude possibilities and motion and only be limited by the imagination. With regard to care, it's even more important for a teacher to enact this than to say it (Noddings,1992). From the very beginning, the teacher needs to communicate (model) that his/her students are expected to engage with and deeply explore a concept, like the energy transfer that takes place during this simulation. This models caring for ideas. Noddings talks a lot about this, too. As we've said earlier, while this is happening, the caring teacher also challenges students, offering constructive critique to keep them moving forward.

Merrie: For my part, I'd like to see our model include features of aesthetic inquiry as described by Pestalozzi (1894), Dewey (1938), Greene (2001), and Eisner (2002)—a participatory and inclusive, sense-based, interactive, connecting, and above all, purposeful engagement with the arts, as we did here with this spatial music simulation. I don't see STEAM achieving deep learning unless the dynamics of the art form are in synch with that of the STEM content being explored. When students perceive the arts as an add-on or a frill or prescriptive craft activity, they don't enter into an appreciation of how it makes it their minds better and stronger. They don't get how rigorous the arts can be, especially when critique is valued and practiced.

Meta: That's huge! For example, making an artbot from a kit can then evolve into designing new versions of artbots that can draw in different ways, which students then share with one another.

Merrie: Exactly. I think today's Maker Spaces can epitomize STEAM learning when explicit connections to both arts and STEM knowledge, skills, and practices are made. There needs to be time for students to notice what's happening and then to improvise with the materials. Within reason and safety measures, they can ask, "What if I try to manipulate a different variable than the one directed by the kit instructions? What will happen then?"

Meta: And then you critique again. The process keeps spiraling, and the performance can better and better as skills are developed through practice (Noddings, 1992).

Merrie: We are clearly both channeling Seymour Papert's (1991) and Kafai's (2006) kind of making, in which dialogue and critique go hand-in-hand as part of the overall practice, establishing the relationship between all the components of the teaching/learning system. His theory of constructionism helps us to see that making a public artifact (a presentation or poem, a musical, theatre, or dance performance, a painting, robot, a new piece of software, etc.) informs and transforms learning.

Meta: Yes. Skills are built and knowledge is constructed. Through practice, understanding emerges as you begin to lean into the learning. For example, if I have to learn to sing a specific run in a song, I have to practice it again and again until it's in me. The muscles controlling my vocal cords and diaphragm have literally learned what to do so that, when it comes time for the performance, I don't have to think about it.

Merrie: I so get what you're telling me. We can't expect our students to be prepared for STEM jobs if we don't give them time to learn and practice STEM skills, which include the ability to creatively and collaboratively identify and solve a problem as well as use the equipment and perform the skills necessary for work. I can remember having to throw away countless electrophoresis gels before I made even a single one that could be properly used in an experiment.

Meta: That's what we mean by practice. Once you've got it, it's yours!

Merrie: So often, we just march through a lab activity, follow the steps, and don't allow time for playing with the materials and seeing what else might happen. The music simulation allowed for the important aesthetic component of taking time for noticing. It's a use of technology that has real educational value. When we use technology, I think we need to ask, "How is it moving the lesson forward? Is it part of the problem-solving process and facilitating deeper learning, or is it just another distraction?"

Meta: Exactly! For the Ives simulation to be a meaningful form of arts integration (STEAM), you can't just play with the controls. Otherwise, it's just another fun technology toy or game. You have to have dialogue about what is happening. For example, what does manipulating the technology signify for the one at the controls? For the student, those jabbering winds could signify anything from a nagging mother to a little brother who won't be quiet. You yourself said they reminded you of the student who ceaselessly interrupted you, possibly to call attention away from the fact that he or she could not read.

Merrie: You're right. That's because my own work is about helping the struggling reader find a way into meaning-making in the science classroom. For me, the matter of signification is really important. I think a lot of what we are talking about resides at the intersection of communication and signification. What kinds of messages are being sent and how are they interpreted? How has the experience been organized so that it promotes inclusion rather than exclusion? Francois Tochon, a world language scholar and semiotician, has stated that the form communication symbols take and the way they are presented is the key to whether the educational system functions as intended and deep learning occurs (2013). All too often, students feel left out of community because they can't understand STEM content languages. These languages are not only foreign, they don't seem relevant to their lives. Thus, there's no incentive to do the work to learn them. The arts can readily serve as a common language and as a bridge to meaning-making and deep learning. They can act as media for both confirmation and affirmation.

Meta: This is both empowering and freeing. AND, you can't take this power away once they have it. What we are talking about gives rise to a pedagogy of possibilities, one that is in constant motion and nurtured, with feeling, by building caring relationships of many kinds. This takes courage and energy.

REFERENCES

Adjibolosoo, S. (1995). *The human factor in developing Africa.* Westport, CT: Praeger Publishers.

Ainsworth, S. E. (2006). DeFT: A conceptual framework for learning with multiple representations. *Learning and Instruction, 16*(3), 183–198. doi:10.1016/j.learninstruc.2006.03.001

Bordieu, P. (2003). The forms of capital. In A. H. Halsey, H. Lauder, P. Brown, & A. S. Wells (Eds.), *Education: Culture, economy, society* (pp. 46–58). Oxford, UK: University Press.

Bresler, L. (2005). What musicianship can teach educational research. *Music Education Research, 7*(2), 169–183. doi:10.1080/14613800500169399

Coleman, J. S. (2003). Social capital in the creation of human capital. In A. H. Halsey, H. Lauder, P. Brown, & A. S. Wells (Eds.), *Education: Culture, economy, society* (pp. 80–95). Oxford, UK: University Press.

Csikszentmihalyi, M. (1997). *Finding flow: The psychology of engagement with everyday life*. New York, NY: Basic Books.

Dewey, J. (1938). *Experience and education*. New York, NY: Kappa Delta Pi.

Duckworth, A. L., Peterson, C., Matthews, M. D., & Kelly, D. R. (2007). Grit: Perseverance and passion for long-term goals. *Journal of Personality and Social Psychology*, *92*(6), 1087–1101. doi:10.1037/0022-3514.92.6.1087 PMID:17547490

Duckworth, E. (2006). *The having of wonderful ideas: And other essays on teaching and learning* (3rd ed.). New York, NY: Teachers College Press.

Dweck, C. (2006). *Mindset: The new psychology of success*. New York, NY: Ballantine Books.

Eisner, E. (2002). *The arts and the creation of mind*. New Haven, CT: Yale University Press.

Finson, K., & Pederson, J. (2011). What are visual data and what utility do they have for science education? *Journal of Visual Literacy*, *30*(1), 66–85. doi:10.1080/23796529.2011.11674685

Greene, M. (2001). *Variations on a blue guitar: The Lincoln Center Institute lectures on aesthetic education*. New York, NY: Teachers College Press, Columbia University.

Halverson, E. (2016). *The art of improvisation in teaching and learning*. Paper presented at the American Educational Research Association Conference, Washington, DC.

Hetland, L., Winner, E., Veenema, S., & Sheridan, K. (2013). *Studio thinking 2: The real benefits of visual arts education*. New York, NY: Teachers College Press.

Kafai, Y. B. (2006). Constructionism. In K. Sawyer (Ed.), *Cambridge handbook of the learning sciences* (pp. 35–46). Cambridge, MA: Cambridge University Press.

Koester, M. (2015). *Science teachers who draw: The red is always there*. Blue Mounds, WI: Deep University Press.

Koester, M. (in press). Deep learning in science: An argument for an aesthetic paradigm of science education. In F. V. Tochon & D. Busciglio (Eds.), *Deep Education Across the Disciplines and Beyond: A 21st Century Transdisciplinary Breakthrough*. Blue Mounds, WI: Deep University Press.

Lederman, N. G., & Lederman, J. S. (2015). What is a theoretical framework? A practical answer. *Journal of Science Teacher Education*, *26*(7), 593–597. doi:10.1007/s10972-015-9443-2

Newmann, F. (1981). Reducing student alienation in high schools: Implications of theory. *Harvard Educational Review, 51*(4), 546–564. doi:10.17763/haer.51.4.xj67887u8715t66t

NGSS Lead States. (2013). *Next generation science standards: For states, by states.* Washington, DC: The National Academies Press.

Noddings, N. (1992). *The challenge to care in schools.* New York, NY: Teachers College Press.

Noddings, N. (2003). *Caring: A feminine approach to ethics and moral education.* Berkeley, CA: University of California Press.

Papert, S., & Harel, I. (1991). Situating constructionism. In I. Harel & S. Papert (Eds.), *Constructionism* (pp. 1–11). Norwood, NJ: Ablex.

Pestalozzi, J. H. (1894). *How Gertrude teaches her children. Syracuse, NY.* W.: Bardeen.

Sartre, J.-P. (1976). *Anti-Semite and Jew* (G. J. Becker, Trans.). New York, NY: Schocken. (Original work published 1948)

Siegesmund, R. (2010). Aesthetics as a curriculum of care and responsible choice. In T. Constantino & B. White (Eds.), *Essays on aesthetic education for the 21st century* (pp. 81–92). Rotterdam: Sense Publishers.

Tochon, F. V. (2013). *Educational semiotics: Signs and symbols in education.* Blue Mounds, WI: Deep University Press.

Trefil, J. (2008). *Why Science?* New York: Teachers College Press.

Uhrmacher, P. B., Conrad, B., & Lindquist, C. (2010). The power to transform: Implementation as aesthetic awakening. In T. Constantino & B. White (Eds.), *Essays on aesthetic education for the 21st century* (pp. 183–203). Rotterdam: Sense Publishers.

Van Sickle, M., & Spector, B. (1996). Caring relationships in science classrooms: A symbolic interaction study. *Journal of Research in Science Teaching, 33*(4), 433–453.doi:10.1002/(SICI)1098-2736(199604)33:4<433::AID-TEA5>3.0.CO;2-T

Wright, P. (2016). *The unmixed question: the spatial music of Charles Ives.* American Mavericks. Retrieved from http://musicmavericks.publicradio.org/features/feature_unmixed_question.html

KEY TERMS AND DEFINITIONS

Aesthetics of Care: Using the arts to develop an appreciation of how learning through them can create multisensory ways into learning. In this work, we try to show how the aesthetics of care is directly linked to an ethic of care.

Aesthetic Inquiry: We employ Maxine Greene's conceptualization of this concept, which features a curriculum that begins with and is sustained by noticing what there is to be noticed. The lesson progresses through a participatory engagement with the arts, as new connections are made and conceptual relationships evolve.

Axiology: The philosophical study of value. It typically involves the discussion of both ethics and aesthetics.

Constructionism: A learning theory developed by Seymour Papert, in which he contends that you have to make it to learn it. Learning is achieved through the making of a public artifact, which is then critiqued. This shareable construction may take the form of a robot, musical composition, paper mache volcano, poem, conversation, or new hypothesis. We believe that this theory holds a key to understanding and developing the potential of what is now being called a Maker Movement.

Critique: A process described by Yasmin Kafai as the ways in which a learning artifact is analyzed and reviewed in a social context. In the process, connections between old and new knowledge can be made.

Dialectic: The exploration of seemingly contradictory ideas and the relationships between them. It employs logical argumentation techniques to search for a unity of opposites.

Dissonance: In music, the sine waves of pitch interfere with each other to the point that the sound may cause the human ear to react with discomfort. As a metaphor for teaching and learning, being off-pitch can have a deadening effect on the lesson.

Dynamics: Musical conductors use dynamics to achieve entirely different effects. Dynamics can include anything from loudness and softness (volume) to fast and slow (tempo).

Ethic of Care: In this chapter, we compare the ethic of care to the Hippocratic oath of doing no harm. An ethic of care always asks the question of what happens in a relationship. When there is an ethic of care, a system can thrive. When students, the primary actors in an educational system, don't feel cared for, they stop interacting, and they stop learning (Newman, 1981). They become alienated from the adults and the ideas. They may even retaliate against the equipment and each other. They'll either act out or shut down.

Resonance: In music, resonance is the compounding of sound waves so that what you get is more than the sum of the parts. As a metaphor for teaching and learning, resonance is the achievement of synergy— when all the parts of the system unite to produce a level of learning not possible if the elements were separated. It often manifests as joy and exuberance. It's a beautiful thing.

Semiotics: The study of the meaning-making process, or semiosis. In Peirce's theory of signs, semiosis depends on the triadic interaction between an object in the world, a sign or symbol that re-presents that object (or phenomenon), and the way(s) a person in a given culture or situation interpret(s) that contextual sign.

Spatial Music: A modern music composition technique that purposefully uses the location of the instruments to achieve different effect. Charles Ives's "Unanswered Question" of 1906 was the first ever 20th century composition to feature spatial separation as a major element.

Studio Habits of Mind: A conceptualization of the epistemology and practices of the artist as they inform learning. The writing on this subject emerged from the Studio Thinking Project, part of Project Zero, at Harvard University (http://www.pz.harvard.edu/projects/the-studio-thinking-project). The investigators established eight studio habits of mind: 1) develop craft, 2) engage and persist, 3) envision, 4) express, 5) observe, 6) reflect, 7) stretch and explore, and 8) understand arts community.

Voice: At a minimum, this is the utterance of the possible sounds a human can make. At a maximum, voice is the ability to express personal ideas.

Section 2
Art Practice First

Chapter 2
Graphic Novels and STEAM:
Strategies and Texts for Utilization in STEAM Education

Alex Romagnoli
Monmouth University, USA

EXECUTIVE SUMMARY

Focusing on the interdisciplinary connections between STEAM education and graphic novels, this chapter first establishes historical and educational contexts for the use of graphic novels in STEAM education. A literature review focusing on the use of graphic novels in a science course as well as how graphic novels have been used in secondary classrooms will be discussed. Literature that is reviewed includes books, articles, and edited volumes. The strategies for implementing graphic novels in STEAM education promote constructivist learning as students are asked to access their intellectual and cultural capital in order to ascertain meaning from given content. Additionally, multimodality and multiliteracies are highlighted throughout the article. Finally, this chapter ends with linking the use of graphic novels in STEAM education to the power of narrative inquiry in educational contexts.

DOI: 10.4018/978-1-5225-2334-5.ch002

Copyright ©2017, IGI Global. Copying or distributing in print or electronic forms without written permission of IGI Global is prohibited.

INTRODUCTION

In *All-Star Superman* (Morrison & Quitely, 2011), the superhero Superman finds himself in a laboratory full of amazing and bizarre science experiments. After having saved pioneer astronauts who were exploring the outer atmosphere of The Sun, Superman tests his strength levels in this futuristic menagerie of scientific extravagance. The head scientist tells Superman, "You're now pushing against the equivalent of 200 quintillion tons, Superman" (p. 20). However, Superman's trip to The Sun to save the astronauts had a side-effect:

Your trip to the Sun exposed you to critical levels of stellar radiation, more raw energy than your cells are able to process efficiently. Apoptosis has begun. Cell death. There can be only one outcome even for You. (p. 20)

Superman's exposure to The Sun is more than he can take, and within one year, The Man of Steel will meet his end. This is the premise of Morrison and Quitely's acclaimed graphic novel *All-Star Superman* which has been used as a text to teach composition (Romagnoli, 2015).

The value of this text doesn't end with the story, though. Looking back at the scenario that was just presented, unmistakable teaching moments have arisen. 200 quintillion tons? Solar atmosphere? Apoptosis? Cells processing solar energy? A science lesson has just occurred.

LITERATURE REVIEW

The STEAM and Comic Connection

Comic books have traditionally been associated with science and technology, and Jones (2004) even traced the connection of science and popular literature to the "pulps" of the early 20th century:

In the archaeology of popular culture, the August 1928 issue of Amazing Stories appears again and again as a pivotal memory of a generation of moviemakers, science fiction writers, cartoonists, astronomers, futurists, and rocket engineers. Against a lucid yellow background floats a man in a skintight red costume, leather pilot's helmet, and sleek black boots, his body prone but angling upward in an at titude of nascent flight. (p. 30)

Most important to note from Jones' historical study of comic books is his implicit connection of various and diverse disciplines: cartoonists, astronomers, moviemakers, and rocket engineers. This is the essence of STEAM education: the idea that the logistics of scientific realities are informed by the aesthetics of the imaginations of the scientists and engineers who are responsible for the creation and implementation of new machines. Most importantly, this has been the case for a long period of time.

The graphic novel, "A comic style narrative, fictional or nonfictional, that is either a collection of individual comic book issues or an extended, original work" (Romagnoli, 2013), has transcended its pulp roots to inform culture, popular media, and education. Tatalovic (2009) has presented a proverbial call to action for both educators and scientists about incorporating graphic novels and comics into science education. This incorporation is to help bridge the gap between those who work in the sciences (engineers, scientists, mechanics, educators, etc.) and those who do not (students, general public). Recognition of the benefits that comics possess for education is becoming more apparent in science education:

from my exploratory research of science comics, it appears that scientists and educators are becoming more aware of the appeal that comics hold for young people and are starting to use them more as a vehicle to communicate scientific ideas. (p. 4)

The benefits of the comics lies in the very structure of the medium which incorporates multiple modes of interpretation for readers to interact with.

The unique combination of visual stimuli with accompanied narration and dialogue allows for students to access multiple modes of interpretation and comprehension that might otherwise be missing in a monomodal text. An interesting way of approaching graphic novels is to think of them as traditional novels that have the setting and the action drawn instead of written. Some educators may view that as blasphemous, but the visual nature of graphic novels provide unseen benefits to students who are becoming increasingly versatile in their abilities to interpret meaning from varying sources, both traditional and academically unestablished.

Even though comics have traditionally been relegated to the proverbial slums of literary canon because of their sophomoric content and/or emphasis on superheroes, their structure can be beneficial to students because of their visual characteristics. Specifically, graphic novels access students' "multiliteracies" (Cazden et al., 1996) which have become a mainstay in literary studies over the past twenty years.

Multimodality, Graphic Novels, and the STEAM Connection

Even though comic books have been a mainstay of popular media since the early twentieth century, it wasn't until 1978 when comic book artist Will Eisner wrote *A*

Contract with God which became what is widely considered to be the first graphic novel. This text informed what became Will Eisner's professional thesis on comic art and storytelling, mainly *Comics and Sequential Art* (1985). Eisner's focus as a creator of what he termed "sequential art" was on the relationship between text and images which he said were, "irrevocably interwoven" (p. 122). The connection that Eisner speaks of is symbiotic as, "the words become welded to the image and no longer serve to describe but rather to provide sound, dialogue and connective passages" (p. 122).

McCloud's seminal work *Understanding Comics* (1993) deconstructed comic storytelling and further highlighted the important connection between text and art. However, McCloud's commentary on this relationship informs Tatalovic's (2009) assertion that comics can help bridge knowledge gaps. McCloud sums up the popular argument against comics this way: "Words and pictures together are considered, at best a diversion for the masses, at worst a product of crass commercialism" (p. 140). The irony of this statement is that as technology evolved, the importance of visuals only increased because, "words and moving pictures have half the world in thrall to their charms…" (p. 141).

The New London Group (Cazden et al., 1996), which was comprised of elite compositionists from around the world, deemed it necessary to coin a new term to address the evolving nature of media and the eventual dominance of the internet and computers as the premiere form of communication and entertainment. "Multi-literacies" was established, and it embodied a new set of views on how audiences interpret media. Traditionally, literacy, "has referred to reading and writing of printed text (Westby, 2010, p. 65). The privileging of printed text over more dynamic and visual forms of media had been powerful in schools, and that prevailing notion of accepting printed text over forms of media is still present. The New London Group saw literacy as being multifaceted especially in terms of pedagogy:

> *A pedagogy of multiliteracies, by contrast, focuses on modes of representation much broader than language alone. These differ according to culture and context, and have specific cognitive, cultural, and social effects. (p. 64)*

For STEAM education, this has profound repercussions. Jolly (2014) states, "The purpose of STEAM should not be so much to teach art but to apply art in real situations. Applied knowledge leads to deeper learning." This mirrors The New London Group's view on how multiliteracies work where readers possess multiple literacies that work in tandem with one another in order to achieve a holistic understanding of a given text.

The majority of research and literature on incorporating graphic novels into educational contexts has been in Language Arts (Carter, 2007; Eckard, 2015; Romagnoli

& Pagnucci, 2013). Social Studies also has become a popular subject in which to utilize graphic novels. However, mathematics and the sciences are just beginning to assess the potential of graphic novels.

Brozo, Moorman, and Meyer (2014) have also outlined ways in which graphic novels can be used in STEAM education. The benefit of using graphic novels in science courses is that "with their blend of dialogue and images, [they] allow students to gain and interpret science concepts and information through a unique and accessible format, opening a window to science for even the most reluctant learners" (p. 77). The "blend of dialogue and images" is akin to "multiliteracies," and it highlights the very importance of the multiple forms of interpretation and creation that graphic novels embrace. With mathematics, the authors offer a unique view of how mathematics can benefit from graphic novels: "We would like to propose a vision for mathematical teaching and learning that moves beyond routine knowledge to something closer to socially constructed mathematical inquiry" (p. 100). Interestingly, Brozo, Moorman, and Meyer are not the only educators who feel this way about mathematics and graphic novels.

Reilly (2015) presents a way in which a middle school mathematics course can utilize graphic novels to help students understand mathematical concepts. Reilly had students in an eighth grade math course create "three comic books with various themes. All of the comic book adventure themes involved heroes working together to prevent some disaster from occurring" (p. 67). Students were specifically tasked with writing a story "which focused on the overall theme of fully developing one's talents in the face of societal pressure to conform" (p. 66). Reilly connects this socially conscious angle of her assignment to the superhero team The X-Men who are faced with fear by humans who do not fully understand them. This assignment provides three significant forms of literacy:

1. **Social Literacy:** Students are required to consider social issues while creating the mathematics themed graphic novel
2. **Mathematical Literacy:** Students are required to fully explain a mathematical concept in the context of their story
3. **Creative Literacy:** Students are required to create a comic book with both text and visuals.

All of the components of STEAM education are present in this assignment and blends the various disciplines into a cohesive and powerful unit for learning mathematics. And, as The New London Group (Cazden et al., 1996) asserts, the students are able to access their multiliteracies to ensure understanding of complicated mathematical concepts.

In science education, the graphic novel has become a means to instill interest in the subject and make complicated scientific principles intellectually accessible for all students. Tatalovic (2009), whose call to action to bridge this proverbial gap was cited earlier, has more concrete commentary on the benefits of utilizing graphic novels in scientific education. The benefits of graphic novels with science relate to issues of accessibility of content and generating interest. Tatalovic asserts that, "from my exploratory research of science comics, it appears that scientists and educators are becoming more aware of the appeal that comics hold for young people and are starting to use them more as a vehicle to communicate scientific ideas" (p. 4). However, Tatalovich also warns of incorporating graphic novels or comics that may be perceived as being too loaded in their agendas: "Similarly, some have even suggested replacing textbooks with comics. But if the text in comics is overly didactic, it may easily lose connection with the story that is being told resulting in a less interesting read" (p. 13). There needs to be a balance in how graphic novels are used in STEAM education.

Reilly (2015), whose mathematics class was discussed earlier, maintains this balance through the students creating their own graphic novels. By creating their own graphic novels, Reilly's participants had intellectual and cultural capitals invested in both the story *and* the content. This is the key to integrating graphic novels into the STEAM education: balancing the entertainment of the story with the scientific/mathematic principles that are expected to be understood given a particular curriculum.

Texts and Activities for STEAM Education and Graphic Novels

When discussing the activities and graphic novels that are useful to STEAM education, the overarching benefit here for students is multiple forms of interpretation for complicated content. The combination of the visuals and text, which graphic novels are revered for, is the prime benefit here for students. Traditionally, much of the content students studied in their courses would be monomodal: one form of interpretation. For example, a traditional prose text would be monomodal because it only employs text to convey meaning whilst an instruction manual could be multimodal because it employs both text and images to convey meaning. In providing multiple ways to interpret meaning from a text, a reader's multiliteracies are accessed. Having established the importance of multiliteracies and multimodality when it comes to STEAM education, the following are examples of powerful texts teachers can utilize for various disciplines. This is by no means an exhaustive list, but it is a significant starting point for STEAM educators.

Biology and Evolution

Evolution can be a startlingly complicated concept for students to grasp, and evolution's place in curriculums around the country is sometimes challenged for socio-political and socio-cultural reasons. These issues notwithstanding, Charles Darwin's *The Origin of Species*, which was originally published in 1859, serves as the premiere text for studying both the historical context for this scientific theory and the science behind the theory.

The original text, which clocks in at 502 pages, may prove unwieldy to some students, and that is understandable given the language of the original text: "It need not be supposed that all varieties or incipient species attain the rank of species. They may become extinct, or they may endure as varieties for very long periods, as has been shown to be the case by Mr. Wollaston with the plants by Gaston de Saporta" (Darwin, 1909, p. 68-69). Citing this language is not to criticize the work of Darwin or downplay the significance of the text. Rather, citing this language presents what students would be tasked with were they required to read Darwin's text. This leads to how graphic novels may help students with access to Darwin's important and seminal work.

Charles Darwin's On The Origin of Species: A Graphic Interpretation (2009) is written and illustrated by Michael Keller and Nicolle Rager Fuller and presents the story of Charles Darwin's original scientific observations with beautiful illustrations that provide multiples forms of interpretation for students to analyze.

One particularly powerful example of the benefits of this book is when Darwin's Laws of Variation are explored. The Laws of Variation, which looks to clarify why certain species vary in appearances and builds, is particularly interesting. This law helps students understand why all animals of a given species have similar characteristics (i.e. birds and their wings) but look decidedly different and possess varying degrees of attributes (i.e. which birds can fly and which cannot). Keller and Fuller present the following panels to help explain this (Figure 1 and Figure 2).

The visual of the canine's fur changing as it goes farther north, coupled with the original narration from Darwin, helps solidify the concept of variation. In short, the canine needs less fur if it lives in a warmer climate. Complimenting this is the next set of panels which elaborates on the effects of climate on the evolution of a particular species.

The multimodality of this section is a highlight of a text that is rife with these multimodal explanations of Charles Darwin's theory of evolution. Intertwined with the evolutionary theory is Darwin's actual story during his study. Charles Darwin's story, which provides even more context for students as they read about his evolutionary theories, encapsulates the very spirit of STEAM education. Incorporating the stories of a given scientist while he/she was developing ground breaking discoveries

Figure 1.

Figure 2.

contextualizes the struggles, both scientific and social, that many brilliant minds faced throughout history.

To complement the study of this text, one activity a teacher could utilize is having students create their own sections on Charles Darwin's evolutionary theories. The graphic adaptation is very extensive, but it is by no means complete. Clocking in at 189 pages, *Charles Darwin's On The Origin of Species: A Graphic Interpretation* leaves parts of Darwin's theory out; this is where the students come in. A teacher could explore the original text on his/her own and provide short excerpts for students to create their own graphic novels about. Not only would this encourage students to interpret a text using multiliteracies, it extends the discussion on the topic of Darwin and evolution through creation. Additionally, the creation of an original graphic novel section informs the very nature of STEAM education, particularly as it relates to the "A." By having students create these original sections, they are simultaneously exploring advanced scientific concepts whilst creating stories through their art.

Mathematics and the Moon Landing

One of the most significant events in American (and world) history was when Apollo 11 successfully landed on the moon and astronauts walked on the surface. This historical event had implications that extended beyond the science and math that was necessary for astronauts to safely make it to the moon and back again. With the backdrop of the Cold War and the Space Race, The United States and The Soviet Union competed for political and military dominance on Earth with the battlefield being beyond the planet.

The scientific lessons that are associated with the moon landing are apparent, and the study of the moon landing serves as the basis for studies of gravity, the moon itself, and how the moon's proximity to Earth affects the planet. However, the math of this human achievement is unmistakably important for students to explore in some high school classrooms and beyond.

T-Minus: The Race to the Moon (Ottaviani, Cannon, and Cannon, 2009) tells the story of the astronauts, scientists, politicians, and mathematicians who were responsible for getting the United States to the Moon with Apollo 11 in 1969. Throughout the graphic novel, there are sequences that outline both the scientific and mathematical concepts that both scientists and mathematicians had to navigate throughout the preparation process.

One of the major factors that needed to be addressed for astronauts and rocket builders was the amount of G force(s) an astronaut could withstand. This becomes both a scientific and mathematical problem for students to explore. Scientifically, understanding how gravity affects the body informs the study of space travel for humans in general. Mathematically, understanding how to calculate G forces addresses the reality of space travel and how understanding mathematics plays into that reality. One poignant scene from the graphic novel is when a cosmonaut is put into a centrifuge for training purposes (Figure 3).

Below the image of the cosmonaut who is clearly distressed by the training is a definition and example of "G-Force" from which students can begin to explore further. However, the image of the cosmonaut in fear does wonders for a student to connect the mathematics of gravity to the insanity of space flight.

One page later, this same cosmonaut is put in a pressure chamber to test his ability to concentrate during a crisis in space travel. This test involves mathematics (Figure 4).

Figure 3. (p. 27)

Figure 4. (p. 28)

The extreme conditions the cosmonaut is put through are intended to mimic what a space traveler would have to navigate if they were actually in space and an emergency arose. Additionally, having to solve math problems in rapid succession during a given crisis in space makes sense should the ship's trajectory need to be adjusted or the cosmonaut would have to make quick, life-saving decisions using data from the telemetry of the ship. However, one major factor that students reading this text in a math class have to explore is the absence of calculators. Not having a calculator to complete arithmetic becomes a discussion point in class which has sociocultural implications. In a math class, where conversation is stereotypically associated with strictly quantitative reasoning, students are able to apply experiences to support assertions concerning the use of a calculator.

Additionally, *T-Minus'* multidisciplinary content encourages cross-curricular exploration of themes. This particular graphic novel has literary, scientific, political, and mathematical themes embedded throughout, and this highlights the symbiotic relationship all of the disciplines share. One of the major disciplines that is cited in *T-Minus* is also present in other graphic novels: chemistry.

Chemistry and the Graphic Novel

The prominent visual for chemistry is the Periodic Table of Elements. The table, which displays the atomic details of all of the known elements that are the building blocks for all matter, is a mainstay in chemistry and science classrooms around the world. Each of the elements on that table is a story unto itself, and how those elements were discovered and what that discovery meant for mankind is as interesting as the element itself. In short, chemistry involves both the technicalities of the elements and the stories behind those elements.

Gonick and Criddle (2005) took the history of chemistry and told the story of its evolution in *The Cartoon Guide to Chemistry*. This graphic novel represents a unique way to explore chemistry with humorous illustrations and comments. Within the first two pages of the graphic novel, Gonick and Criddle summarize the field of chemistry:

Fire – and those other processes – revealed ***hidden features of matter****. If you heat a piece of wood, all you get is a hot piece of wood, at first...but suddenly, at some point, the wood bursts into flame. Where did* ***that*** *come from?* ***Chemistry*** *is the science that answers that question, and* ***chemical reactions*** *are the strange transformations that reveal matter's* ***hidden properties****. (p. 2)*

Accompanying this explanation is a humorous cartoon of two cave people confusingly observing wood burning. *The Cartoon Guide to Chemistry* illustrates complicated concepts whilst explaining the concept in a humorous and accessible way.

One of the concepts that is explored is the three stages of matter. Understanding the differences between solids, liquids, and gasses is a fundamental concept that involves complicated science for students. Identifying solids, liquids, and gasses is easy enough; taking an ice cube and placing it in a hot pan shows the process occur. However, the science of these states is something else entirely. Gonick and Criddle describe the states using Figure 5.

Placing the character in oversized jars and explaining how the states of matter work is an effective visual for students. It also serves as a great discussion piece about why the characters are reacting the way they are. Even before exploring temperature, pressure, and other factors that influences states of matter, students are free to propose their own hypotheses as to how states of matter work.

This is the prime benefit of this particular text: it encourages discussion through the images. More so than the other graphic novels that have been cited throughout this chapter, *The Cartoon Guide to Chemistry* serves as a discussion starter.

Figure 5. (p. 105)

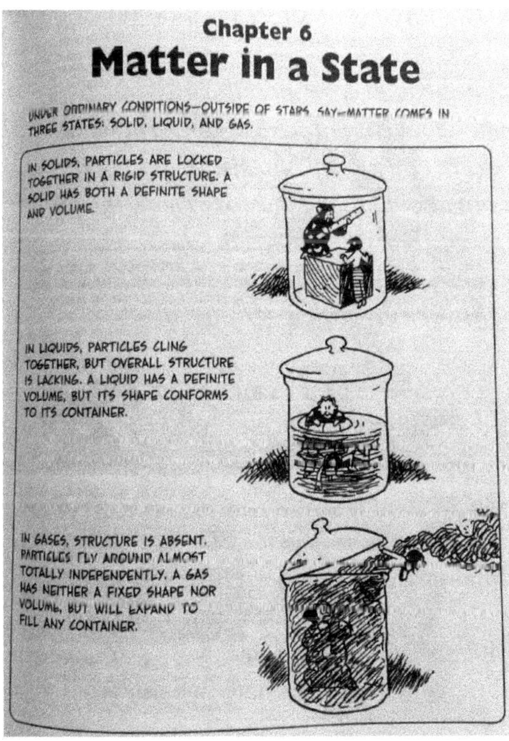

Throughout the text, there are cartoons that visualize chemical processes and allow for class discussions.

One activity that would be beneficial to students is to use this text as a starter for beginning lessons on chemistry concepts. With chemistry, the instructor's expertise and experience in the field serves as the primary resource, but *The Cartoon Guide to Chemistry* is excellent for beginning conversations and accessing students' intellectual capital when it comes to chemistry.

CONCLUSION

When looking back on the graphic novels that have been cited in this chapter, it becomes apparent that all of the graphic novels have one characteristic in common that transcends all of the disciplines; they are stories. While stories and storytelling are often limited to English or history classes, they are the core of human experience, and those experiences account for the scientific and mathematical discoveries that students explore in STEAM education.

For STEAM education, making the content accessible to students of all experiences and background is of paramount importance, and this mentality reflects the reality of what it means to be a scientist. Kakalios describes what it means to be a physicist in the book *The Physics of Superheroes* (2005):

Despite the impression gleaned from popular movies, being a physicist does not require an encyclopedic knowledge of equations and fundamental constants, coupled with the ability to perform complex arithmetic in one's head with robotic speed and precision. Physics is not about having memorized all the answers, but rather about asking the right questions. For when the right question is posed of a phenomenon, either the answer becomes clear, or at least the manner as to how one should go about obtaining the answer is revealed. (p. 16)

In short, science is about experience and sharing those experiences with each other through stories that enlighten, educate, and connect. Graphic novels serve as a positive and multimodal option for educators and students alike.

Pagnucci (2004) has stated that, "Stories connect us to other people, and we have a responsibility to respond to those stories." (p. 76). When students are accessing their multiliteracies and engaging with multimodal texts to explore complicated and significant scientific and mathematical principles, they are taking part in the discovery of these principles. Their own discoveries are occurring through the text and the images, and graphic novels facilitate this access to discovery.

This is an important distinction to mark when it comes to STEAM education and graphic novels. Utilizing graphic novels in classes is about the balance of literacies and recognizing diverse ways to understand concepts; that is all very important. But a pattern has started to emerge where the perceived coldness and rigidity of the sciences and mathematics is being dispelled. Science is not just lab coats, hypotheses, beakers, and experiments. Mathematics is not just equations, proofs, memorized operations, and prime numbers. Science and mathematics are people, and the experiences of people affect the sciences and mathematics. More so than the multimodality of the comic books, the stories with this literature helps to humanize what are sometimes perceived as robotic subjects. There is beauty in constellations, exploding stars, sea life, and cellular reproduction. There is beauty in perfect proofs, statistics, and pi. There is true beauty in all of the people who sacrificed their lives, their fortunes, and sometimes their well-being in the noble efforts to simply inform the world of how the universe works and why the world should care. There is beauty in knowing these stories and reflecting on not just the discoveries, but the lives of these scientists who made it their missions in life to better the world through discovery and experimentation.

REFERENCES

Brozo, W. G., Moorman, G., & Meyer, C. K. (2014). *Wham! Teaching with graphic novels across the curriculum*. New York, NY: Teachers College Press.

Carter, J. B. (Ed.). (2007). *Building literacy connections with graphic novels: Page by page, panel by panel*. Urbana, IL: National Council of Teachers of English.

Cazden, C., Cope, B., Fairclough, N., & Gee, J et al.. (1996). A pedagogy of multiliteracies: Designing social futures. *Harvard Educational Review*, *66*(1), 60–92. doi:10.17763/haer.66.1.17370n67v22j160u

Darwin, C. (1909). *The origin of species*. New York, NY: P.F. Collier & Son Company.

Eckard, S. (Ed.). (2015). *Yin and yang in the english classroom: Teaching with popular culture texts*. Lanham, MD: Rowman & Littlefield.

Eisner, W. (1978). *A contract with God*. New York, NY: Baronet Books.

Eisner, W. (1985). *Comics & sequential art: Principles & practice of the world's most popular art form*. Tamarac, FL: Poorhouse Press.

Gonick, L., & Criddle, C. (2005). *The cartoon guide to chemistry*. New York, NY: Collins Reference.

Jolly, A. (2014). *STEM vs. STEAM: Do the arts belong?* Retrieved from http://www.edweek.org/tm/articles/2014/11/18/ctq-jolly-stem-vs-steam.html

Jones, G. (2004). *Men of tomorrow: Geeks, gangsters and the birth of the comic book*. New York, NY: Basic Books.

Kakalios, J. (2005). *The physics of superheroes*. New York, NY: Gotham Books.

Keller, M., & Fuller, N. R. (Illustrator). (2009). Charles Darwin's on the origin of species: A graphic interpretation. Rodale.

McCloud, S. (1993). *Understanding comics*. Allenspark, CO: Tundra Publications.

Morrison, G., & Quitely, F. (2011). *All star superman*. New York, NY: DC Comics.

Ottaviani, J., Cannon, Z., & Cannon, K. (2009). *T-Minus: The race to the moon*. New York, NY: Aladdin.

Pagnucci, G. S. (2004). *Living the narrative life: Stories as a tool for meaning making*. Portsmouth, NH: Heinemann.

Reilly, E. (2015). Superheroes in math class: Using comics to teach diversity awareness. *Works & Days*, *32*(1 & 2), 61–74.

Romagnoli, A. (2015). Up, up, and away: Superman in the composition classroom. In S. Eckard (Ed.), *Yin and yang in the english classroom: Teaching with popular culture texts* (pp. 123–135). Lanham, MD: Rowman & Littlefield.

Romagnoli, A., & Pagnucci, G. (2013). *Enter the superheroes: American values, culture, and the canon of superhero literature*. Lanham, MD: Scarecrow Press.

Tatalovic, M. (2009). Science comics as tools for science education and communication: A brief, exploratory study. *JCOM: Journal of Science Communication*, *8*(4), 1–17.

Westby, C. (2010). Multiliteracies: The changing world of communication. *Topics in Language Disorders*, *30*(1), 64–71. doi:10.1097/TLD.0b013e3181d0a0ab

KEY TERMS AND DEFINITIONS

Accessibility: The theory that all students are able to read and interact with a given text so as to be able to understand a given and/or participate in the learning process.

Comic(s): A Shorter comic style narrative that serves as a short story or part of an ongoing serial or series.

Graphic Novels: A comic style narrative, fictional or nonfictional, that is either a collection of individual comic book issues or an extended, original work.

Monomodal Text: Texts that contain a single mode of interpretation (novel, poem, etc.).

Multiliteracies: The theory proposed by The New London Group (Cazden et al., 1996) that readers and learners possess multiple literacies that work in tandem in order to ascertain understanding or synthesize meaning from a multimodal text.

Multimodal Text (Multimodality): Texts that contain multiples modes of interpretation that work in tandem to present a story or concept series.

Chapter 3
High-Quality Trade Books and Content Areas:
Planning Accordingly for Rich Instruction

Carolyn A. Groff
Monmouth University, USA

EXECUTIVE SUMMARY

Integrating high-quality children's tradebooks into elementary content areas has long been considered a best-practice (Olness, 2007). When teachers choose to incorporate these texts into content area lessons, they are exposing students to art through the pictures and reaching an array of visual learners. Hoffman, Collins and Schickedanz (2015) state that teachers have difficulty discussing the concepts presented in the books. The goal is to strike a balance between the literacy skills and strategies needed to read the informational text, and the concepts that must be discussed (Hoffman, Collins & Schickedanz, 2015). In order to increase students' understanding, teachers must be able to successfully merge their book selection with a carefully scaffolded lesson plan (Fisher & Frey, 2015). This chapter presents a lesson plan template that assists teachers in planning for integrated instruction.

DOI: 10.4018/978-1-5225-2334-5.ch003

Copyright ©2017, IGI Global. Copying or distributing in print or electronic forms without written permission of IGI Global is prohibited.

CURRENT CHALLENGES FACING TEACHERS

Choosing high-quality informational texts in the elementary grades can be overwhelming and time-consuming. Often curricula materials may come with their own texts which do not involve as much planning. However, these texts can be dull, especially in an artistic sense, and often do not accommodate the range of concepts which need to be conveyed in a unit (Hoffman, Collins & Schickedanz, 2015). While teachers may possess excellent pedagogical skills, their lack of practice in cross-curricular integration may prevent them from wanting to attempt such endeavors, especially given the high-stakes environments in which many teachers work (Hoffman, Collins & Schickedanz, 2015). According to Hoffman, Collins and Schickedanz, "Even highly skilled, experienced teachers found discussion of concepts during read- alouds challenging in practice" (365). Teachers may become too focused on the literacy skills and strategies involved in reading informational text and overlook the possibilities for engaging in rich discussions about the concepts (Hoffman, Collins & Schickedanz, 2015). With many school districts focusing on the preparation of students for standardized tests in literacy and math, other content areas can be overlooked, and teachers are forced to weave this knowledge into literacy instruction time. Therefore, the teachers lack practice in providing full lessons in other content areas such as science.

LITERATURE REVIEW

According to Adams and Pegg (2012), literacy and content area learning are in a constant state of change: "One dimension relates to shifting understandings regarding student learning, and the other dimension involves the relationship between content and literacy" (p. 151). Citing the work of previous researchers, Adams and Pegg argue that discipline-specific discourse can be taught through literacy practices; that is, the way the students talk about math, science, engineering and technology can be learned through literacy practices in the classroom. Moreover, since discipline-specific discourse are used by authors to create informational texts, using such texts can lead to increased understanding in content areas.

Integrating high quality children's trade books into elementary content areas has long been considered a best-practice (Olness, 2004). More specifically, high-quality informational texts can assist students in learning concepts in the STEM areas; using these texts with rich visuals and vocabulary forges a link between the arts and the STEM focus, creating the use of STEAM. Yopp and Yopp (2012) state that students' engagement with informational texts more richly promotes their knowledge of content areas by teaching them vocabulary, text structure (such as compare and contrast), and text features (such as charts and tables). Teachers need to be able to merge their selection of a high-quality informational text and a scaffolded instructional model in order for students' experiences to be successful (Fisher & Frey, 2015). Fisher and Frey (2015) state, "as students invest themselves in these informational texts, look for opportunities to encourage ways to inspire further investigation, discussion, and writing" (p. 529). However, for pre-service teacher candidates and novice teachers, planning for the blend of literacy and other content areas may not come naturally (Hoffman, Collins & Schickedanz, 2015). Citing prior research, Hoffman, Collins and Schickedanz (2015) state the following with regard to science instruction:

Research on effectively integrating science and literacy instruction to develop conceptual understandings uses informational text as a complement to hands- on scientific investigations in the science curriculum, not as a replacement for such experiences (Palincsar & Magnusson, 2001 ; Pappas & Varelas, 2004 ; Varelas & Pappas, 2006). Thus, informational texts are most meaningfully integrated in literacy instruction when they align with the ongoing science curriculum (i.e., the texts are related to the scientific concepts investigated in hands- on experiences). (pp.368-9)

Taking this idea of informational text integration into STEM content area lessons, the lesson plan template provided in this chapter is meant to assist teacher candidates and practicing teachers in preparing for the integration of high-quality informational texts into STEM content area lessons, focusing on the integration of visual information and concept development.

According to Mallett (2010), who draws on the work of Piaget and Vygotsky, there are three principles for effectively teaching and learning with informational texts. First, the teaching of concepts always begins with active exploration (Mallett, 2010). Children are able to become naturally curious about concepts presented to them, and teachers can draw on that excitement by allowing children to actively explore the concept in combination with evidence from informational texts (Mallett, 2010). "Experiments in science, field trips in geography, outings and museum visits to enliven history and English all help to sustain and deepen interest, and all nourish learning from texts" (Mallett, 2010, p. 232). Further, Mallett continues, this exploration is not done in isolation, but rather through dialogue with peers and teachers. This, she contends, is the second principle. Therefore, during instructional time, teachers should allow students to talk to one another, as well as prompt students to share their ideas about the concepts (Mallett, 2010). This communication helps children forge critical links between they are actively exploring in a hands-on manner and what they are reading in the texts (Mallett, 2010). Third, Mallett states that the teacher's role in the use of informational text and concept development is central. Effective teachers understand how to choose appropriate texts, plan effective lessons that promote active engagement, and model how to learn from text and first-hand experiences so that children can emulate these behaviors (Mallett, 2010).

When choosing informational texts, it is critical that those texts be of high-quality. It should be acknowledged that texts come in many different forms, including digital texts, which can be shown on a screen. However, even in traditional printed texts, the layouts task the reader to think about the content in new ways (Hassett & Schieble, 2010). According to Hassett and Schieble, "The computerization of type design and the photomechanical printing technologies available today make it possible for alphabetic text to mingle with graphics, extending the ways in which thought might be represented" (2010, p. 325). Hassett and Schieble argue that the linguistic cueing systems, graphophonic, semantic and syntactic, traditionally used to understand how reading works are no longer enough to navigate these new types of texts. They suggest that "we might begin by updating our reading cueing systems to encompass a greater scope. For example, visual cues clearly extend beyond the sign systems of print and now include images, graphics, and the look of the word on the page" (Hassett & Schieble, 2010, p. 330). Despite the challenges of doing so, the opportunities that this provides in turning STEM lessons into STEAM lessons are many. Choosing books with not only rich vocabulary and complex structure, but also thought-provoking and exciting visuals, will demonstrate to students that art and literacy can be integrated into any content area.

CONTENT INFORMATION

The lesson plan for integrating informational texts in STEAM lessons presented in this chapter is designed to work with the Common Core College and Career Readiness Standards as well as The Next Generation Science Standards. . The specific grade level standards can be found on the following website: http://www.corestandards.org/ELA-Literacy/CCRA/R/. In most situations, teachers may be using informational trade books to expose students to concepts; therefore, it is important for teachers to look for the specific grade level under *Reading: Informational Text* found on the same webpage. Depending on the content area being addressed, teachers also will have to choose grade-appropriate standards in the STEAM areas. Because high-quality informational texts can be used in diverse ways, the standards addressed vary depending on the area of integration. For example, if a book on "how-to" writing is used during a science lesson to teach procedural writing in a visually artistic way, then the corresponding science and English/language arts standards would be invoked at the appropriate grade level.

CASE DESCRIPTION

The following lesson plan format can be used to integrate trade books into STEAM lessons, forging a link between the content areas, literacy and the arts. As Yopp and Yopp (2012) suggest, the way teachers share an informational text with the students is essential to the success of the lesson. They recommend a "before reading-during reading-after reading" format, which is followed here.

Lesson Plan Information

Activation of Background Knowledge and Hands-On Activity

Elementary school children often carry misconceptions about STEAM concepts because they have not developed enough knowledge to understand basic principles (Hoffman, Collins, & Schickedanz, 2015). Therefore, as a teacher, it is important to discover the misconceptions children carry by having them discuss and hypothesize through an activity. Through this activity, an informal assessment can be done of the misconceptions which can be corrected through the use of literacy strategies, especially using trade books. With your STEAM concept that you would like to teach in mind, develop an activity and corresponding questions. For example, a popular activity in science with young children is whether objects will sink or float. Let the students experiment with objects, each time predicting whether an object will sink or float and coming up with reasons why it occurs. If students incorrectly or cannot fully articulate why certain objects might sink or float, take note of their misconceptions. Further, prompt them with questioning such as, "What made you give the reason why that object could float (or sink)?" Observation of the students is extremely important here; document carefully, preferably through video if possible, their responses and dialogue as they go about their hands-on activity.

Strategy Choice/Book Research

From your documentation of the misconceptions and understandings in the hands-on activity and activation of background knowledge, you can begin to address the central focus of your lesson. Think about which literacy strategies would help the students understand the concept through reading trade books and writing. Choose several informational or hybrid trade books on the concept. Make sure that the books provide different levels of challenge to ensure that all readers can access the concept. One of the books should be the mentor text that is shared with the entire class. Align an appropriate reading strategy with your text; for example:

- Making connections (e.g., text to visuals)
- Visualizing
- Inferring
- Asking Questions
- Determining Importance
- Synthesizing

Aligning reading strategies with books gives the students a purpose for listening to the book, especially for the concepts that relate to the central STEAM focus of the lesson. For example, when teachers model how to ask questions while reading a book, and allow students to do the same, they demonstrate how asking questions can lead to a wealth of information discovery that might have been overlooked had not questions been asked.

It is helpful to create a chart that aligns the chosen books with the strategies, as well as the content area standards being covered and the content area(s) to which the books relate:

Using the Books to Foster Understandings

Before Reading: Before you begin the read-aloud of your mentor text with the whole class, review the discussions that took place during the hands-on activity. Remind the students what they predicted and their reasons they gave for why something occurred. For example, in the case of having objects sink or float, a teacher might say, "Group 1, you said that the rubber duck could float because there was nothing weighing it down." After reviewing the discussion and activity, introduce the book by doing a picture walk. In most cases, the book will be an informational book so be certain to point out the text features (photographs, charts, graphs, bold words, captions). Critical to the picture walk is the introduction of vocabulary that students may not have used when describing the experiments or activities. Providing the students with access to that vocabulary will help them better understand the text and its relationship to the concepts at hand. Finally, state and model the chosen strategy which will help them think about the text as you read. For example, if "*Determining*

Table 1. Related books

Book title/author	Reading Strategy	Content Area Standards	Content Area(s)
Example *Sky Tree* Thomas Locker with Candace Christiansen	Making connections between text and visuals	K-ESS2-1 (Weather conditions)	Earth Science Language Arts/Reading

Importance" was chosen, explain that not every detail in the text may be critical to understanding the main idea; therefore, we have to be able to choose the most important details to remember. It is helpful to guide the students in a brief practice of this strategy, using a paragraph from a previously read informational text on a STEAM concept that they have already mastered. Have the students assist you in determining the most important ideas in the paragraph and highlight them.

During Reading: There are key components to this phase. First, check continuously for understanding by questioning and prompting students for comments. Instead of having the students always answer the teacher, they can tell the answer to a friend next to them while you listen. If the text is appropriate for making predictions, allow students to do so and then confirm whether their predictions were correct. Second, continue the focus on the chosen strategy by frequent modeling and having the students assist in using the strategy. Third, be certain to point out the textual features such as pictures, photographs, charts, graphs, bold words and captions that assist in understanding, especially for ELL students. If possible, having enough copies of the text per two students is desirable; if these are not available, then make sure students who need extra support sit closer to you or project the book on a document viewer.

It is at the discretion of the teacher whether to read the entire text or not, depending on the rest of the lesson and the difficulty of the text. In the case where a text is on the reading level of most of the students in the class, a teacher might read half of the book and then send the students in partners to fill out a graphic organizer while reading the rest of the book together. They would then come back to share their findings. Students who need extra support can be provided with easier texts on the same topic so that they can also participate in the discussion. The mentor text and the other books can then be placed in various content area centers so that students can refer to them as they go about their work.

After Reading: When the text has been read, it is time to assess whether or not the reading increased understanding and cleared up any misconceptions. Therefore, return to the discussion held before reading about statements made during the activity and ask students to confirm or disconfirm those statements using textual evidence. Repeat the activity from the beginning of the lesson or a similar activity. Ask students to describe the events using vocabulary and information from the book. If a copy the book is made available to the students, they can write down the page number where they found that information in their observational logs or lab report books.

Post-Lesson Professional Development

It is critical for teacher candidates and practicing teachers to analyze their integration of informational trade books into STEAM content areas, as this integration may not always come naturally. Therefore, provided below are some reflective questions that can be used for professional development purposes.

1. What do you feel were the strengths and weaknesses of your book choices?
2. How could you embed trade books differently in a future content area lesson?
3. What high-quality characteristics did your mentor text have that made you choose it for your whole group reading?
4. How will reading this text help your students master the Content Area standards listed on your chart?
5. How did the text help the students better understand the STEAM concept that was the central focus of the lesson and clear up any misconceptions?

SOLUTIONS AND RECOMMENDATIONS

During clinical practice and professional development, teacher candidates and teachers need practice with the following:

1. Choosing high-quality informational texts that present, both in words and visually, a rich conceptual development of the topics they wish to cover;
2. Creating hands-on experiences that will link to the reading of the text;
3. Forming lesson plans in the STEAM content areas that strike a balance between concept development and the literacy skills/strategies needed to for text comprehension (Hoffman, Collins & Schickedanz, 2015).

In sum, teacher candidates and teachers need to be provided with the time to practice teaching in this manner; to maintain the status quo of only integrating content areas into literacy instruction robs both students and teachers of the opportunities to deeply engage with concepts and make important connections to the world around them.

ACKNOWLEDGMENT

The author wishes to gratefully acknowledge the help of Brian Edgar, Ph.D. for his assistance and feedback during the writing of this chapter.

REFERENCES

Adams, A. E., & Pegg, J. (2012). Teachers enactment of content literacy strategies in secondary science and mathematics classes. *Journal of Adolescent & Adult Literacy*, *2*(56), 151–161. doi:10.1002/JAAL.00116

Fisher, D., & Frey, N. (2015). Selecting texts and tasks for content area reading and learning. *The Reading Teacher*, *68*(7), 524–529. doi:10.1002/trtr.1344

Hassett, D. D., & Scheible, M. B. (2010). Finding space and time for the visual in K-12 literacy instruction. In M. Cappello & B. Moss (Eds.), *Contemporary Readings in Literacy Education* (pp. 325–331). Thousand Oaks, CA: SAGE Publications.

Hoffman, J. L., Collins, M. F., & Schickedanz, J. A. (2015). Instructional challenges in developing young childrens science concepts. *The Reading Teacher*, *68*(5), 363–372. doi:10.1002/trtr.1325

Mallettt, M. (2010). Choosing and using fiction and non-fiction 3-11: A comprehensive guide for teachers and student teachers. New York, NY: Routledge.

Olness, R. (2004). *Using literature to enhance content area instruction: A guide for K-5 teachers*. Newark, DE: International Reading Association.

Yopp, R. H., & Yopp, H. K. (2012). Young childrens limited and narrow exposure to informational text. *The Reading Teacher*, *65*(7), 480–490. doi:10.1002/TRTR.01072

ADDITIONAL READING

Avgerinou, M., & Ericson, J. (1997). A review of the concept of visual literacy. *British Journal of Educational Technology*, *28*(4), 280–291. doi:10.1111/1467-8535.00035

Burmark, L. (2002). *Visual literacy: Learn to see, see to learn*. Alexandria, VA: Association for Supervision and Curriculum Development.

Coiro, J., Knobel, M., Lankshear, C., & Leu, D. J. (Eds.). (2008). *Handbook of research on new literacies*. London: Routledge.

Considine, D. M., & Haley, G. E. (1999). *Visual messages: Integrating imagery into instruction. A media literacy resource for teachers*. Santa Barbara, CA: Libraries Unlimited.

Cope, B., & Kalantzis, M. (2009). Multiliteracies: New literacies, new learning. *Pedagogies*, *4*(3), 164–195. doi:10.1080/15544800903076044

Dondis, D. A. (1974). *A primer of visual literacy*. Cambridge, MA: The MIT Press.

Elkins, J. (Ed.). (2008). *Visual literacy*. New York, NY: Routledge.

Frey, N., & Fisher, D. (2008). *Teaching visual literacy: Using comic books, graphic novels, anime, cartoons, and more to develop comprehension and thinking skills*. Thousand Oaks, CA: Corwin Press.

Keene, E. O., & Zimmermann, S. (1997). *Mosaic of thought: Teaching comprehension in a reader's workshop*. Portsmouth, NH: Heinemann.

Kiefer, B. Z. (1995). *The potential of picturebooks: From visual literacy to aesthetic understanding*. Englewood Cliffs, NJ: Merrill.

Lankshear, C., & Knobel, M. (2003). *New literacies: Changing knowledge and classroom learning*. Buckingham, UK: Open University Press.

Leu, D. J., Kinzer, C. K., Coiro, J. L., & Cammack, D. W. (2004). Toward a theory of new literacies emerging from the Internet and other information and communication technologies. In R. B. Ruddell & N. J. Unrau (Eds.), *Theoretical models and processes of reading* (5th ed., pp. 1570–1613). Newark, DE: International Reading Association.

Messaris, P. (1994). *Visual literacy: Image, mind, and reality*. Boulder, CO: Westview Press.

Schönborn, K. J., & Anderson, T. R. (2006). The importance of visual literacy in the education of biochemists. *Biochemistry and Molecular Biology Education*, *34*(2), 94–102. doi:10.1002/bmb.2006.49403402094 PMID:21638645

Sinatra, R. (1986). *Visual literacy connections to thinking, reading and writing*. Springfield, IL: Charles C. Thomas.

Stokes, S. (2002). Visual literacy in teaching and learning: A literature perspective. *Electronic Journal for the Integration of Technology in Education*, *1*(1), 10–19.

KEY TERMS AND DEFINITIONS

Content Area Integration: The combining of two or more elementary content areas during a lesson plan or unit.

Elementary Content Areas: All subjects taught in elementary grades (language arts, math, science, social studies, world language, music, art, health and physical education).

High-Quality Children's Trade Books: Books with rich vocabulary and complex structure, as well as thought-provoking and exciting visuals.

Linguistic Cueing Systems: Sound-symbol, structural, grammatical and meaning relationships that aid a reader, writer, speaker or listener in processing language.

Mentor Text: The main text used to teach a whole-group lesson to students of varying ability levels.

Misconceptions: Ideas that students hold about concepts which are inaccurate or false, especially in the scientific sense.

Reading and Writing Strategies: The application of purposeful and metacognitive ways of thinking to assist one with language arts tasks.

Visual Learning: Learning through viewing pictorial representations.

APPENDIX: SAMPLE LESSON PLANS FOR GRADES K-2 AND 3-5

Introduction

The goal of these lessons is to help students comprehend scientific and technical concepts by applying strategies that use images to assist in comprehension. In this sample lesson, we begin by looking and noticing. A strategy that can be used with any book is to begin with the title and look at the images on the cover. Book covers are designed with care, and in the case of picture books, artists spend hours drawing, painting, or using computers to consider and design the images on the covers. We can then read the book and make evaluations about how the story or information is represented visually. During the reading and after we have read the book, we can consider how the images contribute to the understanding of the text.

1. Into the Text: Activation of Background Knowledge and Hands-On Activity

To assess prior knowledge of the seasons and any misconceptions held, have the students draw a picture of a tree in the four different seasons. Ask them to describe the changes in their trees (e.g., Why did you draw the autumn tree differently from the spring tree?). Drawings can be displayed or shared in pairs or groups. Students should be asked about the decisions they make with different colors and scenes.

After they describe their changes, ask them why those changes occur. Document their answers in a way that can be displayed for the whole class.

To prepare students for reading the text and introduce the strategy of making connections between text and images, we might ask students to look at the cover of *Sky Tree* and ask them questions such as:

Table 2. Sample Lesson Plan. K-2 Science: Changes in the Seasons. Next Generation Science Standard.

K-ESS2-1.	Use and share observations of local weather conditions to describe patterns over time. [Clarification Statement: Examples of qualitative observations could include descriptions of the weather (such as sunny, cloudy, rainy, and warm); examples of quantitative observations could include numbers of sunny, windy, and rainy days in a month. Examples of patterns could include that it is usually cooler in the morning than in the afternoon and the number of sunny days versus cloudy days in different months.] [*Assessment Boundary: Assessment of quantitative observations limited to whole numbers and relative measures such as warmer/cooler.*]

Book: Sky Tree by Thomas Locker with Candace Christiansen. HarperCollins Publishers, 1995

- What do you notice?
- What objects do you see?
- What colors do you see?

We might begin by noticing the tree. The blue sky between the branches and the clouds that form the boundary. We might notice the grasses, river and mountains in the background. As we look we might see a leave that is formed by the shape of the clouds and formed in the blue sky. We can notice, areas of light and dark. The tree is the center stage and is on a small hill. Other details might stand out: for instance, there are no leaves on the tree. From this observation of a detail we can make inferences about the text: what season is it? What is the weather like during the season?

Next we may ask students to think about the title.

We might ask: What does *Sky Tree* mean?

2. Through the Text: During Reading

As you begin the book, you tell the students that while you read the book, you will be stopping at different points so that they can make connections between the text and the pictures, notice the images and details the same way you modeled with the class by looking at the cover. Explain that making connections the ideas and the pictures in the book is an important strategy that will help them understand the seasons in an in-depth manner.

Model this by using the first page: "The text is telling me that the leaves are fluttering in the summer breeze, and when I look at this picture, I can see that the tree is full of green leaves, which is true of trees in the summer. Moreover, the grass and the bushes are green, too, which tells me the season is one of the warmer seasons." Turn to the next page and have the students assist you in making connections between the text and the picture.

Begin the reading of the mentor text, *Sky Tree*, by reminding the students about their tree descriptions and the reasons that the trees might change through the seasons. Show the students a few of the pages in the book and ask them what they are seeing. Through modeling, encourage them to use vocabulary relevant to the seasons; you can write these words down to display. For example, "sprout" or "sprouting" would be an appropriate word to describe plant life in early spring.

As you read the text, *Sky Tree* will also raise questions about the images such as: how does this make you feel? Or what season is it and so forth? We are doing two things here: one is building skills and strategies in reading images that students can apply to any texts, and, with the case of *Sky Tree,* we are also helping students with the science concepts of the changes of the seasons and the natural cycles of nature.

3. Beyond the Text: After Reading

After you have read the book, you may use your K-W-L chart and ask students: What did you learn? What did you notice? This can be done orally or in writing, think-pair-share or any way that you like. Here is a point to assess the students' understand and then deepen it by developing conceptual knowledge about the seasons and the cycles of the year: The book begins in the fall or autumn? Where does it end? What is the process that occurred?

Once you have finished reading the book, return to the display of the statements the class made at the beginning of the lesson. Read the statements and ask the students to use examples from the text and pictures to confirm their validity. For example, if at the beginning of the class, a student said the leaves fall off the trees when it gets cold, ask the students to show an example from the book.

As a closing activity, students, in pairs or in centers, can match visual depictions of the seasons with vocabulary words or simple sentences by cutting out provided pictures (pre-made) or drawing them. They will then have to orally explain their reasoning for matching a picture with a word or sentence.

1. Activation of Background Knowledge and Hands-On Activity

To assess prior knowledge of coral reefs, have the students in pairs or groups sort pictures (pre-made by the teachers) of animal and plant that could be found or not be found in coral reefs. Ask the student pairs or groups to explain their sort: why did they think that some could survive in coral reefs and some could not? Document their explanations in a manner that can be displayed for the whole class to see.

2. Before Reading

Begin the reading of the mentor text, *Coral Reefs*, by reminding the students about their explanations for why certain life forms survive in coral reefs. Turn to one of the pages in the book and ask them what they are seeing. The two-page spread that

Table 3. Sample Lesson Plan. Grades 3-5 Science: Coral Reefs. Next Generation Science Standard.

3-LS4-3.	Construct an argument with evidence that in a particular habitat some organisms can survive well, some survive less well, and some cannot survive at all. [Clarification Statement: Examples of evidence could include needs and characteristics of the organisms and habitats involved. The organisms and their habitat make up a system in which the parts depend on each other.] (retrieved from http://nextgenscience.org/pe/3-ls4-3-biological-evolution-unity-and-diversity)

Book: Coral Reefs by Jason Chin. Roaring Brook Press, 2011.

shows the child swimming with a turtle is an excellent visual. Through modeling, encourage them to use vocabulary relevant to the topic; you can write these words down to display. For example, “ocean floor” or “food chains” would be appropriate words to describe a coral reef habitat.

Tell the students that while you read the book, you will not be showing them the pictures so that they can visualize the words in their minds. You will show them the pictures after you finish reading the book. Explain that visualizing the ideas book is an important strategy that will help them understand the coral reefs in an in-depth manner. Model this by using the first page which you have previously covered with paper, leaving only the text exposed. “When I read this, I try to imagine in my mind an animal that looks like a plant moving in the water. I am going to try to draw a picture of that on my paper.” Draw a picture and ask students to comment on whether they were imagine the same picture in their minds.

3. During Reading

Continue reading the book and engaging the students in visualizing. Provide each student with a piece of paper so that they can sketch what they are visualizing. You can also have the students “turn and talk” to classmates in order to share their visualizations and sketches. Check continuously for their ability to make visualizations by asking what evidence in the text made them visualize an idea in a certain way.

4. After Reading

Once you have finished reading the book, show the pictures in the book to the students one page spread at a time, and have students explain how their visualizations and sketches compare to the actual pictures. Ask how visualizing helped them understand ideas about coral reefs and make a list of new understandings they gained. Return to the display of the statements the class made about why different plants and animals could or could not survive in coral reefs at the beginning of the lesson. Read the statements and ask the students to use examples from the text and pictures to confirm their validity. Take note of the animals and plants about which students are confused, and tell the students that they will be researching these to confirm whether they are part of a coral reef habitat in a future lesson.

To extend the learning to cover all elements of the lesson standard, students, in pairs or in centers, should fully draw out their sketches of animal and plant only found in the coral reefs and cut them out. They will then make a web of how these organism are related (for example, predator/prey) within the coral reef habitat on a large sheet of paper. They will then have to explain in writing their reasoning for constructing the web they way they did.

Chapter 4
Getting to "Know" STEAM

Merrie Koester
University of South Carolina, USA

EXECUTIVE SUMMARY

This chapter describes the evolution of a personalized, arts-integrated approach to science curriculum inquiry which has been evolving since the 1990s—before even the national science standards, the acronym STEM, much less STEAM, appeared across educational horizons. It reads as ethnography and has been performed in community, in association with the most caring of souls, with the goal of achieving a more inclusive/empowering, aesthetic science education and a deep appreciation of the importance of the creative arts in the learning process. It presents two research-based iterations in STEAM education in practice: 1) the creation of arts-integrated middle school ocean science curricula and 2) the development of a pedagogical tool called the "Know"tation as a way for teachers and students to make learning visible and integrate the languages of science throughout the process of inquiry. The cases described in this chapter apply many features of the STEAM model developed in Chapter 1 of this book.

INTRODUCTION

How can we engage those students who struggle to read and for whom science informational text seems particularly threatening? I myself encountered this same challenge in my very first teaching job, where I was tasked with teaching physics to 35 older ninth graders who had previously failed the class. The football coach, their former teacher, declared that they were his "dummy class." "Don't expect too

DOI: 10.4018/978-1-5225-2334-5.ch004

Copyright ©2017, IGI Global. Copying or distributing in print or electronic forms without written permission of IGI Global is prohibited.

much from them," he said matter-of-factly. I was astonished. Certainly no one in that class was stupid! Yet, they had most definitely had been made to *feel* that way through tracking, neglect, and marginalization. The real kicker was that the more affluent students up the hall in the "gifted science class" had a nice lab with equipment and space to do experiments. We, on the other hand, had 35 desks crammed into rows in a small corner classroom and an old textbook none of my students could read. I was so mad I could spit. But I was also lucky. I had been trained in the arts, and I knew how to draw and to tell a story, and so used that as a way to connect with language we could all understand. Together, we drew and played with funny little cars, which we rolled down those crummy old text books. These kids started passing, and I started to get a sense of what it meant to *design* a science curriculum as an artist might.

What I did not realize then, as I do now from study of Elliot Eisner, John Dewey, Johann Pestalozzi, Maxine Greene, Richard Siegesmund, Liora Bresler, Bruce Uhrmacher, and other arts education researchers is that what I had observed in my students and myself was an *aesthetic* transformation.

Uhrmacher (2010) noted that a student who acquires "aesthetic capital" may feel or act differently – in a good and positive way. I myself, a brand new teacher, was so transformed by the way an artistic pedagogy turned around previously failing students that I set out to change science education. Ah, to be 25 again! I applied to the MEd program at the University of Hawaii, where I took graduate level classes in both the visual and performing arts as well as art and science curriculum theory. I became determined to identify the ways in which science and art share a common language and developed an experimental curriculum I called Teaching Science through the Arts—TSTA – at a time before even the acronym STEM had become popularized (Koester, 1989). I researched, developed, and then field tested curricula for teaching science through the creative arts of drawing, poetry, music, dance, creative drama, and fictional literature. In my master's thesis, I tentatively proposed that virtually any area of K-12 science could be taught through the arts. The lesson planning objective, I reasoned, was to match the art form to the dynamics of the science content being taught. Right off, I recognized that a science teacher who would implement the TSTA model would need to either personally acquire basic skills in multiple art forms, and/or collaborate with an art teacher specialist. Because I had significant arts training, I managed to create and field test about fifteen TSTA lessons as part of my final thesis. However, I did not have nearly enough data to propose a *theory* about best practices for teaching science through the arts. That process has taken nearly two more decades.

What follows is a true story that narrates the transformation of a science teacher into a researcher and the evolution of an idea into a curriculum model that is still growing and developing.

BACKGROUND

Creating As-If Worlds

A few months after completing my masters at UH, I returned to my home state of South Carolina with my hundred page thesis carefully packed in a special box. I had accepted a middle school science position and was eager to field test my arts integration techniques in a sixth grade physical science and eighth grade Earth science class. After only one week, however, my principal said that she had forgotten to tell me that I would be required to take my sixth graders through their first ever science fair projects. Oh, no! Which art form would possibly *match* the teaching of THE scientific method, something our curriculum dictated *all* students were to *learn*. I was full of dismay and fretted for nearly a week, when suddenly, the answer came to me as clearly as I imagined Archimedes' Eureka Moment had occurred to him. Late on a Sunday evening, after I put my two boys to bed, I began writing a "fictional" story about a twelve year old girl, who unwittingly let her father complete her entire science project – an impressive, meter tall electromagnet crane. Even worse, the girl had won *first* prize for work she had not done herself. Her best friend, who *had* done her own project, came in second. The next morning, I shared with my sixth graders the beginning of a story that later became the first of four science education novels, written over the course of two decades, and recently released in their third editions (Koester, 2015 b-e). Almost instantly, my students entered into the shame my young protagonist felt for cheating her best friend out of first place. We reflected at length on the importance of doing your own, honest work. They learned the steps of the experimental process by performing the "Science Rap," which I had written for them. To my amazement, nearly all of them eagerly dug into their own projects, determined to do their own work. That same year, one of my sixth graders, who was so shy that she rarely spoke, won first prize in the regional science fair. She, like the character in my story, later grew up to be a science teacher.

Who among us did not dread doing their first ever science fair project? Indeed, I was that twelve-year-old girl *and* the teacher who was now all grown up and trying to get my students to actually embrace this arduous task. In my creative drama classes at UH, I had discovered the power of creating *as-if* worlds in the science classroom (more on that in just a bit). The effect seemed extra-ordinary if that world came from a fictional trade book. I developed this hypothetical learning equation: Imaginative story (in a trade book featuring science content) + creative drama (with movement) → Learning + Engagement. To test my claim, I developed for my then fourth graders a lesson on digestion. I began by reading with them the trade book, *Gregory, the Terrible Eater,* by Mitchell Sharmat. This allegorical tale is about a goat named Gregory, whose parents think he is a TERRIBLE eater because he wants to eat foods like fruits, vegetables, eggs, and fish instead of the *normal* GOAT diet of shoes, paper, tin cans, coats, pants, etc. My students eagerly responded when I asked them to get on all fours and pretend to eat a pile of shoes like good little goats. But, oh my! I wish you could have seen the Oscar-worthy performances they gave when I asked them to refuse to eat as their parents demanded. The pawing and stomping and mournful bleating were hilarious. Some rolled on their backs and kicked their feet in the air in defiance. To add a musical performance component, we improvised an operatic, call and response exchange between a group of students, pretending to be goat kids (refusing to eat perfectly yummy shoes) and their worried goat parents. The "tune" we used was Beethoven's Fifth—da, da, da, daaaaaaa (Koester, 1989):

Kids: *We will not eat!*
Parents: *Oh, yes you will!*
Kids: *Shoes make me sick!*
Parents: *Don't talk like that!*
Kids: *We're feeling weak.*
Parents: *'Cause you don't eat.*
Kids: *It tastes like junk!*
Parents: *What will we do?*

Andersen (2004) has written that "the methods of drama in education have the potential to create 'as-if' worlds within the classroom that can foster *situated* learning" (p. 284). Andersen was referring to the cognitive psychology studies of Brown, Collins, and Duguid (1989) on the benefits of learning in *authentic* contexts, *situated* within a context with which one could personally relate. Andersen argued that through creative drama and the creation of *as-if* worlds, students could enter into simulations of what *felt* like authentic experiences. Through the vehicle of our imaginations, stories take us places where we might never, pragmatically speaking, be able to go—in spaceships to Mars; in deep sea submersibles to deep ocean hydrothermal vents, where giant tube worms with bright red tips twist and turn in front of the front window; and into science research laboratories, where people, like we may become one day, race to discover a vaccine against *Ebola*. We enter into deep, empathic relationships with a story's characters.

Neuroscientists have determined that there is a scientific basis for the empathy we feel in *as-if* situations. In 1992, a team of researchers at the University of Parma, Italy, working with macaque monkeys discovered that the same brain cells fired when the monkeys performed a task (like picking up a peanut) as they did when the monkeys simply watched a person do the same task. They named these brain cells *mirror neurons* (Ehrenfeld, 2011). Neuroscientists became very excited about this discovery and soon found that human beings also had mirror neuron systems. Were mirror neurons the scientific explanation for empathy? Gallese (2005), an Italian neuroscientist, certainly thought so. He took a lead in the mirror neuron research and discovered that indeed, mirror neurons fire in our brains even when we just *imagine* doing a specific task. For example, we can look at a picture of a hammer, and the motor neurons that we would use to pick up a real hammer are excited. When we see another person smile, the muscles in our own face that control smiling light up on functional MRI. Gallese (2005) had determined that our brain can "model the acting body of other individuals." He called this phenomenon "embodied simulation" and conceptualized what was happening as a kind of *resonance* between bodies, which he theorized could be at the heart of all social cognition (p. 23-48, in Koester, in press).

Rollo May (1985) explained that artists portray their reality through *images*—poetic, aural, plastic, dramatic, and embodied. The processes described by science are a veritable gold mine for *imag*inative, empathic visualization. Their very names conjure wonderful images and possible as-if *situations*. Science words like evolution, metamorphosis, growth, decay, reproduction, birth, and symbiosis are just a few examples of the exciting events described in a life science class. The physical and Earth sciences provide just as many opportunities for artistic, creative expression and interpretation (Koester, 1989). I still take every kind of creative arts class I can. The process is invigorating and enlightening as long as you check your ego at the door. Just recently, I bumbled my way through an improvisational theatre class and came away understanding the power of "Yes, *and*" messages, as opposed to "Yes, *but*," which kills a scene. Every time I take a new art class, I come away with renewed appreciation of the work ethic of the performing artist and the importance of trying, failing, and being critiqued so that you might improve. The process is scary, too. There is no podium behind which to hide. A supportive, caring teacher makes all the difference between whether you will risk ditching your cover and discovering your inner artist (It's there somewhere!), or quit the class for fear of embarrassing yourself.

I am very fortunate to live and work in a state where I have received enthusiastic support for my experimental pedagogy by both science and art education professional organizations, as well as many colleges and universities. Beginning in 1993, I began giving workshops on teaching science through the arts at the South Carolina Science Council (SC2) annual meetings. In the first 1993 session, I shared videos of my students, who had choreographed their version of "The Tectonic Drag." The beginning verse went like this:

On the floor of the seas
are rifts and ridges,
that circle the world
like gigantic zippers.

I had written "The Tectonic Drag" as a poem to be choreographed. It was my way of trying to bring my eighth grade students into the deep, to the ocean floor, to the beginning, possibly of creation itself. Here's the chorus:

Let your arms go up and out and down;
they're convection currents going round and round.
Convection currents are the driving force that
push the Earth's plates along their course.
The plates are doin' the TECTONIC DRAG –
spreading, crashing, diving – making buildings sag.

The Ocean. Can we ever know its effects on us, or more importantly, our effects on it? We should give thanks to marine algae every single day for the oxygen we breathe! David Ballentine, a noted marine phycologist, believes that if all marine algae were to die, we would have less than one week to live (personal interview, 2007). These primitive life forms have been making oxygen for the planet for millions of years. Why do sponges produce natural antibiotics? We don't yet know. How many fish can be caught before their populations collapse completely and entire populations of all sorts of living things collapse? Do we dare find out? In Lloyd Alexander's novel, *Taran Wanderer*, the wizard Morda sneeringly observes: "I have seen enough of humankind and have judged them for what they are—lower than beasts, blind and witless, quarrelsome, caught up in their own selfishness" (in Koester, 2011). Alexander's novel creates an *as-if* world, inside of which students develop understanding of human impact on the environment and each other. Through the elements of narrative— challenge, conflict, and struggle—eventually resolution is achieved. When teachers learn to develop lessons that move through periods of dissonance and struggle, for which no one right solution is apparent, students will make creating some kind of resolution their own responsibility. They key words here are *making, struggle,* and *responsibility*. In real life, resolution is not always achieved. In real *science*, the struggle is to discover what it is we don't know. This theme of persistence in the face of struggle was most definitely the clarion call of Chapter 1 of this book.

DRAWING FROM THE WATER

On June 7, 2006, I was invited to attend the closing session of the first ever Conference on Ocean Literacy (C.o.O.L.), where a consensus of mostly marine scientists and government policy makers had gathered to review the draft of a publication called the *Seven Essential Principles of Ocean Literacy and Detailed Fundamental*

Concepts (http://oceanliteracy.wp2.coexploration.org/). I was very excited about attending this event, as my students and I had been participating in the field testing of National Oceanic and Atmospheric Administration (NOAA) extensive, virtual, and interactive *Ocean Explorer* marine science curriculum, being developed by the education department of their office of ocean exploration (http://oceanexplorer.noaa.gov/welcome.html). My eighth grade students sent real-time questions to NOAA research scientists, as they whisked about in deep sea submersibles. Images of giant tube worms beamed into my classroom. This educational technology was unlike any I had ever experienced as a science teacher, *and* it was free!

At the C.o.O.L gathering, the legendary oceanographer and first female aquanaut, Sylvia Earle, spoke. During our break, I even got to meet this great and gracious scientist and pioneering explorer. John Michel Cousteau, son of Jacques, reminded us that his father had always said that we will only protect what we love and understand. My heart and mind were stirring in the way they always do, right before I get an idea or make an important decision. I have learned to pay close attention when this happens, to be alert for the clue of what to do next. I didn't have to wait long. Suddenly, the lights went out in the conference hall, just as I had thought the meeting was about to close. Our attention was directed to a large screen on a side wall. Flickers of light played across the screen, and then, Woods Hole Oceanographic Institute scientist and deep ocean explorer Robert Ballard appeared. He was speaking with us via telepresence from a submersible on the bottom of the Aegean Sea! Ballard declared that the battle for ocean literacy (and hence the future of our planet) was to be won at the *middle* school because that's when we started "losing" our nation's science students. Wait! *I* was a middle school teacher! Was Ballard talking to me? Exhorting me to become some kind of ambassador for ocean literacy? I was certain he was! In spite of the fact that I had said I did not have another science education novel in me (I had been doing so for nearly 18 years by then), I vowed to write one *last* book, this time about the dire need for ocean literacy.

At about the same time, in Charleston, SC, my home, the SC Maritime Foundation had just launched a tall ship, the *Spirit of South Carolina*, a full scale replica of the 19th century pilot schooner, the *Frances Elizabeth*. She was built as a sail training vessel to provide experiential education to South Carolina youth. While onboard, school students would learn about the math, science, and natural history of the surrounding waters, as well as how to work as a cooperative crew. The schooner was outfitted with the most modern ocean science equipment and technology. I was fortunate enough to join multiple school groups on several sailing experiences in the Charleston Harbor. Many of these children had never once been aboard any kind of boat, much less a nearly 100 foot, three-masted sailing ship! I determined to prominently feature this proud vessel in my new book.

I next applied for and received a research and development grant from the National Marine Sanctuary Foundation (NMSF). With these funds, I set out to research and have vetted the ocean science content for this book, a journey that took me into some of the finest aquaria and ocean science centers in our country—The Shedd (Chicago), The Baltimore Aquarium, The Monterey Bay Aquarium, The Sant Ocean Hall (Smithsonian Natural History Museum), and of course, The South Carolina Aquarium. I traveled to Puerto Rico, where Ruperto Serrano Chaparro, director of the Sea Grant Puerto Rico, introduced me to prominent marine ecologists and researchers, experts all in coral reef diseases and the phenomenon of dead zones. I snorkeled in the waters off the coast of Puerto Rico and also the island of St. John, and saw with my own eyes the devastation beneath the waves—the gray "boneyards" where once great, colorful "gardens" of coral thrived. And then, I spent three days and nights sailing on the *Spirit of South Carolina*, keeping midnight watches, mustering with the crew, and even practicing man overboard drills. I don't think I slept the entire time for fear of missing something!

By now, I was a very active member of both the South Carolina and National Marine Educators Association. I had learned that the breeding ground of the North Atlantic Right Whale, the most endangered of all marine mammals, was very near me at Gray's Reef National Marine Sanctuary, whose staff sent me classroom sets of information packets on this gentle giant.

In 1995, Daniel Pauly, principal investigator at the Sea Around Us Project, coined the term "shifting baselines" to describe changes which occur slowly enough that we barely, if at all, notice them occurring. As a result, our "baseline" of what is acceptable "shifts." For example, tomorrow's coral reef divers will come to see "boneyards" as the norm (https://www.ted.com/talks/daniel_pauly_the_ocean_s_shifting_baseline?language=en). I was fascinated with this concept. Could it be that science, technology, engineering, and mathematics had somehow played a role in the ocean's downfall? I suspected the answer was yes. Before I knew what was happening, my writing and my message took a critical turn. I started looking for what had been left *out* of the history books. Settlage and Southerland (2007) have written that teachers must be vigilant about the contents of the *null curriculum*. Foucalt (1977) declared that critical researchers in science education have a responsibility to show how dominant classes have manipulated truth to their advantage. I felt I surely must do the same with this new work. Indeed, my book would be "fiction," but my motivation for writing it had become deeply personal, rooted in my own repugnance of the racism I had observed growing up in the deep South as well as my passion for preserving and protecting the Ocean, upon which all life forms, from diatoms to despots, depend for their very existence.

Bill Bigelow wrote a chapter for Wayne Au's book, *Rethinking Multicultural Education* (2009) called "Once Upon a Genocide." He cited a need for books for young students which 1) portray what "the Indians might have been thinking about the arrival of the Spaniards" (p. 78) and 2) asks children to "think about how today's world was shaped by the events of 1492" (p. 84). I sought to answer both appeals. In my novel, I lay bare the crimes of imperialism, exploitation, murder, and racism by Columbus and the Spanish *conquistadores* who followed him. My long range action plan was to communicate at the school, community, and professional educator levels the importance of understanding that entire chapters of human history have been rewritten by minds bent on obfuscating the truth about the human and environmental exploitation and destruction set into motion in 1492. Through the medium of Art – using the rhythm of djembe drums, storytelling, and the "painting" of scenes in the mind's eye, where they would not likely be forgotten, I would attempt to tell the story of what really happened on and after that first "Columbus Day." I wrote and wrote and wrote and wrote. When I slept, the plot twisted and turned like sharks twirling inside the barrel of incoming waves. A translation of Christopher Columbus' 1492 journal sat on my bedside table. His scribblings about the Tainos *naïveté* read like a horror story:

They are like the other peoples I have found with the same beliefs, and they believed that we had come from heaven and they gave what they have for whatever they are given without saying that it is too little and I believe that they would do the same with spices and gold if they had any.

Finally, I was ready for the professional development phase of what I called L.A.S.T. Book Project—a Literacy-based, Arts-infused, Science-centered, Technology-driven ocean science curriculum for middle grade learners (and, the completion of my *last* book). I had just enough funding left to provide training and art supplies to two middle school teams. I created an application and selected two schools, whose faculty seemed most committed to entering in a year-long, collaborative, interdisciplinary effort centered on advancing ocean literacy. Both these schools' principals were fully on board, a factor shown to be of significant importance in the success of any professional development project (Loucks-Horsley, 1996). A year later, the South Carolina Science Council invited me to give a keynote about what had happened during the project.What an exciting November day in 2011 that was! Onstage with me was Tiffany, a middle school science teacher from one of the lowest income districts in my state. During our study, she had collaborated with Annie, the school's art teacher. Behind us, filling nearly the entire stage and the wall behind it was the arts installation their students had created. It featured a gigantic ocean creature, with the body of a fish and a human's head. In the first chapter of my book, entitled "The End and the Beginning," a young, twelve-year old girl, walking on the beach against an oppressive, northeast wind and freezing rain, had come across a dead

North Atlantic Right Whale, its gills ensnared with fishing lines and its dorsal fin jaggedly slashed by marks only a ship's propeller could have made. Spelunking into the whale's cavernous rib cage, the girl had discovered a small, beautifully crafted wooden box, inside of which she found a carved and polished stone amulet, with the head of a human and whale's body. It was a 500 year old carved amulet, a symbol of our human responsibility to preserve and protect ocean resources, but the girl did not know that yet (Figure 1).

As I was told later, as soon as she read this part of the story, Annie, the art teacher, sprang into action. "What if," she imagined out loud, "what if our students design and construct a giant replica of this human fish, constructed from the kind of debris that has harmed or killed so much of marine life?" What if, indeed! (Figure 2).

In a joyful act of collaborative making, Tiffany and Annie's students designed and then constructed their "trash" creature. Through their art, they communicated the ever growing threat of marine debris on ocean life. Students transformed other plastic bottles into schools of cheerfully colored fish (Figure 3).

Tiffany, Annie, and I were overjoyed with the results! The school principal arranged for there to be a special school assembly for the "unveiling" (Figure 4).

As a result of participating in this kind of arts-infused exploration of ocean science, students came to understand that humans have a responsibility for protecting and preserving the ocean. They also "got" that not all scientific discoveries result in good, and that the discovery of the Gulf Stream marked the beginning of an era of human conquest of the ocean and of their fellow human brothers and sisters. They engaged in critical thinking about critical matters by examining how their own history intersected with the history of ocean science and exploration (Figure 5).

Figure 1. The human fish, a metaphor for our responsibility to preserve and protect ocean resources

Figure 2. Students constructing a "fish" artifact from potential marine debris

Figure 3. Arts installation of "human fish "and colorful reef "creatures"

This ocean literacy action research project was my first sustained effort to bring science and arts teachers and their shared students *together* in the process of curriculum inquiry. Acting as role models and mentors, these teachers showed their students that deep learning can happen when balanced art/science teaching partnerships are formed. They cross-trained one another in the knowledge, skills, and

Figure 4. Art teacher, university researcher, and science teacher celebrate installation of STEAM marine education project

Figure 5. Atlantic Ocean currents associated with "New World" exploration, colonization, slave trade, and commerce by Europeans

practices of each other's teaching specialty, and each reported being transformed by the collaborative effort. What was happening at this school was a deeply *aesthetic* inquiry of human impact on the environment—a joyful, participatory, sense and arts-based form of learning described by scholars like Pestalozzi (1894/1973), Dewey (1938), Greene (2001), Eisner (1991, 2002), and Siegesmund (2010) as promoting inclusivity, democracy, and freedom.

A similarly inspiring and highly synergistic art/science education collaboration was simultaneously taking place at the second school enrolled in The L.A.S.T. Book Project. Sadie, the eighth grade science teacher, and Shannon, the school visual arts teacher, had received permission from their principal to create a special projects art/science class. Over coffee one afternoon, Sadie, Shannon, and I had hatched the idea to literally *draw* their students into the plight of the nearly extinct North Atlantic Right Whale. Shannon would teach them the elements of design and the art of batik so that they might eventually work together to create a large mural in celebration of this endangered and majestic animal. Meanwhile, in her science class, Sadie would teach about human impact on the ocean, especially our impact on the lives of once thriving ocean animals. The topics in the science curriculum included, among other things, overfishing, shifting baselines, coral reef bleaching, global warming, harmful algal blooms, dead zones, and the power of youth. On my first day meeting with this combined class, we talked about the potential impact of the mural as an art form to communicate profound and important messages. I asked students to contemplate a projected image of Picasso's *Guernica* (1937). This mural displays the tragedies of war through its graphic display of the suffering of innocents, killed by Nazi bombs on the town of Guernica during the Spanish Civil War. As such, Picasso's mural also became a symbolic call for peace (http://www.pablopicasso.org/guernica.jsp). We talked about how artists need to develop deep knowledge of their subjects before they can make powerful art about them. They agreed that Picasso clearly must have witnessed and known a lot of suffering to have so strongly portrayed the devastation of war.

I then showed them an image of a beached North Atlantic Right Whale, brought to an ignoble death by becoming ensnared in commercial fishing tackle and with propeller gash wounds across its dorsal surface. Many exclaimed, "It's the whale in the book we're reading! They're going extinct!" Sadie, who was also in the room with us, jumped up and said, "That's right! And now it's our turn to try to do something to save this whale."

Next, I shared a video which was similar in effect to this newer NOAA release: http://oceantoday.noaa.gov/whalecall/ . The students were thoughtful and quiet. Shannon and Sadie stood up and explained that we were all going to work together on a special art/science project with the goal to learn as much as possible about North Atlantic Right Whale and then to create art that symbolized its *right* to live (pun intended). Next, I told them about Leonardo da Vinci, an artist/scientist who said that to draw something is to know it. Clearly, we could not draw from a live whale, but the staff at Gray's Reef National Marine Sanctuaries (http://www.graysreefnmsf.org/), a breeding ground for these whales, generously sent us a set of huge posters depicting the anatomy of these magnificent mammals. I placed a poster on each art table, and Shannon distributed brand new sketchbooks and drawing pencils, paid for by the NMSF grant. In addition, Gray's Reef specialists made available to us a wide array of educational materials, now accessible at http://graysreef.noaa.gov/education/materials/welcome.html.

Over the next week, in both science and art class, students were challenged to design their own sketchbook journal entries, depicting their growing knowledge about the North Atlantic Right Whale (Figure 6).

We then scheduled a class during which all students rotated around the art room to offer appreciative critique of each other's sketchbook entries. Shannon and Sadie announced to the students that they were now ready to make art about and *for* this whale—art as an act of care. Shannon tasked each student with drawing a mural design in their sketchbooks, which they would soon draw again with white crayon on a 40 x 40 cm^2 of canvas. The paper sketchbook would be their "experimenting" space, where they would practice batik techniques before moving on to the more expensive canvas. Shannon was engaging her students in the art studio habit of mind involved with *developing craft* through practice (Hetland, Winner, Veenema, & Sheridan, 2013). In this stage of inquiry, students engaged in trial-and-error reasoning as they experimented with cause and effect. This stage of "messing around" or "playing with the material" is no different than the tinkering an engineer does as he or she works through the elements of a design process. It's an essential part of the work-in-progress. In this case, the amount of water added to the acrylic paint was a very important independent variable. So was the kind of brush and how tightly the fabric was stretched. Students *had* to know it was okay to make a big mess at first. *"Mess" is as much a part of the nature of art practice as it is of science. Insights evolve as mistakes get made.*

Before they started, Shannon modeled all the techniques she would expect her students to practice and learn. She held up a canvas square for them to see and then drew an abstract design on the cloth with her own crayon. Then she demonstrated

Figure 6. Student sketchbook entries representing the characteristics of the North Atlantic Right Whale

several different kinds of mistakes that might occur in the process, and said that they would likely make them, too. She emphasized that artists need to *engage and persist* at the task at hand (Hetland et. al, 2013). Prior to this stage in the project, we had engaged in many other art studio habits of mind, characterized by such evocative verbs as *observe, envision, integrate, express, differentiate,* and *connect worlds* (Hetland et. al, 2013). I thought, *would that more science classes functioned like art studios!* Shannon's students then entered the *work* stage. I noticed, too, that everywhere were signs reminding the students to take care of their studio and their materials. *Truly, this was a space suffused with caring messages.*

One of the most significant ways that Shannon and Sadie demonstrated care for their students' feelings and made their curriculum inclusive was the decision to invite them to make a quilt, pieced together from *every* student's mural design (Figure 7).

The class voted unanimously on one of their classmate's designs, a gifted artist named Rose (pseudonym), who later earned acceptance to the public magnet high school of the arts (Figure 8).

Rose was also appointed to oversee the final mural's completion as general "contractor" of an artifact that was to measure 3.5 meters in length and 0.75 meters in width. Bringing the project to completion would require students to work collaboratively through issues of scale and proportion and practice making precise measurements.

Rose was a born leader. She expertly organized her classmates into work crews, each with its own sub-contractor. Pictured here are members of the measurement and layout team (Figure 9).

Rose made sure that every single classmate participated in the painting of this mural. Here she is, putting on the finishing touches, just before we presented it at the National Marine Educators Association (NAEA) meeting in Savannah, Georgia (Figure 10).

Figure 7. Student's paper and canvas mural designs (left) and assemblage of class quilt

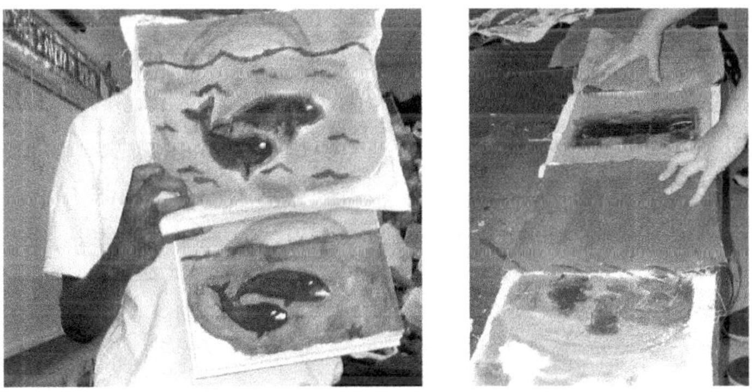

Figure 8. Winning North Atlantic Right Whale mural design on paper and cloth

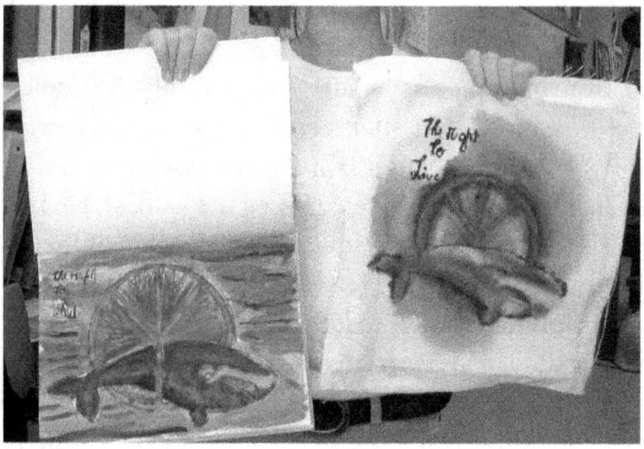

Figure 9. Mural measurement and layout team

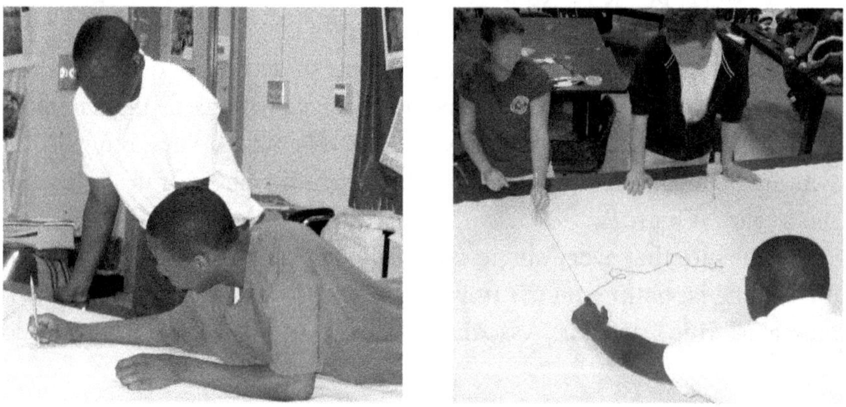

THE MANY FACES OF STEAM

What Is STEAM Anyway?

In 2011 or so, while I was working on this ocean literacy project, I discovered that a pedagogy called STEAM had been "invented" (though I would argue that Leonardo da Vinci had prior rights to that patent). Regardless, I regarded this "newcomer" on the educational landscape as a fantastic harbinger of promising reform. When I began a literature review, however, I grew increasingly concerned that there seemed to be no consensus on just what STEAM was and, most especially, what it was not. My list of questions was long: Which teachers *did* STEAM—art, science, ELA,

Figure 10. Rose puts the finishing touches on the mural, entitled, "The Right to Live!"

social studies, all, some? Did it "happen" after school, or during? If both, how were the versions different from one another? Were specific art forms or science/STEM content areas being privileged over others? What did STEAM look like when it was "working"? What steps were being taken to make sure that STEAM was not just one more task being added to the already overtaxed art teacher's planning book? Were science, STEM, and arts teachers co-constructing project-based lessons? What kind of academic rigor did STEAM pedagogy offer? Further, how were the engineering, technology, and mathematics components being addressed? What kinds of STEAM might be funded? Would competing interests jockey to brand their own versions of STEAM as educational commodities, thus achieving, at best, an isolated impact vs. a collective one? This competitive market strategy seemed counter-intuitive to me. Dewey (1938) challenged us all to consider what conditions have to be satisfied in order that education become a "reality and not just a name or slogan" (p. 116).

Full Disclosure

I myself have never been much for semantics, especially where educational acronyms supposedly signifying some collection of agreed upon skills, knowledge sets, and practices are concerned. Whatever you want to call what it is I am trying to communicate here is fine with me. I am at your service, whether you represent an organization that labels itself as doing science, STEM, STEAM, TSTA, or something similarly catchy. We are all together in this business of educating. I have seen that a collaborative and collective STEAM education supports teachers and students so that both Science/STEM and the Arts actually matter to all stakeholders (Koester, in press). I further believe that STEAM in which the arts are used as "handmaidens" in the service of science/STEM is *not* STEAM. Likewise, when science/STEM content understanding is not an integral part of a more Arts-based version, this is *not* STEAM either. It's something else. The most glorious explosions of STEAM learning I have witnessed emerged when art and science teachers worked together, each celebrating the gifts of the other and joining in the collective goal of effecting deep learning and appreciation of the importance of their respective areas of human endeavor.

Project Draw for Science

Just before I began The L.A.S.T. Book ocean literacy project, I had accepted a position as an adjunct professor at the College of Charleston, teaching science methods for preservice teachers, grades 2-8. I very quickly discovered with a pre-test that most of the elementary pre-service and many of the middle grade candidates lacked deep science content knowledge, especially in the physical sciences. I told them about my first ever teaching job, how I had used drawing to first deepen my own knowledge of physics, and how teaching through drawing had resulted in positive academic turn-arounds. I shared Leonardo da Vinci's claim: *To draw something is to know it.* We studied the maestro's detailed drawings of anatomy, machines, and the movement of water. I then asked these pre-service science teachers to consider the corollary claim: To *know* something is to be able to draw a simple picture explaining it. Immediately came cries of protest, "Not if you can't draw!"

"A valid point," I acknowledged. Let's test my claim *and* your counterclaim. I reached for a bag I had brought with me and withdrew its contents. "Here is a loaf of bread and a jar of peanut butter," I announced. "Using them as a visual reference, can you draw a simple picture that explains how to make a peanut butter sandwich? The drawing only has to be *readable*—not a da Vinci masterpiece," I added. They could all readily complete this task, each in his or her own way. I continued, "Could you argue that you have now used drawing as a kind of *language* to communicate your understanding of how to make a PB & J sandwich? Could I (as your teacher) readily determine from your drawing whether your understanding was well developed (or not)? They agreed with both propositions and quickly surmised that if they did indeed have knowledge about a concept or procedure *and* they had a visual reference, they could produce a simple explanatory diagram—even if they were not gifted artists.

I explained to these science-teachers-to-be that few science teachers whom I had trained had ever been afforded any training in the graphic art of visually organizing text and images on a single page. Often their notes are splattered across their classroom white boards like symbolic vomitus (Figure 11).

Student versions of these notes are often even more confusing and certainly do not invite later study. Why? Because there is no story—nothing to make the scribbles (strange symbols written in a strange language) *mean* something. The result is complete semiotic breakdown (Koester, 2015).

Finson and Pederson (2011) have argued for the importance of "visual data" in science and posited that a "comprehensive understanding of a problem and potential solutions to it are possible only when one knows not only particular skills in visualization, but also in knowing the right questions to ask" (p. 71). I regard this claim as a cause to action (in Koester, 2015a). I ask, "Why is the drawing of science understanding not afforded at *least* equal status with telling and writing, especially in teacher education programs?" Michael Polanyi (1966) has said that we all know more than we can tell. I challenged my pre-service science students to find evidence to support Polanyi's claim throughout their student teaching experience. Over time, I began to imagine an action research project that asked the following question: What if drawing were conceptualized as science language *and* as a legitimate means of formatively assessing science "knowing"? I placed parentheses around the prefix "knowing" to remind one that the very nature of science involves a constant re-testing and revision of what we *think* we "know" based on the evidence we may have at a given time (Koester, 2015a).

In 2012, with funding and support by the University of SC Center for Science Education, I began the Project Draw for Science research with five teachers from the poorest to the most affluent districts in our state, all of whom self-identified as not

Figure 11. Teacher science notes that are visually chaotic in their presentation

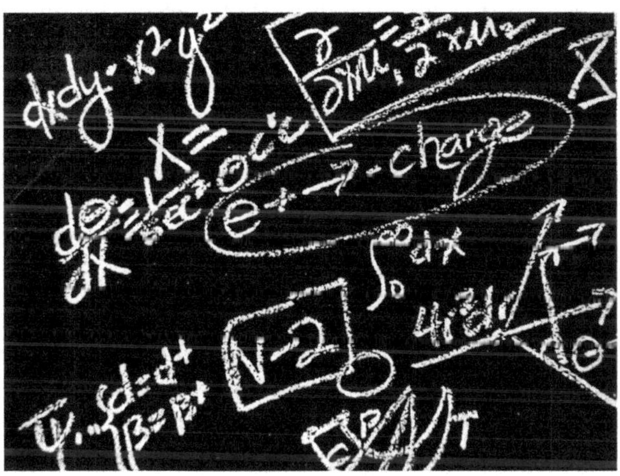

being able to reach their struggling readers and wanted to learn how to teach through drawing. They exhibited deep levels of care for their students and took part in this study with no remuneration at all. After seven months, the students who made the most significant improvement were either struggling readers and/or special needs students. When asked to explain how it was that learning science through drawing affected them, these same students replied with responses like, "It makes me feel like I'm getting help. It helps me think way better than just listening does. It makes me feel joyful and energetic!"

I knew from the years leading up to this new study that the science teachers I had trained felt empowered when they learned the art of visually organizing content for and with their students. They also confessed that creating drawing repertoires as part of their lesson preparation deepened their own content knowledge. The process of teaching science through drawing which I had developed to this point began with first learning basic drawing skills and the elements of graphic design, followed by identifying only the most important science vocabulary words for each concept. Struggling readers, especially those with dyslexia—some 20% of learners (Shaywitz, 2003)— benefit greatly from having new words syllabicated and pronounced out loud as they are written. Inviting students to create illustrated flash cards for new terms is very helpful for English Language Learners. At the "advanced" level of teaching science through drawing, I teach educators how to combine creative drama techniques, scriptwriting, and drawing to effect what I call the Performative Narrative Drawing (Koester, 2015a).

As the beginning of the Project Draw for Science study (http://merriekoester.wix.com/project-dfs), I began an earnest quest to deepen my own understanding of science *language*. My readings took me straight into the important work of Jay Lemke (1999, 2002). Never had I come across a scholar who had had so keenly studied the qualities of science classroom discourse practices and the many ways in which a student could become completely overwhelmed by the multiple and simultaneous de-coding processes required in a science classroom setting. Lemke determined that at any one time in the teaching and learning of science, students were bombarded by not one, but four different languages: WORDS, IMAGES, SYMBOLS, and ACTIONS, all part of a single science *sign system,* which must be integrated if any kind of significant sense making (other than just low level memorizing) were to be achieved (2002).

I was thrilled by Lemke's semiotic portrayal of a science teaching/learning *system*, as it was an important missing ingredient in my own model, which was, at that time, more about "what's" and "how's," but not so much about how those dimensions were integrated. Two epiphanies immediately occurred to me: 1) Lemke's emphasis on *contextualizing* the languages of science was essential practice and 2) we science teachers, myself included, use too many words in isolation from the mathematical symbols, images, and actions associated with the rest of the science inquiry.

The Art and Design of the "Know"tation

I set out, as part of the Project Draw for Science research, to construct and field test a graphic visualization tool for the teaching and formative assessment of science content learning. I visualized a one-page composition which integrated all of Lemke's four science languages: 1) science terminology (the WORDS); 2) explanatory drawings (the IMAGES); 3) the equipment and procedures used during an investigation (the ACTIONS); 4) and mathematical equations, icons, graphs, and tables (the SYMBOLS). I would call this pedagogical tool a "Know"tation. Its purpose was to tell the *story* of situated, contextual, science inquiry as it unfolded in real-time in the classroom. While there would be no one *right* way to design the "Know"tation, the relationship between the four science languages should be readily apparent and the meaning capable of being "read" (Koester, 2015). Depending on the stage of inquiry during which it was created, the "Know"tation might have more words than symbols, but all designs would feature a prominent explanatory and narrative image. Further, an image might serve multiple purposes, such as depicting an experimental procedure (an action) and/or explaining a phenomenon (such as why the sky is blue). Criteria for formative evaluation could be assessed using the developmental terms like Emerging, Proficient, and Mature along criteria that would include scoring the following: Misconception Quotient (out of 5), Graphic Design, Creativity, Language Integration, and Clear Meaning Conveyed. I am still developing the evaluative component of the "Know"tation and enthusiastically welcome all feedback and suggestions!

The readability of a "Know"tation or any graphic design comes from the arrangement of its elements. My favorite book on the subject of visual note design is Mike Rhode's *The Sketchnote Handbook* (2013). I discovered Rhode's work when I enrolled in an online course called "Rockstar Scribe" (https://become-a-rockstar-scribe-at-school-or-work.teachery.co/register) as a way of improving my own graphic visualization skills (before trying to teach it myself). In the Project Draw for Science capacity building training, we began by considering four essential elements of page design: 1) typography, 2) images and their placement, 3) white space, and 4) hierarchy and flow between the elements. We practiced making different kinds of directional arrows and icons for bullets (Figure 12).

We also reflected on the importance of creating a clear visual path through the composition. Rhode (2013) presented several readable "pathways," which I have represented here (Figure 13).

Even teachers with very deep content knowledge readily recognize that unless they employ the elements of design in their graphic representations of content, the result can be rather chaotic (Figure 14).

Figure 12. Elements of graphic design, which enhance visual "flow"

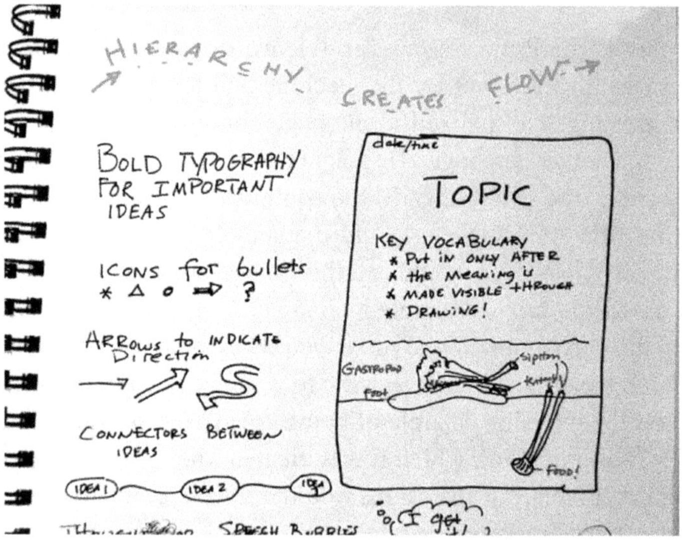

Figure 13. Visually "readable" pathways through a graphic representation

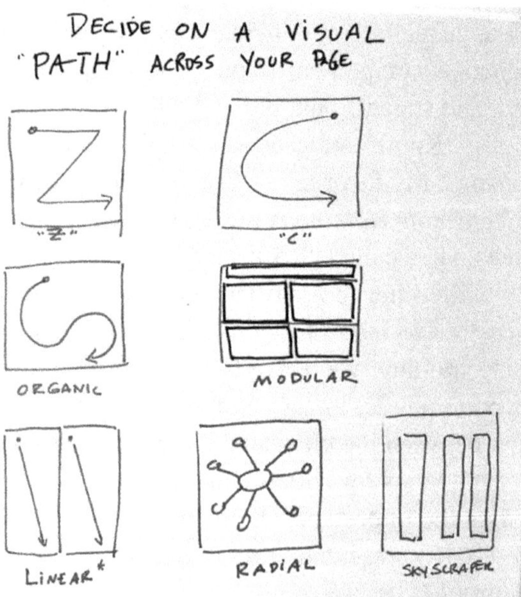

Figure 14. Pre-training attempt by teacher to create a visual explanatory model of climate change

In very short order, teacher participants progressed to making "Know"tations during each stage of 5E science inquiry (Bybee, et al., 1989). One teacher, "Ms. Maya"—most of whose students were reading well below grade level—created *Explain* "Know"tations with her students on large post-it note sheets, which they later used to review for their end-of-year standardized tests (Figure 15).

Unlike *Explain* "Know"tations, which are mostly words and images, with perhaps some symbols, *Explore* "Know"tations feature words, images, symbols, and actions, thoughtfully integrated through the applied use of graphic design elements to tell the story of an actual experiment in progress (Figure 16).

In every way, we came to regard the design of "Know"tations as acts of jointly performed critical making—as public artifacts to be shared and revised as needed and through whose *construction*, students learned to integrate the languages of science.

The Problem of Teacher Content Knowledge

The process of teaching science through drawing and visual thinking is not without its drawbacks. Hashweh (1987) has shown that poor content knowledge significantly impacts teaching, and is a source of misconceptions directly communicated to students (Cochran,1991, p. 12). I have discovered that one of the best ways for teachers to identify weaknesses in their content and procedural knowledge is what I call the "back-of-the-napkin-test." This can be a liminal experience for many teachers.

Figure 15. Large, poster-sized Explain "Know" tation with a large prominent image and only key words

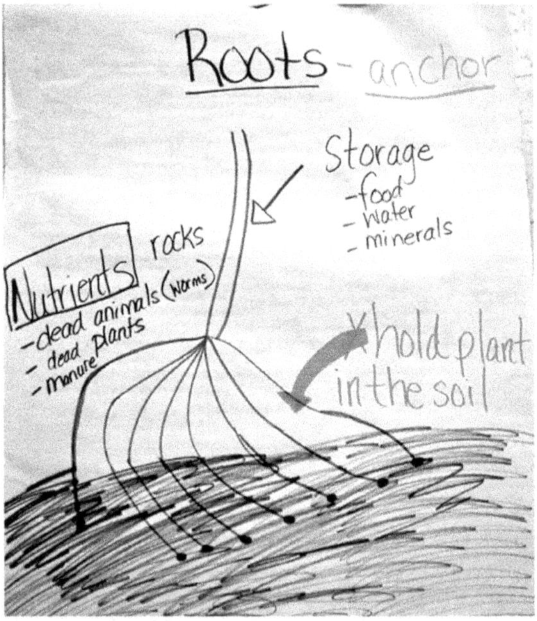

Figure 16. "Know" tation created during the Explore phase of inquiry. Note that all four languages of science are featured here.

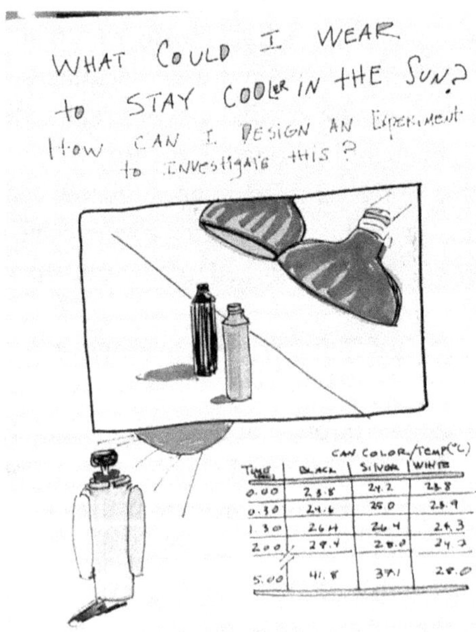

What they quickly discover is that (even if they say they have zero drawing *talent*) they are more than capable of making a simple explanatory drawing of a subject they know well and about which they care. However, when tasked to draw about a science subject which they understand poorly, the napkins quite often remain blank, and teacher faces furrow with scowls. Only a few moments later, they more or less confess that they do a terrible job of teaching about their weak areas. Right away, they also realize that (if they are to actually *teach* science through drawing) they must draw their way into deeper relationship with their most dreaded content area. As was stated earlier, many middle school science teachers have quite under-developed content knowledge about the physical sciences (Figure 17).

Part of the work I do through long-term professional development is to advocate for the creation of standards-based drawing repertoires. Along the way, we work very hard to uncover and reflect upon the misconceptions that many science teachers commonly believe about the content they are teaching (Koester, in press). Critique of student drawings is also essential. The evidence of both learning and *mis*-learning is right there, but can easily be missed by the teacher who becomes distracted by the aesthetic appeal of a student's work. For example, an artistically gifted student produced a visually appealing, explanatory drawing about photosynthesis, incorrectly depicting oxygen being given off by *flower* and carbon dioxide being taken up by the roots. Her teacher was so impressed with the quality of the drawing that he failed to notice the misconceptions. This finding was not an isolated one at all. Accordingly, we are developing a theory of "aesthetic masking" (Figure 18).

Figure 17. The Back-of-the-Napkin Test. Teacher draws a "blank" for physical science concept

Figure 18. Aesthetic masking of student misconceptions about photosynthesis

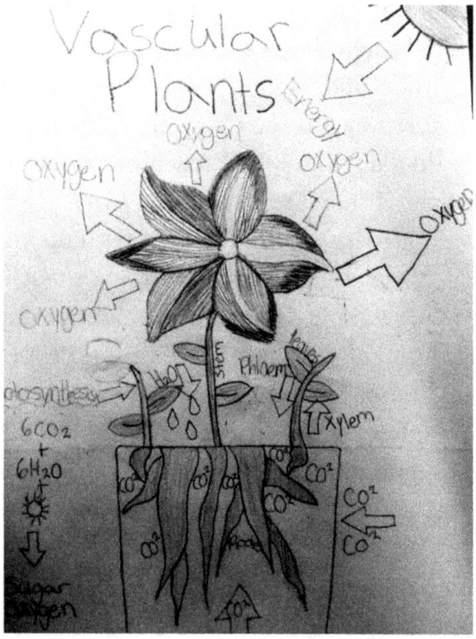

CONCLUSION: C'ING STEAM

In the conclusion of *Science Teachers Who Draw: The Red Is Always There*, I set forth a tentative model for teaching science as caring, non-prescriptive, aesthetic inquiry, one whose goal is achieving a pragmatic level of science literacy for all students, regardless of their reading ability. I believe that an aesthetic paradigm in general and drawing instruction in particular should not only be included in science teacher education programs but become the subject of further participatory action research and professional development efforts. Shulman (1987) concluded that the more we learn about teaching, the more "we will come to recognize new categories of performance and understanding that are characteristics of good teachers, and will have to reconsider and redefine other domains" (p. 13). Lederman's (1992, 2007) reviews of the research on the teaching of the Nature of Science ultimately concluded that the "instructional approach, style, rapport, and personality of the teacher are all important variables in effective science teaching" (2007, p. 845). An aesthetic orientation to teaching implies that teachers can't just script and *direct* student learning, they must, *with* students, jointly *produce* learning *performances*, at the end of which they must also assess student emotions and *feelings* about what just happened (or didn't). Indeed, feedback is crucial, but it can't just be about whether students have memorized the content. Conquergood (1998) has stated that the performance paradigm privileges 'an experiential, participatory epistemology' (p. 27), and believed that only through shared emotional experiences can a performance achieve transformation (in Koester, 2015, p. 186).

I have come a very long way from that twenty-something young woman, whom a class of very bright non-readers inspired to change her way of being a science teacher. They were with me when I developed a pedagogy I called TSTA. They are with me now as I tentatively call what I am doing STEAM. I could not have made this journey without the support of those who believed that what I was doing mattered. All along the way, I have been building tribes. I seek to develop practice-embedded educational research (PEER) projects of the kind that have been called for by Catherine Snow in her 2014 Wallace Foundation Distinguished Lecture for AERA, published in 2015. This PEER work would be characterized by a constellation of C's: *Caring Collaboration* among *Content* specialists, *Cultural relevance*, *Cross-training, Curating*, *Creative process*, *Community*, and the achievement of a *Collective* impact. Participants in this project would collectively build a professional STEAM teaching/learning community characterized by the sharing of lessons that have "worked" for them, as well as obstacles they may have encountered in the process. I am hopefully optimistic about the possibilities!

A democratic and inclusive science education must work to decrease inequalities and achievement gaps. To date, non-dominant culture and low SES students are often tracked into the lowest level classes with the least highly qualified teachers and with consistently low expectations made of them (Oakes, 2005, in Koester, 2015). The result can be complete semiotic breakdown and loss of human capital. Who knows how many of these students might otherwise have actually *enjoyed* the study of science had they also been assigned to experienced educators who established *ways into* meaning-making that did not depend on the written or verbal expression of a "foreign" language. Indeed, I imagine such abandoned students shaking their heads in disgust, declaring, "Forget science! How can I be in your community if I can't speak your language or understand your sign systems?" Who can blame them? The Lakota had much to teach us all: "*Mitakuye oyasin.* We are all related."

REFERENCES

Andersen, C. (2004). Learning in as-if worlds: Cognition in drama in education. *Theory into Practice*, *43*(4), 281–286. doi:10.1207/s15430421tip4304_6

Bigelow, B. (2009). Once upon a genocide: Columbus in children's literature. In W. Au (Ed.), *Rethinking multicultural education: Teaching for racial and cultural justice* (pp. 73–85). Milwaukee, WI: Rethinking Schools.

Bresler, L. (2006). Toward connectedness: Aesthetically based research. *Studies in Art Education, 48*(1), 52–69.

Brown, J. S., Collins, A., & Duguid, P. (1989). Situated cognition and the culture of learning. *Educational Researcher, 18*(1), 32–42. doi:10.3102/0013189X018001032

Bybee, R. W. (1989). *Science and technology education for the elementary years: Frameworks for curriculum and instruction*. Washington, DC: The National Center for Improving Instruction.

Carnine, L., & Carnine, D. (2004). The interaction of reading skills and science content knowledge when teaching struggling secondary students. *Reading & Writing Quarterly, 20*(2), 203–218. doi:10.1080/10573560490264134

Cochran, K. F. (1991). *Pedagogical content knowledge: A tentative model for teacher preparation*. Paper presented at the annual meeting of the American Educational Research Association, Chicago, IL.

Columbus, C. *(1492). Journal of the first voyage (Diario del primer viaje) 1492* (B. W. Ife, Trans. & Ed.). Warminster, UK: Aris and Phillips Ltd.

Conquergood, D. (1998). Beyond the text: Toward a performative cultural politics. In S. Dailey (Ed.), *The future of performance studies: Visions and revisions* (pp. 25–36). Annandale, VA: National Communication Association.

Dewey, J. (1938). *Experience and education*. New York, NY: Kappa Delta Pi.

Ehrenfeld, T. (2011). Reflections on mirror neurons. *Observer, 24*(3). Retrieved from http://www.psychologicalscience.org/index.php/publications/observer/2011/march-11/reflections-on-mirror-neurons.html

Eisner, E. (1991). *The enlightened eye: Qualitative inquiry and the enhancement of educational practice*. New York, NY: Macmillan.

Eisner, E. (2002). *The arts and the creation of mind*. New Haven, CT: Yale University Press.

Finson, K., & Pederson, J. (2011). What are visual data and what utility do they have for science education? *Journal of Visual Literacy, 30*(1), 66–85. doi:10.1080/23796529.2011.11674685

Foucault, M. (1977). *Discipline and punish: The birth of the prison* (A. Sheridan, Trans.). New York, NY: Vintage Books.

Gallese, V. (2005). Embodied simulation: From neurons to phenomenal experience. *Phenomenology and the Cognitive Sciences*, *4*(1), 23–48. doi:10.1007/s11097-005-4737-z

Greene, M. (2001). *Variations on a blue guitar: The Lincoln Center Institute lectures on aesthetic education*. New York, NY: Teachers College Press.

Hashweh, M. Z. (1987). Effects of subject-matter knowledge in the teaching of biology and physics. *Teaching and Teacher Education*, *3*(2), 109–120. doi:10.1016/0742-051X(87)90012-6

Hetland, L., Winner, E., Veenema, S., & Sheridan, K. (2013). *Studio thinking 2: The real benefits of visual arts education*. New York, NY: Teachers College Press.

Koester, M. (1989). *Science and art together again* (Unpublished master's thesis). University of Hawaii, Manoa, HI.

Koester, M. (2011). What's your story? How can we bring students to science through reading and the arts? Keynote address delivered at the South Carolina Science Council meeting, Myrtle Beach, SC.

Koester, M. (2014). *Project Draw for Science*. Retrieved from http://merriekoester.wix.com/project-dfs

Koester, M. (2015a). *Science teachers who draw: The red is always there*. Blue Mounds, WI: Deep University Press.

Koester, M. (2015b). *Agnes Pflumm and the stonecreek science fair* (3rd ed.). Charleston, SC: Read for Science Publishing.

Koester, M. (2015c). *No place like periwinkle* (3rd ed.). Charleston, SC: Read for Science Publishing.

Koester, M. (2015d). *Pond scum and agnes pflumm* (3rd ed.). Charleston, SC: Read for Science Publishing.

Koester, M. (2015e). *Agnes pflumm and the secret of the seven* (2nd ed.). Charleston, SC: Read for Science Publishing.

Koester, M. (in press). Deep learning in science: An argument for an aesthetic paradigm of science education. In F. V. Tochon & D. Busciglio (Eds.), Deep Education Across the Disciplines and Beyond: A 21st Century Transdisciplinary Breakthrough. Blue Mounds, WI: Deep University Press.

Lederman, N. (2007). Nature of science: Past, present, and future. In S. Abell & N. Lederman (Eds.), *Handbook of research on science education Abingdon*. Abingdon, VA: Taylor and Francis.

Lederman, N. G. (1992). Students and teachers conceptions of the nature of science: A review of the research. *Journal of Research in Science Teaching*, *29*(4), 331–359. doi:10.1002/tea.3660290404

Lee, O., Fradd, S. H., & Sutman, F. X. (1995). Science knowledge and cognitive strategy use among culturally and linguistically diverse students. *Journal of Research in Science Teaching*, *32*(8), 797–816. doi:10.1002/tea.3660320804

Lemke, J. L. (1999). Typological and topological meaning in diagnostic discourse. *Discourse Processes*, *27*(2), 173–185. doi:10.1080/01638539909545057

Lemke, J. L. (2002). Enseñar todos los lenguajes de la ciencia: palabras, símbolos, imágenes y acciones. In M. Benlloch (Ed.), *La educación en ciencias: ideas para mejorar su práctica* (pp. 159–186). Barcelona, Spain: Paidos.

Litts, B. (2015). *Making learning: Makerspaces as learning environments* (Doctoral dissertation). Retrieved from www.informalscience.org/sites/.../Litts_2015_Dissertation_Published.pdf

Loucks-Horsley, S. (1996). Principles of effective professional development for mathematics and science education: A synthesis of standards. *NISE Brief, 1*(1). Retrieved from http://eric.ed.gov/?id=ED409201

May, R. (1985). *The courage to create*. Toronto, Canada: Bantam.

NGSS Lead States. (2013). *Next Generation Science Standards: For States, By States*. Washington, DC: The National Academies Press.

NOAA Office of Ocean Exploration. (2016). *Ocean Explorer*. Retrieved from http://oceanexplorer.noaa.gov/

Oakes, J. (2005). Keeping track: How schools structure inequality. New Haven, CT: Yale University.

Ocean Literacy Framework. (2005-2015). Retrieved from http://oceanliteracy.wp2.coexploration.org

Papert, S. (1991). Situating constructionism. In I. Harel & S. Papert (Eds.), *Constructionism* (pp. 1–11). Norwood, NJ: Ablex.

Pestalozzi, J. (1973). *How Gertrude teaches her children*. New York, NY: Gordon Press. (Original work published 1894)

Polanyi, M. (1966). *The tacit dimension*. New York, NY: Doubleday.

Rhode, M. (2013). *The Sketchnote handbook: The illustrated guide to visual notetaking*. Berkeley, CA: Peachpit Press.

Settlage, J., & Southerland, S. A. (2007). *Teaching science to every child: Using culture as a starting point*. Abingdon, UK: Taylor & Francis.

Shaywitz, S. (2003). *Overcoming dyslexia: A new and complete science-based program for reading problems at any level*. New York, NY: Vintage Books.

Shulman, L. S. (1987). Knowledge and teaching: Foundations of the new reform. *Harvard Educational Review*, *57*(1), 1–23. doi:10.17763/haer.57.1.j463w79r56455411

Siegesmund, R. (2010). Aesthetics as a curriculum of care and responsible choice. In T. Constantion & B. White (Eds.), *Essays on aesthetic education for the 21st century* (pp. 81–92). Rotterdam: Sense Publishers.

Snow, C. (2015). Rigor and realism: Doing educational science in the real world. 2014 Wallace Foundation distinguished lecture. *Educational Researcher*, *44*(9), 460–466. doi:10.3102/0013189X15619166

Tochon, F. V. (2013). *Educational semiotics: Signs and symbols in education*. Blue Mounds, WI: Deep University Press.

Trefil, J. (2008). *Why Science?* New York: Columbia University, Teacher's College Press.

Uhrmacher, P. B. (2010). The power to transform: Implementation as aesthetic awakening. In T. Constantion & B. White (Eds.), *Essays on aesthetic education for the 21st century* (pp. 183–203). Rotterdam: Sense Publishers.

KEY TERMS AND DEFINITIONS

Aesthetic Energy: A renewable form of human energy characterized by the presence of openness, care, empathy, emotional connection, sustained acts of noticing and awareness, and flexible, creative ontologies and epistemologies (from Koester, 2015).

Aesthetics: In this book, I have employed Baumgartner's original *sense*-based conceptualization of this term together with Pestalozzi, Dewey, Greene, and Eisner's more ontological characterizations of aesthetics as being purposefully empathic and caring, interactive, connective, feelful, and centered around the practice of keen noticing the particular qualities of something, whatever that may be (from Koester, 2015).

Art/Science Integration: A curriculum and/or creative design task that features time for artistic, personalized exploration and interpretation of science content and phenomena as well as critique of public artifacts that employs the use of the languages of both art and science (from Koester, 2015).

Art: Throughout this chapter, art has been conceptualized as it was *before* it was subsumed by the wave of empiricism that emerged in the 18th century—as any act of making that involves intellectual judgment. Art is the medium through which human beings develop their intuitive and creative potential, regardless of the domain inside of which (like science or teaching) it is practiced (from Koester, 2015).

Deep Education in Science: Is about bringing students into an awareness and appreciation that science can uniquely help them understand three things: 1)what is happening in the world under study, 2) how this happening affects them, and 3) how they themselves affect the happening. Deep learning is enhanced when topics are explored through aesthetic inquiry (from Koester, 2015).

"Know"tation: A play on the word *notation*, I created this term for two reasons: 1) to communicate the reflexive, generative nature of science as a domain of knowledge and practice, and 2) to encourage both teachers and students to visually organize their understandings of science concepts, interactive experiences, or phenomena according to the principles of graphic design. The goal of the "Know"tation is to show on a single page, through an aesthetically pleasing combination of words (typography), drawings, and white space, the story of a teaching/learning experience in science so that an uninformed viewer can "read" that story. In every case, "Know"tations should be collaboratively assessed for the presence of misconceptions and revised until the visual "story" as closely as possible re-presents the most accurate level of "knowing" about that science subject. "Know"tations can be created at every stage of the 5E learning cycle (from Koester, 2015).

Mirror Neurons: These are specialized brain cells that fire whether we see or imagine someone else perform an action *or* do the action ourselves (from Koester, 2015).

S.T.E.A.M School: A participatory action research setting in which teachers with fluency in either S.T.E.M. or the Arts (or both) would form a curriculum building team. First, they would "cross-train" each other and then together create innovative and integrative curriculum for teaching and learning that is congruent with both S.T.E.M. and the Arts standards. Such "schools" could form and function within all levels of K-20 education, including and especially in teacher education programs. An underlying assumption would be that all curricula would be culturally relevant (from Koester, 2015).

Science: A special, usually highly objective way of observing, interpreting, and describing the world, as well as the body of knowledge that results from that study (from Koester, 2015).

Semiosis: The meaning-making process, contextually situated in lived experience. In Peirce's theory of signs, semiosis depends on the triadic interaction between an object in the world, a sign or symbol that re-presents that object (or phenomenon), and the way(s) a person in a given culture or situation interprets that contextual sign. *Semiotics* is the study of meaning-making processes (from Koester, 2015).

Chapter 5
Exploring Simple Machines With Creative Movement

William Paul Lindquist
Hamline University, USA

Martha James-Hassan
Morgan State University, USA

Nathan C. Lindquist
Bethel University, USA

EXECUTIVE SUMMARY

This chapter explores the use of creative movement to extend meaning to inquiry-based science investigations. This process embraces the addition of A to STEM to realize the impact of STEAM. The chapter builds on the import of scientific and physical literacy, interdisciplinary learning, and the power of kinesthetic engagement. Students become active collaborative agents within a dynamic model using creative movement to bring meaning to the science of simple machines. We utilize a Working Words into Movement strategy to help students use their past experiences and motor memory to explore, interpret, and engage with as they seek understanding of simple machines. A Midwest urban elementary school provides the context for a unit plan culminating in a dance performance. The foundational ideas presented within this unit can be enacted within any classroom by creative movement (physical education or dance) specialists, science specialists, or classroom generalists. It follows with a presentation of science content on simple machines exploring the disciplinary core idea of force and motion.

DOI: 10.4018/978-1-5225-2334-5.ch005

Copyright ©2017, IGI Global. Copying or distributing in print or electronic forms without written permission of IGI Global is prohibited.

INTRODUCTION

In the current climate of high-stakes testing and educational accountability extreme pressure is being placed on schools. Career and College Ready is a catchy binary that serves the needs of adults to create order out of the messy process of educating young people. Those young people, however, may be better served by the less structured—and admittedly more challenging to evaluate—framing of educational goals as, "Life Ready." To be life ready Learners must achieve alpha and numeric literacy as well as habits of inquiry, understandings of history and systems, skills to effectively steward, aesthetic awareness, physical well-being, and most of all, self-efficacy and critical thinking. To accomplish this more ambitious goal of education rather than schooling, teachers must build meaningful relationships with students and, from an instructional standpoint, use every minute twice. Interdisciplinary teaching and learning is a critical tool in delivering dynamic and engaging content to students at the same time as it supports reflection, creative thinking, and critical problem solving.

In this chapter, we will explore the richness achieved when providing multiple and diverse representations of scientific concepts brought about with an accompanying focus in creative movement. We begin with a pedagogical background from both science and creative movement. This leads to a block plan for a sample unit in simple machines that can be carried out in the classroom while suggesting a framework for visualizing the means to integrate movement into any lesson. Our goal and hope is that the reader can walk away empowered by the interdisciplinary vision of including The Arts in STEM bringing to life the powerful potential of a STEAM-focused curriculum.

LITERATURE REVIEW

Scientific Literacy

Within an increasingly complex and technological world facing growing global sustainability issues, it is critical our nation's students develop a foundational level of scientific literacy. Whether they become scientists, journalists, or refuse handlers, their participation in today's world requires the ability to think critically and act responsibly based on sound information and judgment. It will be the abilities to think and act that allow us to maintain and improve the quality of life for future generations to come. Through the process of teaching STEM concepts through movement and teaching movement competency framed in STEM content we can empower students in our classrooms to have the agency and self- confidence to become active participants rather than mere observers in the world.

To accomplish this end, students need engagement in the active "doing" of science. Through this engagement, they develop the proficiency to ask questions, design and carry out investigations, and reflect critically on the outcomes. Children learn science best by "doing" science (Martin, 2011). An engaged elementary classroom committed to the integrated needs of a STEAM curriculum needs to begin with a robust commitment to providing hands-on use of science process skills in open-ended inquiries. With that foundation as a base, learning is enriched and broadened throughout all aspects of STEAM – including the use of technology, the analysis of mathematics, meaningful interactions with text (in the broadest sense), and the integration of the arts – in our case, creative movement.

Physical Literacy

The purpose of education is to prepare young people with the knowledge, skills, and experiences to be successful participants in society. In 2002, the United Nations General Assembly passed resolution 56/116: United Nations Literacy Decade: Education for All (United Nations, 2002). The foundation of this resolution was based on the collective conviction that literacy is crucial to the acquisition of essential life skills that enable every child, youth, and adult to address the challenges they face in life. It represents an essential step in basic education and an indispensable means for effective participation in the societies and economies of the twenty-first century. In their understanding of the complexities and unpredictable nature of the knowledge and experiences necessary for current students to apply in the future, members of the General Assembly had the insight to recognize that authentic literacy is not restricted merely to reading and writing. According to the United Nations Education, Scientific, and Cultural Organization (UNESCO; 2003), literacy is concerned with how we communicate about knowledge, language, and culture within our social practices and relationships.

This holistic, comprehensive understanding of literacy supports the development of interdisciplinary STEAM instruction merging scientific literacy with physical and dance literacy. As articulated by Mandigo, Francis, Lodewyk and Lopez (2009),

physically literate individuals consistently develop the motivation and ability to understand, communicate, apply, and analyze different forms of movement. They are able to demonstrate a variety of movements confidently, competently, creatively, and strategically across a wide range of health-related physical activities. (p.7)

Physical literacy includes viewing physical activity, dance, and sport through a lens of creating, analyzing, interpreting, and evaluating movement as text. For an interdisciplinary theme-based instructional unit to be successful, standards within each of the individual content areas appropriate to that learning must have equal weight and representation. For physical literacy, focus needs to remain on the dance and physical education standards as well as on efforts to get and keep students engaged in moderate-to-vigorous physical activity as much as possible during the teaching, learning, performance, reflection, and assessment process. Traditionally, the concept of physical literacy focuses on instruction in physical education. We have chosen to frame the artistic component of this chapter as creative movement for the intentional purpose of opening up space for movement both inside and outside of the gymnasium. Creative movement, as it is expressed in this writing, is an art form whose medium is the human body in motion (Dow, 2010). Aligned with both dance and physical literacy standards, teachers can plan—and, students can engage in—creative movement experiences in physical education, classroom, science lab, or any other learning space within the school community.

Interdisciplinary Teaching and Learning

Interdisciplinary teaching applies methods and language from more than one academic discipline to examine a theme, issue, question, problem, topic, or experience. (Coffey, n.d.). Interdisciplinary methods create connections between traditionally discrete disciplines for the purpose of increasing engagement and the development of rigorous habits of mind around learning, understanding, and application of knowledge to novel situations. Well planned and executed interdisciplinary teaching enables students to see the links between theories, concepts, and subject areas. Heidi Hayes Jacobs (1989) defines interdisciplinary learning as a knowledge view and curriculum approach consciously applying the methodology and language from more than one discipline to examine a central theme, issue, problem, topic, or experience. Further, Barton and Smith (2000) explain that interdisciplinary teaching and learning enables teachers to use classroom time more efficiently while addressing content in depth, allowing students to see the relationship between content areas.

The National Council for Teachers of English (NCTE) stress the importance of interdisciplinary teaching by recognizing that educational experiences are more authentic and of greater value when the curricula reflect the multi-faceted nature of real life — rather than being compartmentalized into neat subject-matter packages (1995). NCTE highlights the benefits of interdisciplinary teaching promoting the natural and logical connections cutting across content areas organized around questions, themes, problems, or projects rather than traditional subject-matter boundaries (NCTE, 1995). Recognizing the greater value of authentic understanding students will be less likely to wonder *when am I ever going to need to know this* or leave schooling with the stunted perception science is something important from 9:00 am-9:50 am on Thursdays.

It can be argued that the industrialized model of education in the United States was designed for the purpose of indoctrinating students with just enough information to remember and repeat in order to create a functional civic society. As education adapts to meet the needs of preparing students to live, work, and flourish in a global society, the impetus exists to move away from learning as recall towards a learning process that is on-going, respectful of diversity, and focuses on being able to find, evaluate, and use information to communicate ideas. The goals of learning in the 21st century include the development of self-efficacy, cultural norms and expectations, and a fundamental understanding of academic subjects that support students in asking meaningful questions while becoming self-sustaining, lifelong learners (Bransford & National Research Council, 2000). Changing language included in the reauthorization of the Elementary and Secondary Education Act of 1965 under the title of the "Every Student Succeeds Act" recognizes the value of and need for students to experience a wide variety of content areas and interdisciplinary study in order to be fully educated as whole learners.

Interdisciplinary teaching and learning help students develop knowledge, insights, problem solving skills, self-confidence, self-efficacy, and a passion for learning. Interdisciplinary teaching and learning can also be included in the classification of culturally relevant pedagogies as it values past experiences, interconnectivity, and context. Further, interdisciplinary teaching is culturally responsive because there is a focus on introducing content from a variety of perspectives that may challenge existing notions, while encouraging students to integrate concepts and ideas into a broader conceptual framework of analysis (Bransford & National Research Council, 2000).

It is the belief of the authors that the "A" in STEAM should be a driving force in the creation of highly engaging, relevant, interdisciplinary teaching and learning in the disciplines of science, technology, engineering, and math (STEM). Our perspective is based on combining research illustrating that Arts-based content integration may be critical to meeting the increased demand for creative thinkers in a global economy and that "art and the process of creation/making promote collaboration and help students synthesize and put into practice what they have learned" (Bedetti, 2015, p. 1). Further, in isolation STEM is missing a key set of creativity-related components that are equally important in developing the creative thinking and problem solving that will likely be essential for success in future job markets (White, 2011). In addition to adding to a student's meaning making ability through the creative process, there is a significant portion of the population whose dominant cognitive processing system is *not* language based and therefore is in need of alternate instructional and assessment opportunities such as those that arts— visual, dance, theater, and music—learning experiences provide in order to reach their full cognitive potential (Groff, 2013). Lastly, arts-based interdisciplinary learning fosters a positive school/community climate and increased self-confidence and academic self-efficacy of teachers and students.

The specific emphasis on creative movement as a platform for learning and expressing understanding of STEM theories and concepts has a number of additional inherent benefits for health, neurological conditions for learning, and social skill development that are worth acknowledging even though they are beyond the scope of this chapter. This chapter will focus on the power of creative movement "as an authentic semiotic meaning-making tool that supports sophisticated levels of embodied thinking, aesthetically oriented expressive communication and empathetic social contagion" (Deans, 2016, p. 55).

Addressing Standards

The integration of arts-based interdisciplinary teaching and learning with authentic and effective STEAM practices requires effort. Haring and Kelner (2015) describe three practices supporting cross- content collaborative teaching: (a) develop a common language, (b) be creative about combining classes, and (c) co-plan creatively. Developing a common language, across all grade levels and content areas, facilitates teacher collaboration and supports students in transferring and applying learning between classes. Creativity about connections in concepts, theories, ideas, and work products need not rest solely with the teachers. Interactions with community experts engaged in the STEAM experience will contribute to making creative connections if the environment is supportive of critical thinking and reflection. Simultaneously encompassing and moving beyond these three practices offered by Haring and Kelner, Spitzer (2013) describes that successful interdisciplinary teaching and learning includes a clear and shared vision of interdisciplinary themes, the theory and practical application of teaching and learning—the didactics—as well as the insight and ability to exploit institutional opportunities. Exploitation of institutional opportunities may include leveraging local expertise, community partners, space, materials, schedules, and staffing. Supporting these additional contributions made by Spitzer, the authors add a fourth practice to support high quality STEAM experiences for students: all instruction and assessment is standards-based.

Moving forward with a standards-based perspective to interdisciplinary STEAM experiences in schools requires a few definitions:

1. ***Encore* Content Areas:** *Encore* content areas are those content areas often marginalized through the implementation of policies such as *No Child Left*

Behind. They are the classes in schools that for the last several years have been considered to be outside the bounds of "core content" areas. Part of the importance of encore content areas is the high engagement levels often associated with them. Classes typically included as encore content areas are the arts, science, social studies, world languages, health, and physical education. Encore content areas can serve either as a primary driver of interdisciplinary learning or as an enhancement for more traditional core content.

2. **Text:** In supporting authentic literacy development for students, Merriam-Webster offer this valuable supporting definition of text: "something considered as an object to be examined, explicated, or deconstructed." In this definition, text is seen as any symbolic representation of meaning. Utilizing this definition in relation to STEAM teaching and learning is beneficial for two reasons: (a) it calls to attention the academic value of observational, interpretational, and aesthetic texts in shaping meaning- making, and (b) illustrates the literacy needs for comprehension beyond phonetic decoding and interpreting of the written word. In this broader interpretation, text includes but is not limited to charts, maps, graphs, movies, music, choreography, sculpture, posture, gesture, tapestry, painting, poems, books, short stories, and informational writings.

The expanded definition of text as symbolic representation of meaning becomes especially critical in the light of the literacy strand within the Common Core State Standards (CCSS) intended to guide alpha-literacy in science, social studies, history, and the technical subjects. In addition to supporting alpha-literacy development and usage, this robust definition of text supports the fundamental understanding that, while alpha and numeric literacy are critical, the primary driver of instruction in encore content areas should be the content Standards for that area of study. As previously stated, relying on Standards is a key factor in implementing an academically rigorous STEAM unit. Therefore, the literacy stands included in the CCSS provide a common academic language and foundational structure for interdisciplinary teaching inclusive of encore content areas.

Quality interdisciplinary teaching should focus instruction and evaluation around content area standards. The "A" in this STEAM project is a creative-movement based exploration and demonstration of student understanding of concepts associated with simple machines. From an instructional perspective, a critical factor in the unit presented is to embrace the relationship between concepts and how they can be metaphorically understood and represented through movement. The texts, or learning products, from this unit may include; visual representations, science lab/observation reports, choreography, written or oral critical reflections, or other aesthetic representations of content, process, or application. Components of each of these texts can be introduced, developed, or assessed based on specific literacy strands in the CCSS (see Appendix A).

Models and Metaphors

NGSS Practice 2 focuses on the development and use of models as "explicit representations that are in some ways analogous to the phenomena they represent" (National Research Council, 2012, p. 56). Modeling of concepts is an integral component of contemporary scientific practice. The use of models as a way of exploring phenomenon is something we need to engage children throughout their P-12 school years. Children possess a native desire for knowledge - a desire that is in place long before they arrive in school. Even very young children are able to utilize the practice of modeling in their quest to understand the world (DeLoache, 2000). Gopnik has shown there is a good degree of scientific thinking present in infants and toddlers (Gopnik, 2012). This provides a strong foundational cognitive structure to build on as students enter school.

The metaphors elicited through the use of models can be effectively used to build understanding. Metaphors are able to tap into the pre-linguistic structures that developed our foundational cognitive schema (Lakoff & Johnson, 1980). Children are wired from an early age to develop these schemata of understanding through their basic bodily experience.

As children move through the early stages of cognitive development described by Piaget, they are reliant on physical engagement with the world as means to develop their basic understandings. The majority of elementary students reside in the concrete operational stage of development. During this stage students need continued access to physical and concrete experience as they more fully develop their understandings. Quoting Thomas Edison, "Great ideas originate in the muscles."

Our approach in this chapter is to engage students as active agents within the conceptual model itself. Models are inexact copies of the phenomenon under investigation with each model providing only a glimpse and the use of multiple models providing an aggregate of glimpses along the path toward greater understanding. Models are "deliberate simplifications of more complex systems" (National Research Council, p. 110). Within this framework of simplified glimpses, it is critical we are careful and deliberate with the modeling in which we choose to engage. We need transparency with our students regarding both the explanatory power and the limitations inherent in each model we utilize. Students must be actively engaged in critiquing the model with a look to how the use of the model adds to new understandings. To this end, the framework we describe here engages students in a deliberate and recursive critique of our kinesthetic models, discuss what worked, what didn't, and look toward ways it might enhance understanding. It is critical we develop a well-worn path between the originating phenomena under investigation, the conceptual understanding attempting to be developed, and the kinesthetic modeling activity. At each stage, students should be engaged in action, dialogue, writing, drawing, and further investigation. It is this practice of multiple representations that can result in the understanding we seek.

This practice of intentional reflection on each product and step in the process is supported by invoking an expanded definition of text inclusive of all symbolic representations of meaning. This expanded interpretation allows each participant in the thematic collaboration to focus on the text most relevant to addressing the academic standards of that content area. Further, this expanded definition of text allows students, with support of peers and the facilitator, to compare and contrast multiple texts on the same topic to verify and strengthen understandings.

Science Standards

Figure 1 identifies two third grade performance standards drawn from the Next Generation Science Standards (Lead States, 2013) within Motion and Stability: Forces and Interactions. Captured as well is the Disciplinary Core Idea PS2.A: Forces and Motion.

CASE DESCRIPTION

The school is a typical elementary school in an upper Midwest urban school district. The gymnasium is bustling with teachers, parents, and students putting the finishing touches on tri-fold boards, suspended mobiles, and live experiments in various stages of composition, eruption, explosion, and decomposition. In the center of this gym-turned-science-fair stands an elevated stage where groups of students will soon share their artistic interpretations of scientific inquiry dance performance to a packed house theater-in-the-round.

In the midst of all this activity, Jason, a sixth-grade student, was anxiously waiting for the other members of his group to arrive. Jason's group had investigated and interpreted the cyclical stages of hurricanes. It quickly became apparent that his group members were not going to make it to the performance on time and could not be reached by phone. As his teacher, I spoke to Jason ensuring him his grade would not suffer if his colleagues did not arrive. Significantly overweight, much taller than his peers, an ethnic minority in this urban school, with a severe speech impediment, looked at me with pleading eyes and through stuttered speech asked, "If I can figure out how to do it by myself, can I please go on stage and dance by myself?" Jason did just that.

Figure 1. Motion and stability: Forces and interactions

Students who demonstrate understanding can:

3-PS2-1.
Plan and conduct an investigation to provide evidence of the effects of balanced and unbalanced forces on the motion of an object. [Clarification Statement: Examples could include an unbalanced force on one side of a ball can make it start moving; and, balanced forces pushing on a box from both sides will not produce any motion at all.] [*Assessment Boundary: Assessment is limited to one variable at a time: number, size, or direction of forces. Assessment does not include quantitative force size, only qualitative and relative. Assessment is limited to gravity being addressed as a force that pulls objects down.*]

3-PS2-2.
Make observations and/or measurements of an object's motion to provide evidence that a pattern can be used to predict future motion. [Clarification Statement: Examples of motion with a predictable pattern could include a child swinging in a swing, a ball rolling back and forth in a bowl, and two children on a see-saw.] [*Assessment Boundary: Assessment does not include technical terms such as period and frequency.*]

The performance expectation above was developed using the following elements from the NRC document *A Framework for K-12 Science Education*:

Science and Engineering Practices	Disciplinary Core Ideas	Crosscutting Concepts
Planning and Carrying Out Investigations Planning and carrying out investigations to answer questions or test solutions to problems in 3–5 builds on K–2 experiences and progresses to include investigations that control variables and provide evidence to support explanations or design solutions. • Plan and conduct an investigation collaboratively to produce data to serve as the basis for evidence, using fair tests in which variables are controlled and the number of trials considered. - - - - - - - - - - - - - - - - *Connections to Nature of Science* **Scientific Investigations Use a Variety of Methods** • Science investigations use a variety of methods, tools, and techniques. **Science Knowledge is Based on Empirical Evidence** • Science findings are based on recognizing patterns.	**PS2.A: Forces and Motion** • Each force acts on one particular object and has both strength and a direction. An object at rest typically has multiple forces acting on it, but they add to give zero net force on the object. Forces that do not sum to zero can cause changes in the object's speed or direction of motion. (Boundary: Qualitative and conceptual, but not quantitative addition of forces are used at this level.) **PS2.B: Types of Interactions** • Objects in contact exert forces on each other. **PS2.A: Forces and Motion** • The patterns of an object's motion in various situations can be observed and measured; when that past motion exhibits a regular pattern, future motion can be predicted from it. (Boundary: Technical terms, such as magnitude, velocity, momentum, and vector quantity, are not introduced at this level, but the concept that some quantities need both size and direction to be described is developed.)	**Cause and Effect** • Cause and effect relationships are routinely identified. **Patterns** • Patterns of change can be used to make predictions.

His mother met him afterwards with a beaming smile, presenting him with flowers picked up at the gas station on her way home from work and a congratulatory embrace. Jason's feeling of accomplishment and his mother's pride in him were readily visible for the world to see.

In strictly artistic terms, Jason's performance did not reach the high-end criteria described on the rubric. Yet the ownership of learning embodied by this young man, not a frequent visitor to School's typical markers of academic success, rose above all other indicators of success. Providing multiple and varied means to explore scientific concepts allowed Jason agency over his learning - creating a deep, memorable learning experience. An effective integration of the arts, including the creative, kinesthetic qualities of dance, provides students like Jason an on- ramp to content.

The preceding example is one of many, many stories accrued over ten years of using creative movement as a tool to introduce, develop, and assess interdisciplinary content learning. Jason's story exemplifies the possibility of both content mastery and the development of efficacy. Following is an outline of another STEAM unit we carried out in a city-wide option urban elementary school in a midwestern city of 300,000 people. The demographics of the district are >75% non-White, approximately 30% non-native English Speakers, and over 70% living in poverty as measured by those eligible for free or reduced meals. The unit brings together the best practices of the interdisciplinary nature of STEAM with a focus on science and the arts, particularly dance and physical education. The unit utilizes planned movement as a living model of scientific concepts related to the core ideas of force and motion embodied in simple machines. Students build off concrete experiences with materials, engage in movement exploration activities, and culminate in a dance performance. While the unit has its roots and outcomes in student mastery of the physics and applications of simple machines and final products that are movement and visual in nature, it is not place-based. This unit could be implemented in a classroom, laboratory, gymnasium, studio, or any learning space where teachers have the courage and students need to learn.

Important to the implementation of this unit is an understanding of the Working Words into Movement Tool utilized in Creative Movement. The Working Words into Movement (WWM) Tool was developed to support the inclusion of movement as a pedagogical tool in introducing, developing, and assessing interdisciplinary content (Figure 2). WWM is based on the work of movement theorist, Rudolf Laban (1879-1958), who developed the Laban Movement Analysis (LMA) taxonomy to generate and describe movement. LMA contains four major categories of movement: effort, body, shape, and space (Konie, 2011). For interdisciplinary use with young learners, the LMA has been translated into the WWM Tool (Figure 2). One purpose of the tool is to help students use their past experiences and motor memory to explore, interpret, engage with, and understand novel vocabulary. The second, and no less important purpose for the WWM Tool, is to serve as a catalyst for choreographing creative movement performances.

Prior to using the WWM Tool students must have ample time and experience in exploring the four major categories to ensure they have an understanding of the relationship to movement of these descriptive classifications.

In the first phase of using the Tool, students begin by brainstorming all of the possible ways the novel vocabulary word might "look" in each of these categories. For example, when brainstorming on the word "forces" as would be included in a unit study on simple machines, in the Effort/Energy category, students may think of words such as: push, pull, hard, light, constant, always, necessary, forward, or down. Encouraging students to include any and all of their thoughts, even if they

Figure 2. Working Words into Movement (WWM) Tool

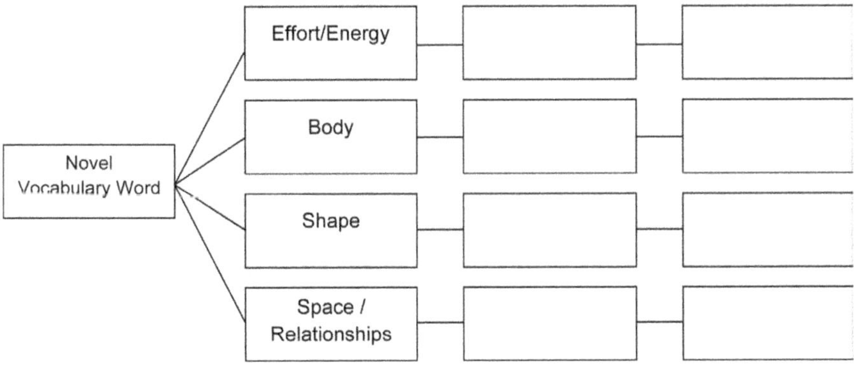

Figure 3. Brief examples of some of the types or qualities of movements that may be included under each heading

Effort/Energy	Body	Shape	Space/Relationship
Weight: firm, light, strong **Flow:** Bound, free, jerky **Emotion/Intent:** Direct, indirect, forceful, angry, joyful **Speed:** Fast, slow, accelerating, sudden, sustained	**Actions:** curl, stretch, twist, support, lead, follow **Locomotion**: run, walk, gallop, skip, hop, jump, leap **Non-locomotor:** twisting, rising, spinning, pushing **Body Parts:** apart, asymmetrical, meeting, above, below	Straight, bent, curved, open, closed, tight, frozen, shifting	**Areas:** General/personal Directions: North, south, east **Facing:** Sideways, backward, forward Levels: Medium, high, low, changing **Pathways**: In air, on floor, direct, indirect, zig-zag, curved, straight **Planes:** Sagittal, frontal, horizontal **Other:** music, walls, props, literature, art, other dancers

think they are only guessing or are using parts of the word in an attempt to decode the whole, can be a very positive tool. After the initial brainstorming, student responses to the WWM Tool can be used as a formative assessment to guide instruction based on abundance or absence of prior knowledge or misinterpretations. For example, if a student included "Yoda" as one of their words related to "force" they may be acknowledged for their humor and wide-range brainstorming but re-directed towards the scientific application sought to be understood through the WWM Tool.

The second phase of WWM Tool use is individual or peer evaluation of brainstorm lists. Evaluation takes place for two purposes. The purposes of this step are to eliminate any words or phrases listed that students do not believe would be easily translated to movement and further engage in dialogue about the vocabulary word being examined. Any words or phrases that are deemed to be irrelevant, inaccurate, or that do not readily appear to have intrinsic movement are simply crossed out. All remaining words are transferred to the next line on the flow chart. This is repeated for each of the four categories. In our example using the word "force" words such as "always" and "necessary" may be eliminated. While there are few "right" or "wrong" answers, it is the individual reflection or group discussion about which words to eliminate that holds the power in helping all learners understand the novel vocabulary word (Figure 4).

Step three of the process can be student or teacher directed, based on instructional purposes and intended learning products. In Step Three the list of remaining words in each of the four categories: effort/energy, body, shape, and space are selected for application to a final movement problem. Step Four, represented by the final area in the WWM Tool includes only the words from Step Three that the individual or group would like to use in their movement sequence. The choices students make from Step Three for Step Four should take into consideration the parameters established by the teacher, co-created by teacher and students, or determined by students for solving the *movement problem*. In this case a movement problem is the pre-determined criteria for a final movement performance. The final movement performance can range from a quick, unpolished informal performance in class to a very formal, highly rehearsed performance with an external audience. In the example of the word "force", students would select the words from each category

Figure 4. Working Words into Movement (WWM) Tool Step One

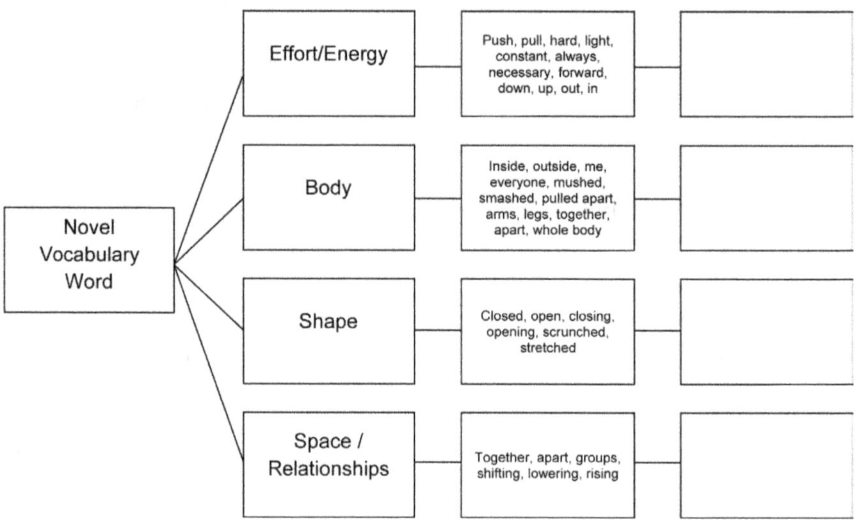

(Energy/Effort, Body, Shape, Space/Relationship) that they believe, and can justify—either orally or in writing—best support their solution to the movement problem.

It is anticipated this unit may be put into practice by the dance or physical education specialist, science specialist, classroom generalist, or a collaborative of engaged educators. In this case study, our context was a collaborative effort between the creative movement specialist and a classroom generalist. The classroom generalist carried out an active 'doing-based' science program in her classroom, while students expanded their understanding through engagement in movement as text, metaphor, and modeling in physical education. While working with movement may be easier in open spaces that are bounded enough for teacher and peer engagement throughout the experience, having a gymnasium or dance studio is not essential. During implementations of our collaboration we explored using classrooms, common areas, and courtyards. The space used had little impact on the student's comprehension or final products. In addition to finding effective work spaces throughout the school, we found this kind of collaboration between the generalist and specialist offered the ideal opportunity for students to fully engage in a robust exploration of content. The collaborative partnership allowed for alternative means of knowing and expressing knowledge—for deep processing of the models and metaphors that emerged within the student performance.

An Explanation of Simple Machines Through Movement

Prerequisite knowledge, skills and experiences to ensure success in this unit:

- Clear, consistent rules, expectations, and behavior management in place in all collaborating classrooms.
- Mastery of benchmarks and outcomes identified for English language arts (Common Core), science, dance, and physical education through grade two.
- Experience with Working Words into Movement (WWM) framework.

As in the use of any models, the critical reflection following the experience is able to make the transformation from students having fun to deep rich meaning in science, arts, and self. Following are some of the kind of questions that might be put to good use:

- What did you notice as you prepared and carried out your performance?
- What did you notice about your movements?
- How did your movement connect to force and motion?
- How was your movement like simple machines?
- How well did your performance 'model' force and motion?

Table 1. An exploration of simple machines unit plan

Day One	Day Two	Day Three
INTRODUCTION/WARM-UP: moderate-to-vigorous physical activity (MVPA) in the form of a low organized tagging game suitable for the space **FOCUS:** Inquiry stations: • Students rotate between 8-15 stations with a variety of common & novel simple machines for open exploration • Science notebook: o Describe how objects move or can be used o Compare/Contrast notes to co-create classification of all materials into like- categories **CONCLUSION/COOL-DOWN:** Exit slip: At least one question generated during exploration or collaboration	**INTRODUCTION/WARM-UP:** How many categories did you co-create yesterday? **FOCUS:** • Domain specific, rigorous vocabulary sorting race • Naming: each group reviews vocabulary related to each category and determines a name and function for that particular simple machine **CONCLUSION/COOL-DOWN** 5-minute yoga w/intention on "where in your body do you find some similarities to simple machines"	**INTRODUCTION/WARM-UP:** 10-min walk/jog **FOCUS:** Direct instruction • Review naming to ensure proper vocabulary • Co-create definitions for each simple machine and generate examples list Movement Exploration Individual movement problem: • How can you represent the actions of a simple machine through locomotor and non-locomotor actions? **CONCLUSION/COOL-DOWN** 1/3 of students at a time share their solutions, audience identifies which machines are represented, repeat
WORK PRODUCT/ ASSESSMENT:	**WORK PRODUCT/ ASSESSMENT:**	**WORK PRODUCT/ ASSESSMENT:** Visual recording of individuals presenting with group
Day Four	**Day Five**	**Day Six**
INTRODUCTION/WARM-UP: Repeat mvmt solution from Day Three, "trying on" mvmt observed in others • Teacher loudly reinforce vocabulary through observation **FOCUS:** Forces • Direct or inquiry-based instruction into forces, as related to simple machines	**INTRODUCTION/WARM-UP:** MVPA-low organized tagging game **FOCUS:** Working Words into Movement • Groups of three (not partners from yesterday) • Assigned vocabulary words to work into mvmt • Create and rehearse one 30 second mvmt sequence to	**INTRODUCTION/WARM-UP:** Open space jog while playing video of Rube Goldberg machine **FOCUS:** Final Mvmt Problem introduced • Co-determine a problem to be solved in school or community that a Rube Goldberg-like machine could tackle.
• Mvmt exploration: Mvmt problem: (In pairs), • ABA of simple machine o action (A), o acted upon by simple machine (B), o simple machine action (A) **CONCLUSION/COOL-DOWN** Presentation of mvmt problem solution • Each pair evaluates another	reflect one of the words **CONCLUSION/COOL-DOWN** • Presentation of mvmt problem solution • Each group evaluates another	• Review of group work norms/ expectations • Each group is to create a repeatable 16 count mvmt sequence (approximately 10- 15 seconds) representing their assigned simple machine as a part of solving the community/school problem **CONCLUSION/COOL-DOWN** Fist-to-Five reflecting personal responsibility in the group
WORK PRODUCT/ ASSESSMENT: Peer evaluation exit slip	**WORK PRODUCT/ ASSESSMENT:** Peer evaluation exit slip	**WORK PRODUCT/ ASSESSMENT:** None

continued on next page

Table 1. Continued

Day Seven	Day Eight	Day Nine
INTRODUCTION/WARM-UP: High cardio/low organized tagging activity with assigned group member teams **FOCUS:** • Continued group creation/rehearsal of 16 count sequence • As groups complete rehearsal, send "spies" to give feedback to each other's groups • Groups each determine how they need to be acted on to initiate and to pass along action **CONCLUSION/COOL-DOWN** Put it all together, one 16 count sequence after another	**INTRODUCTION/WARM-UP:** • 5-10 min walk/jog while watching prior day's video • What can make your sequence more accurate & more aesthetically interesting **FOCUS:** • Refining individual 16 count sequences • Put it all together • Structure: o Initial resting shape (held by each member of each group until it is time for that group to begin their 16 count sequence for the first time and between executions) o First verse: Each group 16 count, one group at a time, following the previous group, holding resting shapes when complete) o Second verse: The second verse begins when the final group concludes the first execution of their 16 count sequence and the dance starts over with the first group. This verse is an ABAB add-on structure: A=16 count movement sequence, B=16 count rest, repeat. The first group begins verse two with their 16 counts, group two then begins as group one rests. Then groups one and three execute their A while all others hold resting shapes, then groups two and four move, and so on until every group is participating in the 16 count move, 16 count rest. o Third verse: Continuous repetition of 16 count sequence, culminating in an 'overheating' or 'explosion' action leading to the... o Final resting shape **CONCLUSION/COOL-DOWN:** Yoga sequence with intention of appreciating group contributions	**INTRODUCTION/WARM-UP:** Socratic Jog Topic: What makes for a quality dance performance? **FOCUS:** Rehearse and refine performance (Performance should have an audience and be recorded if, that is 'in class' a tenth day should be added for drafting of the blueprint prior to performance so that they may also be displayed. If limited time, collaboration with visual arts or classroom teachers valuable here. **CONCLUSION/COOL-DOWN:** Oral reflection on learning, process, and product
WORK PRODUCT/ ASSESSMENT: Video recording of initial attempt	**WORK PRODUCT/ ASSESSMENT:** None	**WORK PRODUCT/ ASSESSMENT:** o Video recording of performance o Rube Goldberg illustration/ blueprint of Simple Machine Dance solution to school or community problem

- How did working with movement help you understand the principles of force and simple machines?
- What were some limitations of your model?
- How did your experience in modeling through movement help you better understand force and motion?

Science Content

This section is meant to provide the physics content for the educator to both understand and be able to explain forces, motion, and simple machines to the target student audience. Examples are given to link these scientific terms and ideas to the WWM tool and the teaching unit.

Forces and Motion

All the motion we observe can be explained with Newton's Three Laws. Newton's first two laws of motion may be summarized as follows: Motion that is constant requires no explanation and all changes in motion are explainable by forces (that is to say, a "push" or a "pull" as discussed in the WWM tool above). Here "constant" motion means either no motion or motion in a straight line without changes in speed or direction. Importantly, Newton's First Law identifies the concept of inertia, or the property of all mass to resist changes in motion. Newton's Second Law then describes how changes in motion do occur, namely, that there must be an external force applied to an object. This can get it to change its speed or direction. Finally, Newton's Third Law, while often more mysterious, is another way to say that you can't lift yourself up by your bootstraps, or, that there are no "free" forces.

Unfortunately, the persistent presence of friction can lead to some misconceptions. While it may seem that a *constant* force must be exerted on an object for it to stay moving at a *constant* speed, this is not true since friction is *also* a force that must be taken into account. Consider, for example, pushing a desk across the floor. The force of the push and the force of friction are in opposite directions. If they balance the desk moves at a constant speed. If the force of the push is greater than the force of friction, the desk will speed up; if it is less, the desk will slow down accordingly. This speeding up or slowing down is called "acceleration." Newton's Second Law provides a quantitative summary of that change via an equation:

$$Acceleration = Force \div Mass$$

What this is saying is that if you apply more force, the object's motion changes more abruptly and the motion is no longer constant. Students could also engage the concept of "constant motion" by considering whether or not forces are at work when a book is simply sitting on a desk, e.g. gravity pulling down and the force from the desk holding the book up. Since the forces balance, the zero motion remains constant at zero motion.

Finally, Newton's Third Law provides a summary about how objects can exert forces on each other. In effect, there are no "free" forces. By pushing on something, I exert a force on it, and can get to accelerate. But, that object also pushes back equally hard on me. So I have to push against the ground to get more force on me to get the object moving. But the ground pushes back equally hard against me. What does the ground push against? Well, in fact, it is the earth itself. But the earth has nothing to push against in empty space, so, in effect, the earth changes its motion ever so slightly in an equal and opposite manner. Of course, since the earth is so large, this change is imperceptible. So taken together, Newton's three laws tell us that: (1) Mass has a property called inertia that resists changes in motion; (2) To get a massive object to move you must apply a force on it; (3) There are no "free" forces and all forces of interaction are equal and opposite.

Forces and Rotation

Forces can also cause objects to rotate. Consider the situation shown below in Figure 5. The two boxes are being pushed from opposite sides with two equal forces. Since the forces balance, the two boxes will not be accelerated to the left or to the right. However, when these balanced forces are *offset* from each other, they can cause the object to *rotate*. Offset forces produce what is called "torque," or a force that causes rotation. If students are asked to brainstorm words for how they get things to rotate, perhaps terms like "twist" or "spin" may arise. It is also likely that the same terms "push" or "pull" would get discussed, providing a direct link between forces and torques. Indeed, rotation is also a form of motion, so forces *must* be involved. So just like unbalanced forces cause an object to accelerate to the left or right, an unbalanced set of *torques* can cause "rotational acceleration" in either a clockwise or counterclockwise direction. The amount that the force or forces are offset from the point of rotation tells us how much torque, and therefore how much rotational acceleration, those forces will provide. So again, in Figure 5, the different placements of the two pushing forces will either cause no rotation, some rotation, or quite a bit of rotation. Mathematically, the amount of torque provided by a force is proportional to both the force and to the "lever arm" or distance from the point of rotation. This is described as.

Torque = Force × Lever Arm

As with forces pushing objects along a path, we can now calculate the torque produced by each separate force, add them up, and therefore predict whether or not there will be any rotation. If the total net torque is non-zero, the object will start to rotate. In terms of creative movement, the instructor might connect why your arms rotate downward from your shoulders when held straight out: they are rotating due to the torque produced by the force of gravity. The muscles in our shoulders are providing the balancing torque to keep the arm held straight out. If there is another force, say, someone placing a weight in the outstretched hand, there is now another torque wanting to rotate the arm downwards and the shoulder muscles need to work even harder. If the weight is placed instead at the elbow, the "Lever Arm" is less, and the torque provided by the shoulder doesn't need to be as strong.

Forces and Work

When we exert a force on an object and the object starts to accelerate, it will move a certain distance. The relationship between the *distance* the object moves and the *force* applied to that object is described with the concept of *work* or energy spent. I have to do work on an object (expend energy or exert effort) to move it a certain distance by applying a force. You can understand this concept by again considering the effort of pushing a desk across a rug. To describe this, students would likely think of words like work, effort, hard, try, power, strong, or difficult. But when considering how to move the desk from here to there, students would also naturally understand that you need to either "push" or "pull" the desk, e.g. apply a force. The instructor might then ask if it takes more effort to push the desk a farther distance, and of course it does. So mathematically, the concept of work is expressed as follows:

Work = Force × Distance

Figure 5. Forces and Rotation. Even when two forces are balanced, so that the boxes don't accelerate to the left or to the right, an offset force can cause an object to start to rotate. An offset force that causes rotation is called a "torque."

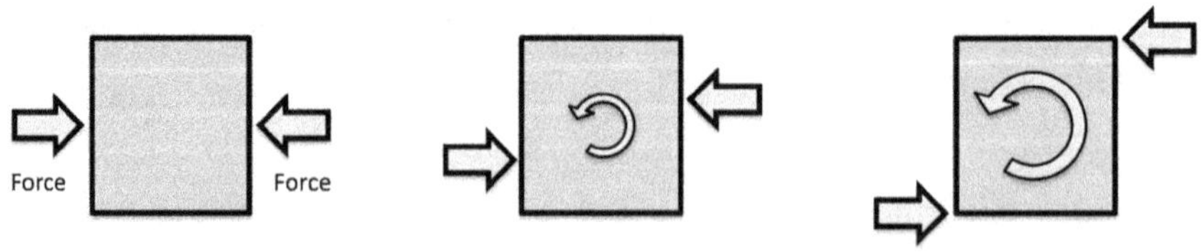

To move a massive object, we require more force, and therefore exert more effort. To push an object a farther distance, we also need to exert more effort or energy.

Work and Simple Machines

With this physics background, we can now understand how simple machines do what they do. A simple machine is a mechanical arrangement that allows us to do work on an object — e.g., move it a certain distance by applying a certain force. However, using our understanding of how work (or energy or effort) is related to the force and the distance, we can make the mechanical arrangement so that —*for the same amount of total work* — the force we apply is much less. This can only happen if the distance over which the force operates is much longer. For example, students could consider two scenarios: pushing the desk across the rug, or getting it on some wheels. While they will understand that the wheels (a simple machine) make it a lot easier, the instructor could ask them if eventually, after pushing for a long time over a much longer distance, they might still get just as tired. So for the same amount of work or energy spent, we can either choose to apply much more force (hard) over a small distance or much less force (easy) over a larger distance. Of course, in using a simple machine, the amount of effort we put in (small force / large distance) must equal the amount of work done by the machine (large force / small distance). This is a statement of "Conservation of Energy." Ignoring any negative effects of friction, the amount of work in will equal the amount of work out, or "Work In = Work Out."

The amplification of the output force relative to the input force provided by a simple machine is called the "Mechanical Advantage" or MA. Depending on the particular mechanical arrangement, simple machines can be characterized by one or more of the following:

1. Amplify the magnitude of an input force.
2. Amplify the distance (or the speed) of an input force.
3. Change the direction of an input force.
4. Transfer an input force from one location to another.

Classically, there are six simple machines: (1) the lever; (2) the pulley; (3) the inclined plane; (4) the wheel and axle; (5) the screw; and (6) the wedge. Many of these are related (e.g. the screw is a twisted ramp and a pulley is similar to a wheel and axle but with ropes). Therefore, the following sections will provide examples and the physics background for only three of these.

1. The Lever

Consider the situation illustrated below in Figure 6, lifting a heavy box by using a lever. The point of rotation is called the fulcrum. In this case, a small force down on the lever over a larger distance is able to lift with a large force over a smaller distance. The product of each of the "Force × Distance" on both sides of the lever is the same but the force exerted by you, the lifter, is much less than lifting the box directly. Unfortunately, the box isn't lifted nearly as high. That is the trade-off with all simple machines using this principle — we can lift things as heavy as a car with our own hands and a long lever, but only a few inches. Notice the characteristics of this simple machine: (a) The magnitude of the force that you exert is magnified; (b) The speed at which the box lifted was slower than the speed that you were pushing down (over the same amount of time, the box went a shorter distance up than your hands pushing down, thereby moving more slowly); (c) The direction of the force was flipped from pushing down on the lever to pushing up on the box; (d) The force was transferred from one end of the lever to the other.

Considering the geometry of the lever, we can see that the MA of the lever is: *L1 / L2,* or the ratio of the lengths from the fulcrum. This is labeled in the figure. We can also understand the mechanism of the lever by using and balancing *torques.* Recall that to determine whether something will rotate (i.e. whether or not the lever will lift by rotating around the fulcrum) we need to consider each force and how much torque that force is producing. This is diagrammed in Figure 7. By using our

Figure 6. The Lever. Using the concept of forces acting over a distance allows us to design a lever that requires less force, but only lifts the box a small distance. The same amount of work is done in two cases: using the lever (small force × large distance = large force × small distance) or lifting the box up that same small distance directly (large force × small distance) without the lever.

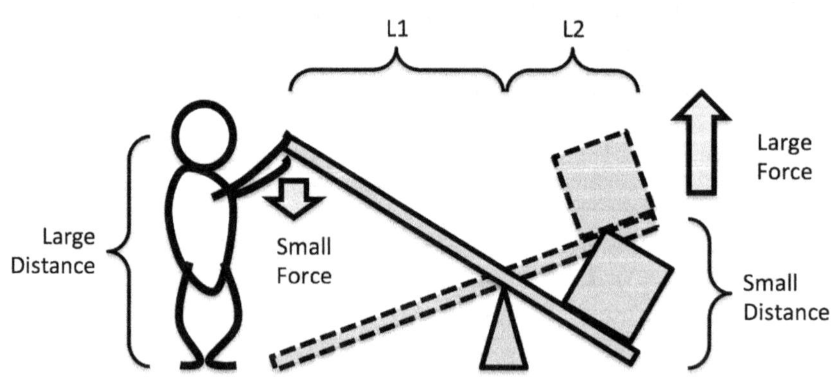

Figure 7. The balanced lever. By considering each force and the torque produced by those forces, it is possible to determine whether or not an object, in this case a lever, will rotate or remain balanced. Recall that torque is force times lever arm. Therefore, a large force with a small lever arm can be balanced (or overcome, if you are trying to lift something heavy!) by a small force and a large lever arm.

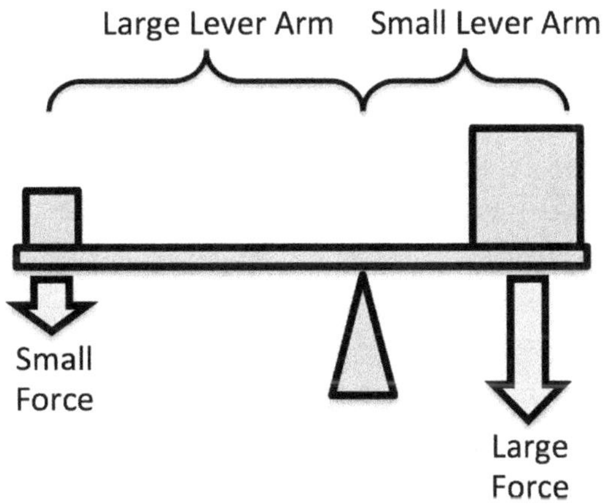

equation that torque is proportional to the force and the lever arm, we can tally up the clockwise torques and the counterclockwise torques to determine the net torque. If the net torque is balanced, the lever will not rotate. If we can somehow push a bit harder on one end, we have upset the balance of torques, and the lever will rotate, lifting the object up. In terms of body movement, students can consider the simple machine of their arm as a lever. With an elbow on a table, a student can try to lift up a heavy book. The instructor would ask the students where the fulcrum is (the elbow) and what is providing the input force (the tendon and bicep muscle). But by considering how much the tendon actually moves (short distance) compared with the book (long distance), we see that the elbow simple machine actually has a MA less than one. So the force that the tendon is applying to the bone is actually quite a bit greater than the weight of the book.

2. The Pulley

A second type of simple machine is the pulley. Here, we are concerned again with reducing the amount of force required to perform some work (i.e. move) some heavy object. Pulleys also change the direction of a force, depending on how many pulleys are used. A single pulley offers no mechanical advantage, as shown below in Figure 8, because the force is only changing direction and not acting over a different distance. If, however, two pulleys are used, you have to pull the rope over a greater

Figure 8. The Pulley. The single pulley system on the left only changes the direction and location of an input force. There is no mechanical advantage because the force acts over the same distance. A double pulley, however, does offer mechanical advantage because to lift the heavy object a small distance, you must pull the rope (e.g. exert a force) over a longer distance. The mechanical advantage is equal to how many portions of the rope are pulling vertically on the hanging weight. The more pulleys, the higher the mechanical advantage, but the farther the rope needs to be pulled.

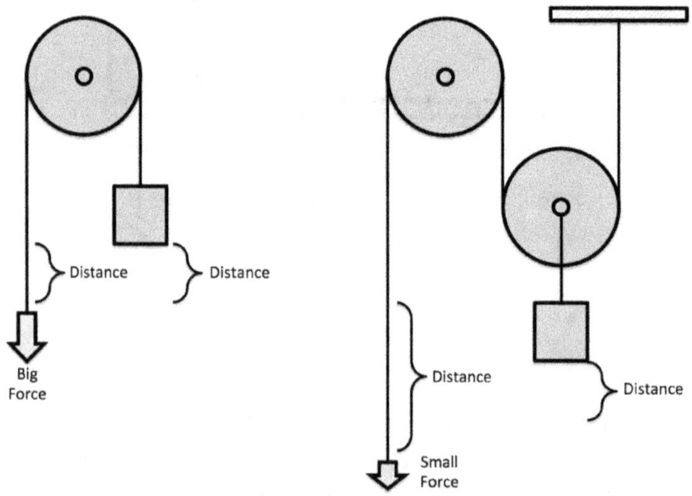

distance than before to lift the object the same amount as shown on the right. This means that the force of your pull is less than before since the same amount of work (force lifting the weight up the same distance) is performed. For example, a double pulley offers a mechanical advantage of "2" because twice as much rope has to be pulled out (by you) to move the far end of the rope (the object you are moving) by the same distance as before.

One rope can also be tied around a different sized pulley wheel, meaning that you have to pull out more rope over a longer distance to move the heavy object over a shorter distance, thereby again gaining mechanical advantage. This is similar to how a "wheel and axle" gives mechanical advantage. Since the inner pulley and the outer pulley each rotate the same amount, say a full 360 degree turn, the rope attached to the outer pulley will unwrap a much longer amount compared to the rope wound around the inner pulley. This is shown in Figure 9. This is also understood in the sense that the rope you are pulling on now has a larger *lever arm*, so for the same amount of force, more torque is produced and the pulley will now begin to rotate. Another example is imagining trying to open a door with a doorknob that

Figure 9. The Pulley as Wheel and Axle. Wrapping a rope around a pulley / wheel with a different diameter offers mechanical advantage because the force both (1) has a larger lever arm and can produce more torque and (2) the force must act over a much longer distance to turn the pulley / wheel one full revolution. The mechanical advantage here is the ratio of the two diameters of the two wheels.

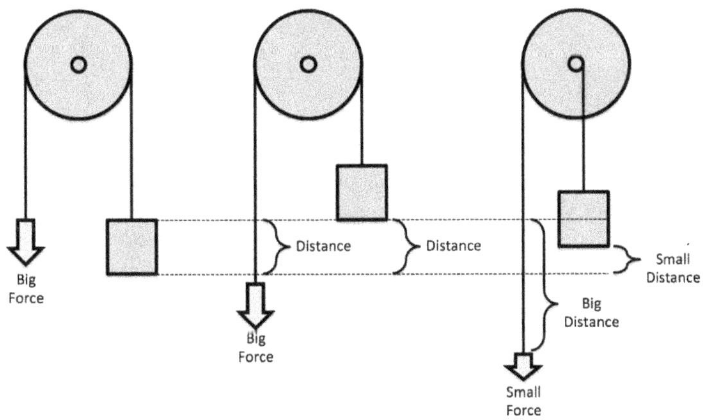

was only a thin rod, not a bulging handle. The mechanical advantage is the size of the door handle to the size of the actual axle that releases the lock.

3. The Inclined Plane

To lift an object to a certain height, you must do work against the force of gravity. Therefore, the amount of work that is performed is equal to the "Force of Gravity (the weight)" *times* the "Distance Lifted." If done directly, i.e. straight up, you must exert a force equivalent to the weight of the object for the full height. This is difficult. If, however, you can spread the force of your push over a longer distance while still lifting the weight to the same height, you have performed the same amount of total work, but with much less force. This is the advantage of the inclined plane, shown in Figure 10. Since the weight is lifted upwards, "Work Out" for the ramp is set to be "Work Out = Weight × Height." The amount of work we put in is "Work In = Force of Push × Ramp Length." Since "Work in = Work Out" we find that the mechanical advantage of the ramp is "Ramp Length ÷ Height." Therefore, the longer the ramp, the greater the mechanical advantage less pushing force is required. However, you need to push for a longer distance. Connecting this concept to creative movement or dance could touch on jumping up in the air. Why can't we jump to the second floor of a building but can easily climb stairs or a ramp to get there? The amount of force required to jump straight up is more than our legs can give us, so the tradeoff in taking the stairs is that we'll need to walk a much longer distance to get there.

Figure 10. The inclined plane. You can perform the same amount of work by either: (1) lifting a weight directly against the force of gravity; or (2) pushing with less force over a longer distance but still getting the weight to the same height. The mechanical advantage of a simple ramp is the Ramp Distance divided by the Vertical Distance. The longer the ramp, the greater the mechanical advantage, but you have to push over a much longer distance to lift the object up to the same final height.

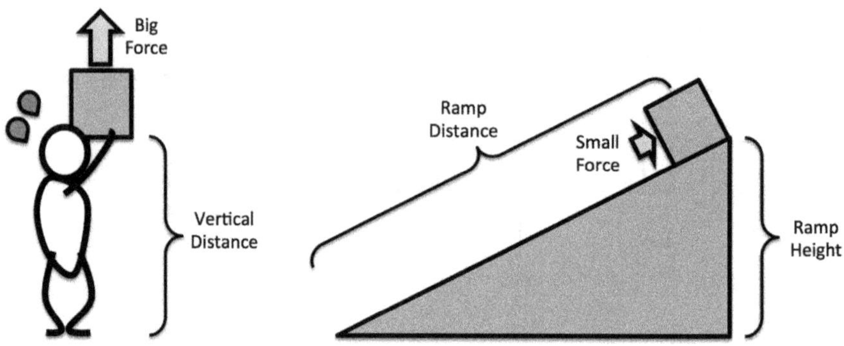

CONCLUSION

Combining the scientific concepts of force and motion with human movement is an organic collaboration that simply makes sense for students. Deep and lasting understandings on the theory, concepts, and skills develop in the students. Further, students routinely share that, not only do they understand simple machines, they also get along better as a community and feel pride in themselves for being able to express their understandings through a movement performance. For the teachers involved, each time we have co-facilitated the unit, we learn something new about our strengths and needs as teachers. We have learned to take ownership and release control. Over the years, we have invited other content area teachers to join in the project, adding their content area expertise and learning from our process. While the unit structure always starts the same, each experience is different and perhaps better, than the one before. From each experience we have learned how to make the collaboration and the learning experience better, stronger, more effective, and more fun. Because, life—and certainly life teaching in urban schools—is way too much work to not also be fun!

REFERENCES

Barton, K., & Smith, L. (2000). Themes or motifs? Aiming for coherence through interdisciplinary outlines. *The Reading Teacher*, *54*(1), 54–63.

Bedetti, G. (2015). Collaborative college playwriting and performance: A core course "Trespassing" onto the dramatic arts. *International Journal of Education & the Arts*, *16*(19).

Bransford, J. (2000). *How people learn: Brain, mind, experience, and school.* Washington, DC: National Academies Press.

Coffey, H. (n.d.). *Interdisciplinary teaching*. Retrieved on October 3, 2016 from http://www.learnnc.org/lp/pages/5196

Deans, J. (2016). Thinking, feeling and relating: Young children learning through dance. *Australian Journal of Early Childhood*, *41*(3), 46–57.

DeLoache, J. S. (2000). Dual representation and young children's use of scale models. *Child Development*, *71*(2), 329–338. doi:10.1111/1467-8624.00148 PMID:10834468

Dow, C. B. (2010). Young children and movement: The power of creative dance. *Young Children*, *65*(2), 30–35.

Gopnik, A. (2012). Scientific thinking in young children: Theoretical advances, empirical research, and policy implications. *Science*, *337*(28), 1623–1627. doi:10.1126/science.1223416 PMID:23019643

Groff, J. S. (2013). Expanding our frames of mind for education and the arts. *Harvard Educational Review*, *83*(1), 15–39. doi:10.17763/haer.83.1.kk34802147665819

Haring, D., & Kelner, T. (2015). Why we got serious about interdisciplinary teaching. *Educational Leadership*, *73*(4), 68.

Jacobs, H. H. (1989). The growing need for interdisciplinary curriculum content. In H.H. Jacobs (Ed.), Interdisciplinary Curriculum: Design and Implementation (pp. 1 11). Alexandria, VA: ASCD.

Konie, R. (2011). *A brief overview of laban movement analysis*. Retrieved on April 27, 2016 from http://www.movementhasmeaning.com/wp-content/uploads/2010/09/LMA-Workshop-Sheet.pdf

Lakoff, G., & Johnson, M. (1980). *Metaphors we live by*. Chicago, IL: University of Chicago Press.

Lead States. (2013). *Next generation science standards: For states, by states*. Washington, DC: National Academies Press.

Mandigo, J., Francis, N., Lodewyk, K., & Lopez, R. (2009). *Position paper: Physical literacy for educators*. St. Catherines, ON, Canada: Brock University. Retrieved May 10, 2016 from https://www.phecanada.ca/sites/default/files/pl_position_paper.pdf

Martin, D. J. (2011). *Elementary science methods: A constructivist approach* (6th ed.). Belmont, CA: Wadsworth Publishing.

Merriam-Webster. (n.d.). *Definition of text*. Retrieved Oct 4, 2016 from http://www.merriam-webster.com/dictionary/text

National Council for Teachers of English. (1995). *Position statement on interdisciplinary learning, pre-K to grade 4*. Retrieved May 10, 2016 from http://www.ncte.org/positions/statements/interdisclearnprek4

National Research Council. (2012). *A framework for K-12 science education: Practices, crosscutting concepts, and core ideas. Committee on a Conceptual Framework for New K-12 Science Education Standards. Board on Science Education, Division of Behavioral and Social Sciences and Education*. Washington, DC: National Academies Press.

Spitzer, H. (2013). Introduction of interdisciplinary teaching: Two case studies: Commentary on Teaching Science, Technology, and Society to Engineering Students: A Sixteen Year Journey. *Science and Engineering Ethics*, *19*(4), 1451–1454. doi:10.1007/s11948-013-9475-y PMID:24085355

The Common Core State Standards Initiative. (2016). Retrieved on September 1, 2016 from http://www.corestandards.org/ELA-Literacy/introduction/key-design-consideration/

United Nations. (2002). *56/116. United Nations Literacy Decade: education for all. Resolution adopted by the General Assembly, Fifty-Sixth Session*. Retrieved on May 10, 2009 from: http://www.undemocracy.com/A-RES-56-116.pdf

United Nations Educational, Scientific and Cultural Organization. (2003). *Literacy, a UNESCO perspective*. Retrieved on May 10, 2016 from: http://unesdoc.unesco.org/images/0013/001318/131817eo.pdf

White, H. (2011). Our education system is not so much "broken" - as it is totally outdated! In *STEAM*. Retrieved September 16, 2016, from http://steam-not-stem.com/articles/our-education-system-is- not-so-much-broken-as-it-is-totally-outdated/

KEY TERMS AND DEFINITIONS

Creative Movement: Objective driven activities through which learners explore movement for the purpose of developing skills, building efficacy, or stimulating imagination with the intent of communicating knowledge of facts, concepts, principles, ideas, and/or feelings.

Encore Content Areas: Encore content areas are those content areas taught in schools that are often marginalized through the implementation of policies such as No Child Left Behind that focus nearly exclusively on content and direct instruction in reading and math courses. They are the classes in schools that for the last several years have been considered to be outside or or adjacent to the bounds of "core content" areas. Examples of encore content areas are social studies, the Arts, health, science, physical education, career/technical education and world languages.

Force and Motion: A push or a pull that can accelerate, rotate, or otherwise alter the motion of an object. The relationship between applied forces and the resulting motion is the subject of Newton's Laws of Motion.

Physical Literacy: The ability to move with competence and confidence in a variety of physical activities in multiple environments that benefit the healthy development of the whole person.

Scientific Literacy: The knowledge and skills necessary to participate in scientific decisions impacting personal, local, and global affairs.

Simple Machines: A mechanical arrangement used to apply a force. Typically used to amplify a small input force into a larger output force. Examples include a lever, pulley, or an inclined plane.

Text: Symbolic representation of meaning. Text includes but is not limited to charts, maps, graphs, movies, music, choreography, sculpture, posture, gesture, tapestry, painting, poems, books, short stories, graphic organizers and procedural manuals.

Working Words into Movement: A process and support tool to bring to light the organic movement embedded in vocabulary words.

APPENDIX A: CCSS STANDARD BENCHMARKS THAT MAY APPLY TO A CREATIVE MOVEMENT/ SIMPLE MACHINES UNIT FOR GRADE 3

Table 2.

3NCAS-Dance Standards that *may* Apply to Simple Machines Unit Plan, Grade 3	
Strand Identification	Standard
Artistic Process: Creating. Generate and conceptualize artistic ideas and work.	
DA: Cr1.1.3	b. Explore a given movement problem. Select and demonstrate a solution.
Artistic Process: Creating. Organize and develop artistic ideas and work	
DA: Cr2.1.3	b. Develop a dance phrase that expresses and communicates an idea of feeling. Discuss the effect of the movement choices.
Artistic Process: Creating. Refine and complete artistic work	
DA: Cr3.1.3	a. Revise movement choices in response to feedback to improve a short dance study. Describe the differences and changes made in the movement.
Artistic Process: Performing. Select, analyze, and interpret artistic work for presentation	
DA: Pr4.1.3	a. Judge spaces as distance traveled and use space three dimensionally. Demonstrates shapes with positive and negative spaces. Perform movement sequences in and through space with intentionality and focus.
DA: Pr4.1.3	c. Change use of energy and dynamics by modifying movements and applying specific characteristics to heighten the effect of their intent.
Artistic Process: Performing. Develop and refine artistic technique and work for presentation	
DA: Pr5.1.3	a. Replicate body shapes, movement characteristics, and movement patterns in a dance sequence with awareness of body alignment and core support.
DA: Pr5.1.3	b. Adjust body-use to coordinate with a partner or other dancers to safely change levels, directions, and pathway designs.
DA: Pr5.1.3	c. Recall movement sequences with a partner or in group dance activities. Apply constructive feedback from teacher and self-check to improve dance skills.
Artistic Process: Responding, interpreting intent and meaning in artistic work.	
DA: Re8.1.3	a. Select specific context cues from movement. Explain how they relate to the main idea of the dance using basic dance terminology.
Artistic Process: Connecting, synthesize, and relate knowledge and personal experiences to make art.	
DA: Cn10.1.3	b. Ask and research a question about a key aspect of a adance that communicates a perspective about an issue or event. Explore the key aspect through movement. Share movements and describe how the movements help to remember or discover new qualities in these key aspects. Communicate the new learning in oral, written, or movement form.

SHAPE America Physical Education Standards that *may* Apply to Simple Machines Unit Plan, Grade		
Strand Identification	Standard/Outcome	Application to Simple Machines Unit Notes

continued on next page

Table 2. Continued

Standard 1. The physically literate individual demonstrates competency in a variety of motor skills and movement patterns.	
S1.E6.3	Performs a sequence of locomotor skills, transitioning from one skill to another smoothly and with- out hesitation.
S1.E7.3	Balances on different bases of support, demonstrating muscular tension and extensions of free body parts.
S1.E8.3	Transfers weight from feet to hands for momentary weight support
S1.E11.3	Combines locomotor skills and movement concepts (levels, shapes, extensions, pathways, force, time, flow) to create and perform a dance.
S1.E12.3	Combines balance and weight transfers with movement concepts to create and perform a dance.
Standard 2. The physically literate individual applies knowledge of concepts, principles, strategies and tactics related to movement and performance	
S2.E3.3	Combines movement concepts (direction, levels, force, time) with skills as directed by the teacher.
Standard 3. The physically literate individual demonstrates the knowledge and skills to achieve and maintain a health-enhancing level of physical activity and fitness.	
S3.E2.3	Engages in activities of physical education without teacher prompting.
Standard 4. The physically literate individual exhibits responsible personal and social behavior that respects self and others.	
S4.E4.3a	Works cooperatively with others.
S4.E4.3b	Praises others for their success in movement performance.
CCSS ELA Standards that *may* Apply to Simple Machines Unit Plan, Grade 3	
Strand Identification	Standard
Key Ideas and Details	
CCSS.ELA-LITERACY.RL.3.1	Ask and answer questions to demonstrate understanding of a text, referring explicitly to the text as the basis for the answers.
CCSS.ELA-LITERACY.RI.3.3	Describe the relationship between a series of historical events, scientific ideas or concepts, or steps in technical procedures in a text, using language that pertains to time, sequence, and cause/effect.
Craft and Structure	
CCSS.ELA-LITERACY.RI.3.4	Determine the meaning of general academic and domain specific words and phrases in a text relevant to a grade 3 topic or subject area
Integration of Knowledge and Ideas	
CCSS.ELA-LITERACY.RI.3.9	Compare and contrast the most important points and key details presented in two texts on the same topic.
Writing	
CCSS.ELA-LITERACY.W.3.2.B	Develop the topic with facts, definitions, and details.
CCSS.ELA-LITERACY.W.3.7	Conduct short research projects that build knowledge about a topic.

continued on next page

Table 2. Continued

CCSS.ELA-LITERACY.W.3.10	Write routinely over extended time frames (time for research, reflection, and revision) and shorter time frames (a single sitting or a day or two) for a range of discipline-specific tasks, purposes, and audiences
Speaking and Listening: Comprehension and Collaboration & Presentation of Knowledge and Ideas	
CCSS.ELA-LITERACY.SL.3.1.B	Follow agreed-upon rules for discussions (e.g., gaining the floor in respectful ways, listening to others with care, speaking one at a time about the topics and texts under discussion).
LITERACY.SL.3.1.C	Ask questions to check understanding of information presented, stay on topic, and link their comments to the remarks of others.
CCSS.ELA-LITERACY.SL.3.1.D	Explain their own ideas and understanding in light of the discussion.
CCSS.ELA-LITERACY.SL.3.2	Determine the main ideas and supporting details of a text read aloud or information presented in diverse media and formats, including visually, quantitatively, and orally.
CCSS.ELA-LITERACY.SL.3.6	Speak in complete sentences when appropriate to task and situation in order to provide requested detail or clarification. (See grade 3 Language standards 1 and 3 here for specific expectations.)
Language: Conventions of Standard English	
CCSS.ELA-LITERACY.L.3.3.B	Recognize and observe differences between the conventions of spoken and written standard English.

APPENDIX B: STEAM LESSON PLAN TEMPLATE

Table 3.

<table>
<tr><td colspan="3">Lesson Title:</td><td>Sequence in unit:</td></tr>
<tr><td colspan="2">Scientific Big Idea(s):</td><td colspan="2">Encore Big Idea(s):</td></tr>
<tr><td colspan="2">Target Grade level:</td><td>Lesson Date:</td><td>Duration & # Ss:</td></tr>
<tr><td colspan="4">Academic/Scientific Standards Addressed</td></tr>
<tr><td>Introduced</td><td>Developed</td><td>Assessed for Mastery</td><td>Included for repetition/extension/ transference</td></tr>
<tr><td colspan="4">Academic/Encore Standards Addressed</td></tr>
<tr><td>Introduced</td><td>Developed</td><td>Assessed for Mastery</td><td>Included for repetition/extension/ transference</td></tr>
<tr><td colspan="4">Academic/CCSS Addressed</td></tr>
<tr><td>Introduced</td><td>Developed</td><td>Assessed for Mastery</td><td>Included for repetition/extension/ transference</td></tr>
<tr><td colspan="4">Standards-based Performance Objectives and/or I can Statements (cognitive, affective & psychomotor)</td></tr>
<tr><td colspan="4">Key Vocabulary</td></tr>
<tr><td colspan="4">Space, equipment, materials:</td></tr>
<tr><td>Instruction</td><td>What</td><td>Who is responsible?</td><td>Organization of students</td></tr>
<tr><td>Engaging message/ Instant Activity</td><td></td><td></td><td></td></tr>
<tr><td>Lesson Focus</td><td></td><td></td><td></td></tr>
<tr><td>Assessment</td><td></td><td></td><td></td></tr>
<tr><td>Culmination/closing/ wrap-up/cool down/ processing</td><td></td><td></td><td></td></tr>
</table>

Chapter 6
Theater as the STEAM Engine for Engaging Those Previously Disengaged

Paul C. Jablon
Independent Researcher, USA

EXECUTIVE SUMMARY

Connects the use of creative dramatics with engaging students previously disengaged with STEM classes. Elucidates a variety of reasons how creative dramatics allows STEM teachers and their students to succeed. Each of the assertions is backed by citations of research studies, classroom practice, and details of the theoretical underpinnings. Also included are detailed descriptions of three effective classroom methodologies using theater in STEM classes, along with specific examples of each that include student and teacher interactions.

INTRODUCTION

Skilled inquiry-based elementary teachers who engage students in daily materials-based STEM investigations maintain students' interests throughout their elementary school experiences. However, for much of the older school age population, science, mathematics and engineering are seen as "cold" subjects that are lacking in emotion

DOI: 10.4018/978-1-5225-2334-5.ch006

Copyright ©2017, IGI Global. Copying or distributing in print or electronic forms without written permission of IGI Global is prohibited.

and humanity (Heaverlo, 2011; Burke & Mattis, 2007), and by the time students reach early adolescence they have pretty much become disengaged from these STEM subjects (USDOE, 2006; Osborne, Simon, & Collins, 2003).

Why Do So Many Adolescents, Who Previously Enjoyed Elementary School STEM, Flee From These Subjects as They Get Older?

In order to answer this question it is important to investigate what is relevant to early and later adolescents. Perhaps a better way to ask the question is "what matters to teenagers?" Before you read on, please take one or two minutes to write down your top two answers to this question.

Over the years I have done this exercise with thousands of teachers and parents and in virtually every case the consensus has been the same. I have also done it with hundreds of teenagers and they have also come to the same conclusions. Everyone pretty much agrees that they care most about being *accepted by a group of their pee*rs who matter to them. Secondly, they care that they can *demonstrate their competence* in things important to their peers and important to members of the community. This is different than the former need in that in the latter they are *respected* for who they are and what they can accomplish.

If a teacher can meet these needs while engaging students in STEM learning then it is a natural, organic process. It is what students want to do, not something that they must be forced to do. They will come and do it even if they aren't required. Let us use the humanistic psychologists' explanation of human needs, which is supported by our current understanding of brain neurobiology, to put this in a better conceptual perspective.

In the 1950s and 60s, Abraham Maslow (1968) and his colleagues studied individuals whom lots of people recognized as the most mentally healthy individuals they knew. The people they wish they could be. From this study the researchers deduced a hierarchical set of human needs. Until one need is reasonably met, the others are not of consequence to the person. After testing this with thousands of individuals around the world, researchers have come to realize that this sequence of needs is *genetically programmed* into all humans. This matches well with our knowledge of neurobiology where the frontal cortex and other reasoning parts of the brain surround the amygdala, the most ancient part of our evolutionary brain.

Figure 1 (page 2) illustrates the hierarchy of needs (Maslow, 1968). From the figure you can see that the survival and more emotional aspects of the hierarchy are the foundation and, therefore, always override the more rational aspects up the triangle. Once people have met their physiological—food, clothing and shelter—needs, then they have to focus on their safety needs. Homeless children keep slipping back

Figure 1. Maslow's hierarchy of needs

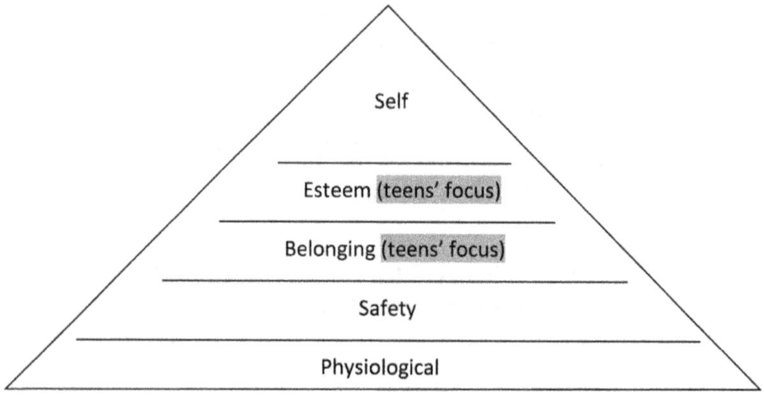

into this category and can't much focus readily on anything happening in school. Safety needs to come next. This is the ability to regularly predict what will happen in your life. Not just your physical safety, but also your psychological well-being. It is why kids who come from abusive, dysfunctional households can't focus on school. It is why divorces that upset family routines and create constant uncertainty send students into a tailspin. For the teenagers where these needs are fulfilled, they live most of their lives within the two aspects of adolescence we all have identified that teens care about most. The next two levels of belonging (acceptance in a group) and esteem (respect for competence) are where *genetically* teens must be. We keep asking them to be at the top level, self-actualization, where you can function once the other levels are sufficiently fulfilled, but of course adolescents don't have these levels filled. Nevertheless, schools generally ask them to do things because they will need it in the future, or not to worry about what their friends think. Yet, these are all things that are virtually impossible for teenagers to do.

If we can run our classrooms so that what we do meet these needs while students are engaged in explorations that uncover concepts and provide practice for their thinking skills, then students will rush to be there. They will put more effort into learning than most teachers have had the opportunity to experience. Examples of how their needs can be met while studying STEM concepts will be demonstrated in detail in the next section.

Thirteen years into my public school teaching career, I turned to my colleague Terry and pointed at the classroom and said, "Look at *that*. I never imagined it could be this self-sustaining. They are so intellectually and socially engaged. It is beyond anything I could imagine."

The students had engaged in two role plays earlier in the class about an underlying theme in a piece of literature that they were reading related to biological concepts they were investigating, and were now working in small groups to create some two minute theater pieces that would demonstrate their current insights and understandings of the relationship between the science and social science concepts of *evolution and revolution*. What was so amazing was the complete self-motivation of each group as they first intellectually argued with one another about the intricacies of these complex concepts, and then worked equally hard to dramatically convey their intended meaning. The previous term a number of these students were cutting their didactic transmission style classes. It boggled my mind (Jablon, 2014).

Therefore, as students progress into early adolescence, their daily attention centers around the drama of personal inter-relationships, their struggles with the unfairness of the world, and their confrontation with rules and regulations and ethical issues. The pre-teen and subsequent "teen in angst" rarely finds relevance in anything outside of their daily drama, and the previously engaging mysteries investigated in inquiry and materials-based science, engineering, and math fall to the wayside (Kuenzi, Matthews, & Mangan, 2006; Bennett & Hogarth, 2005). Until these inter-relationships are brought into the STEM classroom, as Maslow noted, most adolescents cannot deeply engage. Using drama as a tool in the STEM classroom can assist teachers to accomplish this task.

EFFECTIVE MODELS FOR USING THEATER IN STEM CLASSROOMS

Three approaches to using theater, and creative dramatics in particular, are outlined in this section. There are other ways that theater can be utilized in STEM classrooms. One of these alternative approaches is where students role-play physical STEM systems such as the example mentioned in a latter section about students physically representing the chemicals in a redox equation. This can easily be done, as can representing parts of mathematical algorithms. However, it is not nearly as powerful a tool in deeply engaging students in STEM understandings, especially those previously disengaged, as are the three models explained below.

Each approach is explained in step-by-step detail so that teachers or university faculty can understand how they might utilize it in their classrooms, followed by an example that has worked effectively in actual classrooms.

ROLE PLAYING OF ACTUAL STEM PROFESSIONAL SITUATIONS

These are role-playing scenarios where students actually role-play various constituencies and the STEM professionals that are involved in government, legal, medical, engineering design, or other professional situations where decisions need to be made. In these role-plays, students are given criteria for their characters and they then improvise these situations. Some students play the STEM professionals utilizing their subject area knowledge, while others play lay people who interact with the professionals where both utilize their decision-making skills. These role-plays are usually brief, lasting fewer than 5 minutes. However, the follow-up analysis by the rest of the students in the class can last for a more extended period.

Although some of the situations in these involve teenager characters, they usually do not. Rather, they are adult situations where you have STEM professionals in governmental hearings or local community meetings. Another common scenario is having individuals who are in conversation with personal advisors such as health professionals.

This approach is the most commonly used by STEM educators and is a good starting point for those who have never utilized creative drama in their classrooms. Because it has teenagers playing adults, rather than teenagers playing teenagers making decisions about issues they deal with in their daily lives, they are not as engaging to the students. Nevertheless, it is still a powerful tool.

These can be used as a motivational tool to introduce a new topic by engaging the students in a controversial application of the STEM knowledge that encourages them to be curious about the content. The same role-play can be utilized by the teacher in the middle of a unit where the discussion of the role-play can be a formative assessment tool, or used at the end as a summative assessment tool. At the beginning of the term students are told that everyone will be asked to improvise and act out a role explained to them by the teacher. The teacher tells each character separately how they will react and what they believe. The characters don't know what the others have been told. The rest of the class and the two or three role players are given a bare bones outline of the situation and the role-play begins with the students making up, improvising, the dialogue. The rest of the students are asked not to say anything at all to anyone while during the role-play and to try not to laugh unless something is very funny since it makes it hard for the role-players to focus.

The rest of the students are told that although they will each get credit for doing the role-plays when it is their turn, most of the credit will come from the whole class discussion that will follow the role-play. So, it behooves them to pay close attention to what is going on. The teacher watches how the role-play progresses and notices if it does not hit upon enough of the issues and STEM content. The teacher then asks the role-players to stop for a minute and asks if the viewers think a particular aspect is "real" to try to move the dialogue in a more useful direction. Surprisingly to teachers who are novice to using creative dramatics, this almost always works and the role-play resumes in a more useful way.

However, most of the time the scene goes on by itself for two or three minutes fulfilling its goal. The teacher then asks students what is real about this situation. This allows the rest of the class to note what they perceived as real human or societal interaction or as accurate math, science, or technological content. Then the viewers have a chance to speak to what they see as not real, or unusual, or inaccurate. They then can suggest other solutions not presented, alternative decisions to those made by characters, or perhaps comment on societal situations that they think are unfair and should be altered. All of these need to be supported by evidence from the content of the class or things they have experienced in their lives. Sometimes the teacher will ask the student who says there is another way for a character to deal with the situation to take over the role of that character and the improv continues.

Most of the time the teacher thinks of these situations ahead of time when they are writing their lesson plans. Many times there are STEM societal scenarios related to the content being studied written in books for students just to read and comment on. These are great sources for teachers to use as the basis for a role-play. There are a few on the NSTA website and many more elsewhere. However, once teachers become proficient doing this they can make up their own role-play situations to fit given content. I have seen teachers only have students read the scenarios and have discussions, and I have seen the same scenario first role-played and then discussed. The level of participation, passion, deeply involved, complex thought, and application of STEM content is always exponentially higher when the students have experienced the dramatic improvisation version in class.

Example #1: A Role-Play of an Actual STEM Professional Situation Focusing on Sickle Cell Anemia Testing on a College Campus

The teacher asked for two volunteers for a role-play. One of the volunteers was told that the character they would be playing was Frankie Fulcher, an African-American college student. About a month ago Frankie and his friends had been walking across campus when they saw a sign advertising free sickle cell anemia testing in the student center. One of his friends convinced the others that, "This is really important." They all had the blood test. Frankie had pretty much forgotten about it when s/he received a letter from a laboratory saying that his/her test results had come back and s/he could set up an appointment in the student center to get the results. Even though s/he didn't know much about sickle cell, s/he made an appointment. The teacher also told the student that Frankie is engaged to be married in the near future. S/he is told that the genetic counselor might say s/he is a carrier of sickle cell. S/he should act skeptical since s/he is a great athlete and is completely healthy. S/he should keep referring to that. If she suggests that he tell his fiancé s/he should be hesitant and then say s/he's probably not going to do that because s/he might think s/he is defective and then won't want to marry him/her.

The other student is told that s/he is a genetic counselor, Pat Nash, who interprets genetic test results and assists people in making decisions by explaining the situation, odds of occurrence, etc. Since it is the beginning of the term the teacher has to coach the student on what to say. The teacher explains that sickle cell is a disease that only manifests itself if the zygote is homozygous for the trait. S/he explains that if you don't have two of the genes for the trait you are a carrier, but you don't have the full-blown disease, which he explains is a really debilitating disease. Therefore, both mother and father need to be carriers, and then there is a one in four chance of the baby having the disease. That is why when Frankie is told that s/he is a carrier for sickle cell, s/he must be asked if s/he is married. If s/he plans on having a child, the partner should be told and tested.

The role-play of Frankie coming to talk with the genetic counselor begins. Frankie plays awkwardly astonished when s/he is told that s/he is a carrier, and protests that s/he is not only healthy, but buff. Pat, the counselor, does a pretty good job of explaining the genetics that s/he has memorized. When s/he asks about being married and Frankie explains about his/her fiancé, she suggests s/he tell him/her and s/he rejects the idea. She talks about the fiancé loving him/her and how she will respect him/her for being so mature, but Frankie is afraid of rejection, and says it is only a one in four chance and his fiancé is rich and probably isn't a carrier, equating wealth with health.

At this point, the teacher stops the improv, and the class claps as they do for all role-plays. The teacher then asks what is real. The students talked about how they have heard something about sickle cell, but like Frankie they don't know much. They talked about how his/her being scared of losing his fiancé could be real, and they talk about themselves and their friends and their failed relationships. They say that the genetic counselor tries to explain things, but the scientific terms make it not so easy to understand.

The teacher then asks the class if they would want to marry Frankie given s/he doesn't want to tell his/her fiancé? This raises a heated discussion, but also has many students in the class asking questions like, "How can you have the gene, but not have the disease?" or "How can this counselor know that there is a 25% chance of the child getting the disease?"

The teacher says that if you really want the answers to these questions to really pay attention to the investigations we will start tomorrow with Fast Plants and then later some Fruit Flies. As the class engages in genetics investigations over the next few weeks, both the teacher and the students keep referring back to Frankie Fulcher as they study Punnett Squares, recessive and dominant traits, and how a disease like sickle cell anemia can actually be a positive thing in a race or ethnic group genetic pool if Malaria is common in the habitat.

In this case the role play was done as a motivation for students by creating a scenario that they get pulled into and therefore want to know the science, math, and technology necessary to fully understand it in order to make informed decisions. Finally, here is a valuable multicultural lesson that doesn't have to do with ethnic food fairs. Sickle Cell Anemia is not limited to individuals of African descent. As a positive adaptation to fight malaria, it is most common among people of African descent but is found in any population with more than a few generations of living with malaria exposure.

The multidimensional ethical issues here are abundant and students need a deep understanding. This scenario can lead to continued discussions about genetic testing as an emerging and informative field. There is a great deal of misinformation being perpetuated related to socially constructed groups and genetic correlation (Roberts, 2011). This is a grand opportunity not only to wrestle with the NGSS disciplinary core ideas of *Inheritance of Traits, Variation of Traits, Adaptation, and Biodiversity and Humans,* but also a grand opportunity to practice the Core Skills of *Asking Questions and Defining Problems*, and even more so *Engaging in Arguments from Evidence*. Likewise, as the students engage in the ethical decision making that this scenario demands, they will also touch upon the core STS idea of *The Influence of Engineering, Technology and Science on Society and the Natural World*

ANALOGOUS INTERRELATIONSHIP ROLE PLAYS

These are role-plays created by the teachers that utilize typical teenage situations, but are analogous to governmental or corporate ethical and economic dilemmas. Many times when looking at a large-scale adult governmental or corporate problem they can get students to understand the complex interplay of ethics, economics, and subject area knowledge when they could not previously comprehend it.

For the past forty years I have observed adolescents in a wide range of classes. For some of these classes I was the teacher but since 1990 I have been blessed with the opportunity to be an outside observer of a cross-section of classes in various disciplines, in both urban and suburban locations, and in small non-traditional schools and large fairly traditional schools. In each of these classrooms there were at least three or four students who were completely disengaged from any learning activities no matter how much inquiry learning and interdisciplinary format was used. If we multiply these students by the number of classes in the school, this number becomes substantial. In many urban schools this number was much higher.

During the twenty-one years—1970-1990 and 2001-2002—that I taught adolescents, together with my colleagues and our students, we created an approach to truly engage these disengaged students in learning, and in becoming a full participant in our community of learners. Inquiry-based learning was a major component but it was not sufficient by itself.

Action-based classes that connect academic subjects to societal issues appeal to the innate desires of young people *to make a difference* in their world. Giving attention to social conscience issues enables students to act upon their decisions about the issues through letter writing, direct community involvement, or some other action. However, many times when teachers introduce societal issues such as solid waste management, genetic engineering, and such, they are surprised to find that many students are not motivated by such matters. And for those students who initially demonstrated an interest, their attention either is sporadic or wanes completely.

Yet there are some programs utilizing societal issues that have the ability to sustain a deep student interest throughout the semester. Some of these programs have utilized societal issues with great success with students who have been labeled "at-risk" (Jablon and Born, 1988). If we look at the nature of human beings, and the nature of adolescents in particular, some of the underlying reasons for this success in engaging students with societal issues become clarified.

As outlined earlier, most adolescents, but especially at-risk students, do not have their security, love, or self-esteem needs met—as defined by Maslow's hierarchy of needs (1968). Yet most of what we do in schools is to ask students to work within the realm of self-actualization, saying that these things they are asked to study they will need some day or will use for the betterment of society. Rarely are they engaged in any activities in the academic classroom that will fill their immediate needs for these lower level foundational elements. These needs when not addressed, disallow the student from engaging in the self-actualized activity we present. Many students not finding any of these more emotional needs being addressed in their academic classes seek them in their anti-school, negative peer groups (Pulkkinen, 1982).

Underlying these societal issues are psychological and sociological relationships. Over the years, teachers found that most of these big ideas and concepts that they wish students to acquire, whether in social studies, science, technology, or in the relationships of characters in novels, are universal relationships between individuals and groups, individuals and other individuals, or between groups and groups. They are in many ways based on the same psychological and sociological foundations as the relationships these students are having with their peers and their parents, the two primary areas of concern to adolescents. They are *analogous.*

Once these needs are met, the students will not only engage in study of the analogous societal relationships but will delve into the deepest esoteric questions that are posed. In addition, they acquire conflict resolution skills and coping mechanisms that they can utilize in their everyday interpersonal relationships. The personalization of the subject matter that comes from having students investigate these *analogous interrelationships* between their own personal interrelationships and those that exist at a societal level is one of the most powerful tools that I have ever experienced in engaging students previously disengaged from schools (Jablon, 2010). If the psychological needs of the students are met as they are engaged in activities with the other students in the STEM class, then a positive alternative peer group is created that students can go to as an alternative to their negative neighborhood peer groups where they normally turn in their search for fulfillment.

Teachers facilitate the role-plays the same way as with the previously described STEM professional role-plays, but the result is far more effective in engaging these emotionally needy students. I could not think of a more powerful pedagogical tool to utilize in a STEM classroom.

Example #2: An Analogous Interrelationship Role Play focusing on the Greenhouse Effect and the Cutting of the Brazilian Rain Forest

In a biology class it was a few weeks into studying the effects of cutting the Brazilian rainforest in order to sell hardwoods and to create grazing land for cattle to be sold as beef. The students had done a myriad of investigations with stream tables in order to understand the relationship between planted and unplanted soil and water erosion. They had done greenhouse effect experiments with jars and thermometers, and done a number of bromothymol blue photosynthesis experiments. They really had a handle on the science and had been especially engaged because of the societal issues raised about the plight of those living in the rainforests, and how Brazil was up against a wall in need of exports to avoid inflation and the World Bank calling in loans. However, they were constantly perplexed that given what ecologists knew as the long-term bad effects on the local as well as the world ecosystem, how adults in Brazil just didn't stop the devastation. Teachers' explanations didn't seem to help with these complex real world interactions.

In order to have students see the interrelationships in both the country of Brazil and in the world groups with whom they had a relationship, the following role-play was created and then done in class the following day.

Three students assume the roles of the following characters:

- Marie Sussman, a fifteen-year-old high school student who must work long hours to add to her family's income because a parent has recently died.
- Ms. Pape, Marie's English teacher who is upset by the drop in quality of the student's work in her class.
- Marie's mother, Diane Sussman, who is caught between wanting the best for the student and the rest of the family's welfare.

Both the student playing the high school student and her mother are given the information about their roles. Marie is the oldest child, having a four-year old brother and an 8-month-old brother. Up until the summer her mother worked part time spending many hours at home taking care of the kids. Her father worked full-time and earned a decent salary. Unfortunately, her father died during the summer and being a "private" family they felt no obligation to notify the school. Her mother went back to work more hours, mostly on the weekends. The family is on food stamps, but having grown up in a welfare family the mother will not put her family through that "same humiliation." In September Marie was doing fabulous in Ms. Pape's class. She loved the novels and the class discussions, and did great on homework, tests, and quizzes. But her mother came to her and said the family wasn't making enough money to pay rent, buy food, and keep her siblings in Pampers. Thus, she needed to get a job to assist with these expenses. They have a family friend who gets Marie a job doing mostly filing work at a doctor's office from 4 to 8 every weeknight. Marie's grades have been dropping radically. By the time she gets home she's exhausted and she has six subjects' homework to do. She is way behind in everything, including English class. On weekends she babysits her siblings while her mom is at work. She and Ms. Pape have spoken before about her grades dropping, her poor quality of work, and the lateness of her assignments. Marie has assured her that she plans on doing better. Marie has even asked the doctor if she can cut back some hours, but the doctor says that he is already doing her a favor and he can't have more part-time employees so no cutback in hours. Ms. Sussman is told that if she gets a call from any of Marie's teachers, to be huffy thinking that these teachers just don't get how difficult a position she is in. Marie is told that in her family, the family's business is kept within the family so she would be hesitant to talk about it when asked by teachers. But that when pressed she can let out a little bit, and then finally the whole story.

The student playing Ms. Pape has not heard any of this. She is told that Marie was a great student in her class in September but her work has dropped off radically, such that she is on the brink of failing. Occasionally an amazing piece of writing still comes in, so she suspects there is some reason that Marie's not keeping up. She has spoken to Marie before, who keeps assuring her she will get in the work and improve, but that she is just busy. She is also told that she has asked Marie to come after class and she will really confront her with how close she is to failing. She is also told she can call Marie's mother if she can't resolve this with Marie.

The class is simply told that Marie is a ninth grade student in Ms. Pape's English class and Ms. Pape has asked Marie to see her after class. They are also told that Marie's mother is at home and can be reached by phone.

Marie doesn't tell the teacher much to start with, although Ms. Pape is probing. Finally, Marie admits that she has a job and it is hard for her to do homework when she gets home at almost nine every night. When pushed further by Ms. Pape, Marie explains about her father's death and her need to help support the family. Ms. Pape is helpful, asking if she can cut back hours (no), asking what about the weekends (babysitting siblings), and offering to stay after school to help Marie (can't because need to travel to work). Finally, Marie needs to go to work. Ms. Pape calls Marie's mother, Diane Sussman, who is at home with her other children. At first Diane is pleasant and saying she is also concerned about Marie's grades, but as Ms. Pape continues to push speaking about Marie's future and how she is almost failing, Diane pushes back. She says she is responsible for her whole family and she tries to give Marie as much time as possible to do her schoolwork, but they don't eat and pay the rent if Marie does not hold down this job.

The teacher stops the role-play at this point, and asks the class what is real. Most hands in the class go up with comments about how the teacher is caring but only cares about grades. Other comments are about it is real that parents don't realize how much pressure there is on high school kids, or that it is true that there are lots of kids like Marie who are hesitant to talk about their family issues with teachers. When asked what is not real, a student says that the situation where Marie really has to work, not just to earn spending money, isn't real. Three students all fire back that either they, or friends, or brothers or sisters are in that situation right now. Others say that teachers really don't care that much, and many students give examples of teachers who bend over backwards caring, or teachers who wouldn't even go so far as calling Marie's mother. Other kids talk about Marie needing to stand up more for herself with her mother, and the class gets into a long and serious discussion about relationships between parents and teens and the power differential and other issues.

The teacher asks the class to think for a few minutes before answering (wait time) the next question,

"Who is to blame?" At first there is a lot of back and forth discussion and disagreement about Ms. Pape being the one to blame and she should give Marie less homework and a break on her tests. Then the same thing happens about the mother or Marie being the one to blame. Finally, after about five minutes a student puts forth the notion that, "None of them are to blame." When asked by others in the class how can that be, someone needs to be wrong, the student replies, "The system in this community, city, or state that they live in is to blame. Each of them are doing the best they can do at that moment." The teacher asks if any students can think of a time when they were caught in something like that, and one example is given, and then another that is not so clearly the fault of the situation.

Up until this point this is about social issues that teenagers are grappling with in their everyday lives, with teachers and parents. However, now the teacher asks the whole class to use wait time before they answer, "Think back to the cutting of the rainforest in Brazil. Who here represents, is like, the ecologist? Who is like the government of Brazil with the World Bank on their back? Who is like the folks living in the rainforest and cutting it down to get by?

There is a very long silence and then various opinions are given with evidence from the role-play and from what they studied about the greenhouse effect and cutting the rainforest to back them up. There were these "Aha!" moments as students said things like, "Ecologists only care about the future, just like Marie's teacher, Ms. Pape, who kept talking about her future in college."

This role play allowed this class of students studying about the cutting of the rainforests to understand the complex relationship between the poor native inhabitants cutting the rainforests (the student), the government of the third world country who is encumbered with a large international debt (the parent), and the environmentalists trying to help both the native and the future world ecological situation (the teacher). It was a clear *Analogous Interrelationship*. Students not only had a unique insight into world economics but into their own personal relationships. Therefore, such analogous interrelationship role play causes students to identify positively with their academic classes.

For days following this long role-play and discussion students were deeply engaged in any of the science related to this topic. Why? Their teen Maslowian needs were met with serious discussions about their daily dilemmas with parents, teachers, and responsibilities, with many coping mechanisms suggested along the way. Although it helps all students, for those students who are most emotionally needy, the more of these types of role plays done, the more they will not cut class and the more they will attend to the content as well.

This example described an actual classroom experience that I encountered in my teaching career and how analogous interrelationship role-plays that were created helped solve some student lack of understanding of a complex topic. However, if a teacher was just to introduce this in a generic deforestation/ecosystem unit, it might be more effective and create less of a them-versus-us mentality if the teacher was to use local or US national examples such as the deforestation of the Pacific Northwest, or golf courses and the impact they have on the water table in the desert southwest, etc. Using the Brazilian, or an Indonesian, example, there is the potential for teaching that deforestation and other topics impacting the environment are something that happen "elsewhere" which may limit student ownership and agency as well as reinforce a narrative of "us: good, smart, stewardly" and "them: unaware, abusive, ignorant, destructive."

Just as in example #1, here is another grand opportunity not only to wrestle with the NGSS (2013) disciplinary core ideas of *Global Climate Change, Ecosytems: Dynamics and Resilience,* and *Interdependent Relationships in Ecosystems*, but also a grand opportunity to practice the Core Skills of *Asking Questions and Defining problems*, and even more so *Engaging in Arguments from Evidence.*

WRITING SCRIPTS AND PRODUCING PLAYS UTILIZING THE SUBJECT AREAS STUDIED

This third model for using theater goes well beyond creative dramatics and role-plays, although they both are utilized in the model. It usually involves other subject area teachers working with one of the STEM teachers, is much more complex and difficult to maintain, and is best explained by citing an example. Therefore, a brief description will be given here and the model will be elucidated in greater detail in *Example #3* where a full description of a program is given.

After a certain amount of subject area material has been uncovered through inquiry teaching approaches, the students as a class decide on the important STEM and societal or ethical issues they want to have others consider. Then as a class, they write, rehearse, and produce a dramatic piece that creatively embeds these issues in the plot. This piece is then performed for a public audience. This process is usually done in collaboration with other subject area teachers.

The various subject area teachers working together with the same group of students usually select some powerful universal theme to help integrate the various cross-disciplinary content, which simultaneously assists students in making sense of complex world issues and decisions. Such themes as "power over versus power with" or "the right of the individual versus the right of society" allow students to investigate some of the greatest of human dilemmas and ideas, while pulling together seemingly unrelated subject areas.

Example #3: Using Student Written Theater Productions to Engage Previously Disengaged Students

Nellie McCaslin, the grande dame of American creative dramatics for over sixty years, highlighted a particularly successful program in her *Creative Drama in the Classroom and Beyond* (2006, pp. 264-5). What follows is a detailed description of that program in a public high school in New York City for students who were either regularly truant or previously had dropped out of school. It shows how an interdisciplinary program with its drama components had these same students attending daily, succeeding in science and other subjects, and finally graduating at amazingly high numbers. A ten-year mixed method study demonstrated how having these students use role plays addressing STEM-Societal topics, utilizing interdisciplinary team teaching, and most especially writing original full length plays based on the subject area content and important themes taught throughout the term completely turned around the academic and social-emotional achievement of a vast majority of the students enrolled. This BONGO program started by Terry Born, an English teacher and myself, a science teacher, has been replicated in many other schools in many other states, but this particular term described below was from one of our group of students at the original Middle College High School.

The program was three periods of a seven-class day where students studied rigorous subject matter in three state-mandated subjects. The students who completed whole class dependent projects with public presentations—many with social advocacy themes—were enrolled for one semester. However, since the mix of subjects changed each semester, many students re-enrolled during a subsequent year for a semester when the classes they needed were available as they realized their need for the *positive alternative peer group* they had previously experienced.

The students in this semester were enrolled in a Biology I, physical education, and English III class as part of BONGO. During the term they read Orwell's *1984*, Huxley's *Brave New World*, and More's *Utopia*.The students also studied a bit about American government structures and engaged in physical education where small bicycling experiences twice a week led to an understanding of bicycle safety. Simultaneously, students were engaging in *doing* biology by designing and completing daily inquiry lab investigations related to genetics, cellular mechanisms, and ecological relationships. Genetic engineering, and its related ethical issues, became a lens through which the students studied various biological concepts and connected these to the novels they were reading. Students also employed many of these biological concepts to make sense of the physiological demands on their bodies during bicycling.

For the first two-thirds of the term, small cooperative teams completed many small interdisciplinary projects and daily *analogous interrelationship role-plays* that used and demonstrated mastery of science, English, and social studies knowledge and skills. These role-plays always involved adolescent characters interacting with other teens, their parents, teachers, or other significant adults in their lives. The personal relationships in the role-plays were analogous to the larger relationships being studied in an academic perspective in science, history, or literature. Students had an opportunity to simultaneously confront and discuss the adolescent issues that made their lives so unpredictable and themselves so needy while studying important concepts in academics.

During the final six weeks, students began to synthesize some of the larger concepts they had identified as being *vital driving questions* that they themselves were wrestling with during the term and that they found of great significance. As adolescents they identified the most important social conscience issue they encountered as the term progressed so that they could then engage their peers, parents, and others who would see the play they were about to write together as a whole class. Like the little projects all term long, the final project they created was group dependent for success. As a class, they outlined, wrote, revised, and produced a two-act play that synthesized complex ethical, scientific, and legal issues into a compelling drama that spanned four decades in present and hypothetical future history. The first act of the play written by the students in 1983 took place in a hypothetical 1983 United States where a constitutional convention had been called to change the constitution so that a law could be passed that allowed the government to override the rights of individuals found in the Bill of Rights. A court scene takes place where a couple refuses to obey the law that would affect their unborn baby. The second act occurs in 2020 where the students devised a culture based on the extension of the idea of the government making decisions for individuals. The baby of the original couple, now a young man who has been hidden away in the mountains, emerges into this new culture with the ethics of pre-1983 Americans. This play was toured to surrounding schools and also presented in a local college theater (Jablon & Born, 1988).

During the thirteen years that Terry and I facilitated the BONGO Program we were astounded by what our previously disengaged students accomplished. It was beyond my wildest dreams about what teaching and learning could be. We got to watch one class reinvent existentialism over a two-day period. The students created photographs and sculptures that were displayed in college galleries. They won the educational award from the National Hunger Project for a play they wrote and toured about world hunger and the politics involved. Four times they won first prize for the young adults' playwriting contest from the Dramatist Guild in New York City. Groups of two or three BONGO students have taught over fifty classes, with adults only as observers, in other public schools before they presented their original plays the next day.

That term was six years into a three-period-a-day program that Terry and I co-taught and directed for thirteen years and that in some form or another has been continued by teachers around the country. A ten-year quantitative and qualitative research study of the program uncovered some fairly amazing results. Despite the almost universal agreement among the students in the program that the coursework was the most challenging they ever encountered in school, a ten-year study of the program demonstrated a remarkable success with a set of previously disengaged students. This structure accounted for some fairly compelling results with a population of urban African-American, Latino, and white students who almost all had a history of at-risk behavior. The results showed:

- Ninety-two percent had no cutting of classes during the time they were in the BONGO Program. Seventy-seven percent of those students had a history of cutting classes before they were in the program, which dropped to zero cuts while in BONGO.
- Attendance also increased, both for students who were mostly truant, and for students who attended school fairly regularly, but had significant absences. Sixty-five percent of the students experienced increased attendance, thirty-one percent showing a massive increase. For any of these students to have above ninety-five percent attendance was unbelievable to their parents, their other teachers, and themselves. The number of students who showed either perfect or ninety-five percent attendance doubled (perfect attendance went from three to five percent; ninety-five percent attendance went from twenty-one to forty-five percent).While students were in BONGO, fifty percent had ninety-five percent attendance or better.
- Eighty-nine percent of the BONGO students either increased or remained constant in their academic performance with fifty-one percent showing a major increase. Thirty-six percent failed science before BONGO but passed during. Fifty-four percent had an increase in their science grade.
- The most startling finding was that thirty-two percent of the students went from passing no courses, or one or two courses, to passing *all* of their subjects, including those classes outside of the three BONGO periods.

The Watson-Glaser critical thinking test administered to BONGO students demonstrated that sixty percent of those students significantly raised their *critical thinking abilities*, which was four times greater than their peers in control classes. The Process Skills and Science, Society, and Technology (PSST) test of process skills in science and abilities in understanding Science, Technology, and Society (STS) relationships revealed that forty-one percent of the BONGO group increased their science reasoning skills and STS awareness over the course of one term as compared to nine percent within the control group (Jablon, 1989).

One final thought about this approach to teaching STEM subjects interconnected with other disciplines using drama as a thread throughout the program: Of these students who almost all had a history of dropping out of another high school, 97% graduated. Although this is a much more complicated version of the use of theater in teaching STEM subjects, the compelling statistics about how it turned around the lives of countless hundreds of students should motivate educators to both implement similar programs in their schools, and university STEM education professors to include it in preparation and professional development programs.

THEORETICAL UNDERPINNINGS AND A REVIEW OF THE LITERATURE OF DRAMA IN THE STEM CLASSROOM

For some students, the utilitarian application of designing useful things for our everyday needs utilizing scientific and mathematical knowledge brings STEM subjects closer to relevance. However, for most early adolescents, this practice of engineering and technology is devoid of the optimization and ethical drama that accompanies real world applications. It is ironic that many of the societal issues raised by technological innovation are analogous to the issues that affect these very adolescents in their daily emotional lives, but these parallel emotional, social, and ethical issues are rarely at the core of the technology and engineering programs.

When some modestly effective STEM programs have infused these social and ethical issues into the curriculum, many times they are introduced through readings about international environmental situations with countries or large corporations being the units of conflict. Even when the issues are locally based, all too often the societal issues are simplistically seen as one dimensional, black and white, right and wrong without the complex ethical and monetary dimensions being introduced. Some students can deal at this theoretical and formal thinking "someday I will need to know this" level, but most are still focused on their daily interpersonal interactions as they try to navigate their way to adulthood.

For decades, teachers have found that using creative dramatics in their classrooms has allowed them to tap into these emotional needs of their students in order to engage them in the previously unlearned subject area conceptual understandings (McCaslin, 1981, 1996; Bolton & Heathcote, 1995; Arieli, 2007). A vast majority of the teachers who utilized this effective approach did so with social studies or language arts (Salisbury, 1986; Duveen & Solomon, 1994; Mardirosian & Lewis, 2009). Science, math, engineering, and computer science teachers have been loath to use these creative dramatics techniques (Kelner, 1993; Dorion, 2009), perhaps because of the way we as adults have experienced theater in our lives. Most of us are familiar with going to the theater, or reading famous plays in school. Unlike theater, creative dramatics in the classroom focuses on those who are participating, rather than on an end product or a final performance. Creative drama is defined by the Children's Theatre Association of America as "...an improvisational, non-exhibitional, process-centered form of drama in which participants are guided by a leader to imagine, enact, and reflect upon human experiences (McCaslin, 1981).

It is almost always an improvised performance where lines are not written and not memorized. Many members of the group get to play various parts. Participants are guided by a teacher, and no decorations, costumes, or special equipment are needed.

Using drama in the classroom builds community, engages students in content, enhances focus and concentration, and often creates meaningful experiences that help students understand abstract concepts.

Creative dramatics, a highly effective method for integrating arts education into core curriculum, produces a positive and lasting impact on student learning, in terms of creative and critical thinking, language development, listening, comprehension, retention, cooperation, and empathy and awareness of others. Creative dramatics not only has the power to bring curriculum to life, but also to stimulate active involvement in the development of conceptual understandings. (Pieczura, 2013)

Well-planned, engaging experiences with creative drama guides and supports student's problem solving and decision-making skills, while at the same time, works at increasing awareness of how to solve issues at hand. Instead of the STEM classroom just being a place where students are being taught and told what to think and feel, drama turns this into a deeper experience in thinking, further motivating students to question, respond, and explain what they themselves are feeling and thinking (Ødegaard, 2003). Creative drama can be a useful tool for all stages of the learning cycle, providing the STEM teacher an authentic assessment measure, as well as an excellent tool for engaging, explaining, exploring, elaborating and evaluating (BSCS, 2005; Llewellyn, 2004).

STEM subjects are still generally taught in didactic, teacher-dominated classrooms by teaching methods that do not consider students' Piagetian cognitive developmental stages, learning styles, or multiple intelligences (Goodnough, 2001; Weld, 2000). Creative drama activities can help students bring abstract concepts into a concrete context by exploring and constructing an understanding of abstract ideas during improvisational interaction (Bailey, 1993; McCaslin, 1996). Students are motivated to build concepts, negotiate ideas, and to solve problems when doing so is relevant and meaningful for them (Dockett, 1995). Effective learning occurs when students construct their understanding by active learning and by building on their prior knowledge (Yager, 1993). "Students who are usually shy, quiet, self-conscious, or seem troubled, "wake up" or "come to life" when they are free to use their imagination. Teachers are often surprised at the knowledge and the vocabulary that students had internalized but had not shared through traditional forms of learning and assessment" (Arieli, 2007, p. 57). Participation in drama activities provides the teacher with another way to assess the student. Through dramatic play, students reveal how they organize ideas, solve problems, work in a group, deal with conflict, and use their imagination. Observing how students dramatize an event offers valuable insight into how they perceive, interpret, understand, and analyze the material at the core of the lesson (Caldwell, 2011).

A student can, if only for a few moments, become another, explore a new role, try out and experiment with various personal choices and solutions to very real problems-problems from their own life, or problems faced by characters in literature or historical figures. This can happen in a safe atmosphere, where actions and consequences can be examined, discussed, and in a very real sense experienced without the dangers and pitfalls that such experimentation would obviously lead to in the 'real' world. (Buchanan, 2016)

Although the use of creative dramatics to tap into the emotional connection of subject matter to students' daily lives is the most powerful use of theater in schools, it is also useful to simply have students physically role play various complex physical or biological scientific or technical systems to give insight into these interactions and concepts related to these systems. In addition to using creative dramatics, it can also be an effective practice to actually have students in classes write and produce short or full-length plays that are final public projects that allow a STEM class to express their science, technology, and societal understandings to those of importance in their own personal lives.

It is also important to note that the integrative thinking that occurs in the right frontal cortex that is predominant in the arts, including theater, is also the same type of thinking necessary for novel insights in all STEM fields. Far too much of STEM subjects taught in schools focus on using linear, logical thinking devoid of any practice of using a Gestalt approach to creating novel insights utilizing concepts and data from seemingly unrelated science and engineering fields. There are a whole set of students who are drawn to the arts for this creativity. Many can actually do well in science, math, or engineering, but choose not to pursue a STEM related career because they never get to see this side of these professions. However, when drama and STEM subjects are intertwined a number of those previously disinterested in STEM professions realize a possible compatibility.

VARIOUS REASONS TO USE THEATER, ESPECIALLY CREATIVE DRAMATICS, IN STEM CLASSROOMS

The following is a list of reasons demonstrating that creative dramatics is an effective tool to be used in STEM classrooms. Although each of these reasons stands on its own, it is the synergy among these schemas that makes creative dramatics such a powerful tool.

Gives Students Insight Into the Scientific, Mathematical, Social, Economic, and Ethical Issues of Users of Technology

Students may not be intrinsically interested in codons, transcription, or mitochondrial RNA, but when asked to role-play, or watch a role-play, about near future parents, religious leaders, and doctors asked to make decisions about DNA testing and genetic editing they become almost instantly intrigued by the ethical dilemmas in a way that reading about such issues does not engender. Once they are hooked on debating the "right and wrong" of issues, they begin to want to know the science and technologies used in order to make sense of the difficult ethical dilemmas. The use of drama to teach about STEM interactions within society has been noted to improve pupils' empathy and identification in socio-political situations of science (Cobern, 1993). In South Africa lessons have been evaluated where role-plays helped raise the conceptual insight and level of argumentation about the ethics of trade in organs for xenotransplantation (Braund, Lubben, Scholtz, Sadeck, & Hodges, 2007). Hence, creative drama can assist in the development of argumentation in science by anchoring scientific debate and concepts in pupils' real world experiences (Duveen & Solomon, 1994).

Gives Students Insight Into the Social, Psychological, and Economic Dimensions of the Design Process

My first teaching job was over 40 years ago in a "street academy" in EL Barrio, Spanish Harlem, created by the Young Lords, a group who were trying to bring the young adult Puerto Ricans into places of power, knowledge, and authority. By the time they created this school they were college graduates, but formerly many were members of the Puerto Rican street gangs made famous by the movie *West Side Story*. In addition to the school, another young Puerto Rican leader who had a degree in architecture ran a studio where a number of Columbia University graduate architecture students chose to do their internships. This gave these mostly white, middle class architects-in-training an opportunity to work on designs for real projects to be built in the neighborhood. They would interview the people who would use the future community center or housing project and then create designs to be evaluated by the staff in the studio. Invariably these students would create designs that were aesthetically beautiful to them and their Columbia faculty, while having met some of the criteria set by both the engineers and the local residential users. However, most of their designs were not utilized because they could not hear the mothers who asked for enclosed courtyards in public housing so their children could be safe outside playing while their mothers cooked in kitchens that faced these courtyards. They couldn't hear that the fathers asked that Spanish Colonial architectural components be included in the designs for this matched their cultural aesthetics. The city engineers said thirty-story buildings were more affordable, but then mothers couldn't keep track of their kids like they could in an eight-story building.

Therefore, it is vital to have STEM students role-play community board meetings, with various students taking on the roles of parents, local physicians, engineers, politicians, local store owners, and perhaps epidemiologists. The board is having a hearing on some design for a local structure that allows students to hear the passion that each representative brings to the table. Mathematics, science, and engineering when thrown into this cauldron of everyday public debate bring the subject matter involved into the emotional realm of everyday human living and the importance it plays in effectively planning our *designed environment*. This approach comes from a long history of educational drama and related approaches to education (e.g. Boal, 2002; Freire, 2000; Heathcote, 1984; O'Neill, 1995; Grady, 2000; Miller & Saxton, 2004). These strategies support a variety of learning styles that keep students actively engaged in the learning process, while also allowing for differentiated instruction for diverse learners, such as those who are English Language Learners or who have disabilities.

These creative drama techniques are useful, helpful, and effective in many curriculum areas, including STEM subjects (Bailey, 1993; Kase-Polisini & Spector, 1992; McCaslin, 1986, 1996; Cottrell, 1987). Science researchers and educators, in particular, have promoted and researched this approach (Butler, 1989; Duveen & Solomon, 1994; Kentish, 1995; Steinert, 1993; Stencel & Barkoff, 1993). Butler (1989) and Webb (1980) see a major benefit of using creative drama in STEM classrooms because it enables students to learn openly from and with each other. Their studies demonstrate that drama is a stimulating alternative to class discussions, lectures, or question-and-answer sessions. STEM become more than learning a collection of isolated facts when drama is involved, where application and synthesis of conceptual understandings occur naturally.

These creative drama approaches do not only exist in secondary school settings. Kelly, Carlsen, and Cunningham (1993) strongly recommend that the sociological aspects of a genuine STEM culture should be included in school curricula. Early childhood science education seems as an ideal place to begin this incorporation through sociodramatic play. It is evident that children utilize play to clarify their scientific thinking (Bailey, 1993; McCaslin, 1996; Smilansky, 1968, 1990). They mentally play with the concepts they are dealing with, incorporating them into an imaginary setting. They use play in order to try to make sense of ideas that are puzzling and stimulating to them and of ideas that emerge from contexts that are familiar to them. The book *Dramatic Science - Inspired ideas for teaching science using drama ages 5–11* by Debra McGregor and Wendy Precious (2015) provides teachers with a range of tried and tested techniques to use drama as a support and aid to the teaching of science to young children. All the strategies in this book have been shown to work effectively in a wide range of primary schools. There are three studies that look closely at successfully using drama in elementary STEM classrooms (Kamen, 1991; Metcalfe, Abbott, Bray, Exley, & Wisnia, 1984; Varelas, M., et al. 2010). This last one demonstrates enhanced STEM learning in urban first through third grade classes using drama enactments.

Gives Teachers Insight Into the Decision-Making Criteria of Their Students and the Origins of These Criteria From Their Student's Lives

In the engineering portion of STEM, it is an overt part of the curriculum to actually teach the decision making process (Aurum & Wohlin, 2003; Gomes, 2016). However, if science and mathematics is taught using effective inquiry teaching methodologies (Jablon, 2014) then a major focus of the pedagogy is also to work on students' decision-making models in their process of doing math and science. In order to be successful at having students metacognitively be aware of their thinking within a STEM subject, teachers need to focus on uncovering the students' decision-making models that they use in their everyday lives. Teachers and students can then compare and contrast the student's everyday decision-making models with ones used in the math and science communities. As students in STEM classrooms engage in STEM-societal role-plays, their improvised roles overtly demonstrate the way they make decisions by the interactive dialogue they create for their characters. Not only does it become clear to the teacher and the other students watching the role-play, but also the student questioned about the role they created gets insight into how they utilize STEM understandings to make decisions in their lives.

In classrooms where teachers effectively facilitate learning through the use of creative drama, students actively co-construct their learning process. They take on the "mantle of the expert" through the scaffolded safety of drama-based activities (Cawthon, Dawson, & Ihorn, 2011), trying on new language (first modeled by the instructor) both in verbal and written form (Bolton & Heathcote, 1995). This dynamic process demands higher-order thinking skills and increases emotional intelligence. In some cases, this shift in roles and knowledge making through drama-based work can illuminate where socially constructed meanings exist and create space for self-efficacy to emerge (Thompson, 2003). Kentish (1995) notes that when students use creative drama strategies, they develop greater ownership of their learning, which is reflected by their development of problem solving and decision-making skills and abilities.

In addition, creative dramatics not only gives insights into students' decision-making and problem-solving skills, but also can teach, or at least reinforce, the same STEM skills in parallel. Similarities exist between STEM and creative drama processes. In both processes students are active, solve problems, ask questions, learn through inquiry, and construct knowledge built on prior experience and information (Bailey, 1993; McCaslin, 1996; Kase- Polisini & Spector, 1992). When integrated into teaching STEM, creative drama can assist students in developing STEM skills as well as affective skills, such as cooperative work, empathy, communication, listening, and reasoning (Christofi & Davies, 1991; Hodson & Reid, 1988; Kase-Polisini & Spector, 1992).

Gives Teachers Insight Into Their Students' Conceptual Understanding of Subject Area Conceptual Knowledge as They Apply it in Dramatic Role-Plays

Just as with their decision-making skills, creative drama helps learners to make their STEM conceptual understandings *explicit* (Onwukwe, Akuezuilo, & Ijioma, 2014;Venkateswaran, 2004; HMI, 1989; ArtsCouncil, 1992). Certainly role-plays are generally more spontaneous and more flexible than end-of-the-chapter questions or discussions led by the teacher (Steinert, 1993).

Drama is an alternative way to assess by observing. When teaching new lessons, we always depend on prior knowledge. We start with what students know, which further guides us with the next step to take in our teaching. It is difficult for some of our students to make sense of specific things, which is simply where drama fits in. Drama can be used to preview or review a lesson; further allowing teachers to assess what students already know or have learned. (Cornett & Smithrlm, 2001)

Whether the role-play has students playing roles in social decision making situations related to STEM subject matter or the students are actually role playing the parts of a scientific or technological system using their bodies, the teacher and students still see and hear a physical analogy manifestation that can then be discussed concretely, and possibly then moved into a more theoretical realm. In *The Drama of Science* (2009), Bernhard Sturm offers a selection of drama-based activities to act or re-enact science in the chemistry and physics classroom. For example, he suggests a method to visualize the exchange of oxygen between different metal atoms within the redox series (after Lavoisier). Each student wears a cap in one of three colors, representing atoms of oxygen and two different metals (there should be equal numbers of each color). In groups of about 8-10, the students enact simple redox reactions with 1:1 stoichiometry, such as $CuO + Fe \rightarrow Cu + FeO$, and afterwards present them to the entire class. Students often find creative ways to represent both the activation energy and the release of energy in these reactions. Although these role-plays of actual physical reactions are not as powerful a way to engage students in the science as are social decision making situations, they are nevertheless still somewhat useful to uncover students understandings, and misunderstandings, of STEM subject matter.

Role-play not only allows students to demonstrate their understanding, but also to explore their views and develop deeper understanding of phenomena by combining new facts and concepts with previously acquired knowledge and applying it to solving real-life problems (Butler, 1989). Role-play permits students to verbalize abstract science ideas in terms that they can understand. Role-play has the potential to assist students in developing and creating their own mental models. Gardner (1991) suggests, "in trying to understand science, students draw on available mental models" (p. 157). Often in science these mental models create images of a microscopic world, which is inferred from observations of theories while trying to explain phenomena (Aubusson, Fogwill, Barr, & Perkovic, 1997; Aubusson & Fogwill, 2006). Creative drama is an effective strategy to teach abstract concepts. According to Stencel and Barkoff (1993), drama is very useful in teaching abstract concepts, especially those at the molecular level, that are often difficult for students to understand. Bachelis et al. (1994) demonstrate that students can best understand an algorithm if they can see the process tangibly at work, and creative drama activities are suggested as a means of accomplishing this goal.

One of the great gurus of drama education, Dorothy Heathcote, went even further than just seeing drama as a milieu in which students can demonstrate and modify their understandings of STEM concepts. She asserts that the challenges involved in creating drama stimulate left-brain activity and hence scientific and mathematical functioning (Wagner, 1979). Evidence from neurological research into brain function supports the view that creative (right hemispheric) activity contributes to general cognitive ability (Gazzaniga, 1989)

Gives Teachers Insights Into Issues Affecting the Emotional Lives of Their Students

As educators we all know that the most defining criteria for successful learning to take place in a classroom is the relationship between the teacher and the student, followed only by the relationships between the student and other students in the class. For over three decades I have utilized role-plays related to the STEM content and skills in my classes and as a result of students improvising their roles, or of other students speaking about the improvised situations, have learned much about events taking place in students' lives that have an impact on their engagement as learners. This is one of the reasons that my colleagues and I have had such success with students who were previously disengaged from school and STEM classes (Jablon, 2014).

In addition to uncovering emotional issues in students' lives, creative drama can also directly enhance students' psychological well-being. Under different characters, students can express their true feelings or sense of personality without fear of being judged or criticized. They can work on personal issue or solve personal problems while in character, which can simply help their overall wellbeing. Essentially, what this does is allow students to get things off their mind, further releasing emotion and tension and allowing students to be who they are (Cornett & Smithrim, 2001).

Meets Many of the Emotional Needs of Their Students (Maslow) so These Students Can Focus on the Concepts and Thinking Skills Involved in Their STEM Subjects

If students are deeply engaged in school lessons where they are actively negotiating ideas and creating concepts, then deep understanding is the result. Therefore, it is simple. We just need to get students to come to school everyday, then come to STEM classes everyday, and then when they are there be completely focused on thinking and meaning making. For we know that student engagement is tied to improvements in student behavior (including both an increase in pro-social behavior and a decrease in antisocial behaviors), improved learning outcomes, and greater respect for the rights of others (Covell & Howe, 2001; DeCoene & De Cock, 1996; Howe & Covell, 2005). High levels of student engagement can also reliably predict lower high school dropout rates (Archambault, Janosz, Fallu, & Pagani, 2009). Student engagement has positive effects for teachers, as well; Covell et al. (2009) found that low engagement is a contributing factor in teacher burnout and that increasing engagement can both reduce teacher burnout and improve teacher self-efficacy.

In STEM classes in middle and high school, most of the topics studied don't immediately appear relevant to students' lives, and especially their inherent psychological needs. The study of acceleration, subduction, or osmosis doesn't appear immediately relevant. Calculating centrifugal force seems so unimportant to almost every sixteen year old. Likewise, uncountable numbers of students have asked when will they ever need to use a quadratic equation. Yet each of these topics, if embedded in the right explorations, contextualized in challenging local projects, with the appropriate large conceptual objectives and effective themes can be made engaging and deeply relevant to many of the learners' lives. However, for the most disengaged students whose lives are unpredictable, unless what occurs in the classroom can meet their immediate psychological needs, they cannot focus on any of the STEM content.

Meets the Needs of Those Students Whose Way of Learning STEM is Through Creative Engagement in the Arts

Twelve years into my university career as a science and interdisciplinary education professor I had an opportunity to spend a year again teaching in an urban high school. This particular school was fairly representative of the traditional pedagogy practiced in American schools in the early twenty first century. However, in my classroom I engaged students in daily inquiry investigations, working in small research teams, focused on large conceptual understandings, problem solving, and students owning their learning. Although this was a whole new classroom paradigm for these students, many adapted well and had a higher level of success than in their other classes (Jablon, 2013). However, there were a number of students who were not thriving, and others who were not succeeding in this STEM class. That is, until I started using creative dramatics' role-plays on a regular basis. There are two students I remember in particular. One boy was there everyday, but rarely participated in his group and didn't bother to study any of the material we were investigating. The other, a girl, attended sporadically, and when present feigned indifference to all that many others were engaged in creating.

One day the class was in the middle of a mini-unit on issues dealing with solid waste in our communities when I suggested we were going to do a role-play of a community board meeting about a proposal to build a solid waste incinerator. Within two days both of these students became animated, taking lead roles, and became engaged in learning the science concepts and technology applications. They not only intuitively understood the societal dilemmas, but also led other students to understand the complexities of the ethical, economic, psychological dimensions of the community roles they played. They were naturals at learning, and helping others learn, within this milieu. They physically attended class everyday, but more importantly mentally attended to the material even when we did regular inquiry investigations – as long as we did some role-playing now and then as part of the pedagogical mix.

This was one of a multitude of examples from my STEM classes where, "drama does engage multi-dimensional learning styles including verbal-linguistic, interpersonal, intrapersonal, kinesthetic, spatial, and logical" (Ashton-Hay, 2005). As teachers, we understand that our students learn differently and at different paces because of their biological and psychological differences (Reiff, 1992). Learning styles not only comprise the cognitive domain, but also the affective and physiological domains (Oxford, Hollaway, & Horton-Murillo, 1992; Kinsella, 1996). Strong, Silver, and Perini (2001) divide verbal-linguistic intelligence into four learning preferences including mastery (knowing), interpersonal (connecting with people; social skills), understanding (discovery and reasoning) and self-expressive (creativity). Using drama in a classroom naturally matches the learning styles of a variety of students, thus their motivation, performance, and achievement are more likely to increase making the STEM content readily accessible to them.

SOME FINAL THOUGHTS

Do I believe that teachers and professors reading this will be so informed and empowered that they will begin using this in their classrooms and professional development workshops starting next term? Actually, for most, they probably will not. However, for some, what has been presented here makes so much sense, and has enough studies to demonstrate its effectiveness, that they feel compelled to try to reach the rest of their STEM students that they are not currently deeply engaging. Therefore, what is a good path for them to take?

If you are currently a STEM teacher, find someone in your school or school district who already does creative dramatics in their classroom. Ask if you can attend their class on a day when they will be doing some role-plays. Watch everything they do closely. Then ask if you can come again, but if the next time you could facilitate a role-play in their class, even if it is an English or social studies class. You will want to meet with them beforehand for a couple of minutes so you know the purpose of the improv and where they expect it to go. That way, if it doesn't go exactly as planned when you are facilitating in their class, they are there to help you out if absolutely necessary. With this success under your belt, ask if they could meet with you after school or on a weekend and help you think up one or two scenarios that will work in your STEM class. Now it is time to jump into the pool. Try it in your STEM class. Pick kids who you think will love to do this and ask them the class before. You will most likely be successful and the students will be engaged. You will also get better at it as you go along. If you are a STEM professor at a university, find the arts in education faculty and do the same protocol with them.

I have always taught these approaches to creative drama in my science methods classes and convinced colleagues in various university education departments to have a required class where these play a prominent role for all of our education majors. I have then seen many first year teachers use these very successfully in their classes, and now years later have many who teach others in their schools how to use creative dramatics. I also run workshops in school districts throughout the country and at national conferences with STEM teachers at every grade level from Pre-K to graduate school, so there are a cadre of STEM professionals utilizing these techniques in their classrooms.

Before you have done these regularly in your classroom you might think there will be a lot of resistance from a substantial number of students. There are some who initially bristle at the thought of doing it, but invariably after the class has been doing them for a month or so, most everyone is regularly volunteering when given the opportunity. You know it is working because students are eagerly asking if the class is going to be doing role-plays today. It also really changes the culture of the STEM class in a very positive way bringing humanity right into the core of everything studied. Students are more creative, think more deeply, and are usually truly engaged in most everything the class attempts to solve.

Now it is your turn to raise the curtain on learning!

REFERENCES

Archambault, M., Janosz, M., Fallu, J.-S., & Pagani, L. (2009). Student engagement and its relationship with early high school dropout. *Journal of Adolescence*, *32*(3), 651–670. doi:10.1016/j.adolescence.2008.06.007 PMID:18708246

Arieli, B. (2007). *The integration of creative drama into science teaching* (Doctoral dissertation). Curriculum & Instruction College of Education, Kansas State University, Manhattan, KS. Retrieved from https://krex.k-state.edu/dspace/bitstream/handle/2097/531/BrachaArieli2007.pdf?sequence=1

Arts Council. (1992). *Drama in schools: Arts council guidance on drama education*. London: The Arts Council of Great Britain.

Ashton-Hay, S. (2005). Drama: Engaging all learning styles. In *Proceedings 9th International INGED (Turkish English Education Association) Conference*. Economics and Technical University.

Aubusson, P., Fogwill, S., Barr, R., & Perkovic, L. (1997). What happens when students do simulation-role-play in science? *Research in Science Education*, *27*(4), 565–579. doi:10.1007/BF02461481

Aubusson, P. J., & Fogwill, S. (2006). Role play as analogical modelling in science. In P. J. Aubusson, A. G. Harrison, & S. M. Ritchie (Eds.), *Metaphor and analogy in science education* (pp. 93–104). Dordrecht, The Netherlands: Springer. doi:10.1007/1-4020-3830-5_8

Aurum, A., & Wohlin, C. (2003). The fundamental nature of requirements engineering activities as a decision-making process. *Information and Software Technology*, *45*(14), 945–954. doi:10.1016/S0950-5849(03)00096-X

Bachelis, G. F., Maxim, B. R., James, D. A., & Stout, Q. F. (1994). Bringing algorithms to life: Cooperative computing activities using students as processors. *School Science and Mathematics*, *94*(4), 176–186. doi:10.1111/j.1949-8594.1994.tb15651.x

Bailey, S. D. (1993). *Wings to fly: Bringing theatre arts to students with special needs*. Rockville, MD: Woodbine House.

Bennett, J., & Hogarth, S. (2005). *Would YOU want to talk to a scientist at a party? Students' attitudes to school science and science*. University of York, Department of Educational Studies Research Paper 2005/08.

Boal, A. (2002). *Games for actors and non-actors*. New York: Routledge.

Bolton, G., & Heathcote, D. (1995). *Drama for learning: Dorothy Heathcote's mantle of the expert approach to education*. Portsmouth, NH: Heinemann.

Braund, M., Lubben, F., Scholtz, Z., Sadeck, M., & Hodges, M. (2007). Comparing the effect of scientific and socio-scientific argumentation tasks: Lessons from South Africa. *The School Science Review*, *88*(324), 67–76.

BSCS - Biological Science Curriculum Study. (2005). *BSCS Science and technology: Investigating physical systems*. Dubuque, IA: Kendall Hunt.

Buchanan, M. (2016). Why teach drama? *Child Drama*. Retrieved June 3, 2016. Retrieved from http://www.childdrama.com/the-essay.html

Burke, R. J., & Mattis, M. C. (Eds.). (2007). *Women and minorities in science, technology, engineering and mathematics: Upping the numbers*. Northampton, MA: Edward Elgar Publishing, Inc. doi:10.4337/9781847206879

Butler, J. (1989). Science learning and drama processes. *Science Education*, *73*(5), 569–579. doi:10.1002/sce.3730730505

Caldwell, K. (2011). Drama as a teaching tool. *Teaching Through the Arts*. Retrieved June 3, 2016. http://teachingthroughthearts.blogspot.com/2011/07/drama-as-teaching-tool.html

Cawthon, S. W., Dawson, K., & Ihorn, S. (2011). Activating student engagement through drama-based instruction. *Journal for Learning through the Arts*, *7*(1). Retrieved from http://www.escholarship.org/uc/item/6qc4b7pt

Christofi, C., & Davies, M. (1991). Science through drama. *Education in Science*, *141*, 28–29.

Cobern, W. W. (1993). Contextual constructivism: The impact of culture on the learning and teaching of science. In K. Tobin (Ed.), *The Practice of constructivism in science education* (pp. 51–69). Washington, DC: American Association for the Advancement of Science.

Cornett, C. E., & Smithrim, K. L. (2001). *Arts as meaning makers: Integrating literature and the arts throughout the curriculum*. Don Mills, Canada: Pearson Education.

Cottrell, J. (1987). Creative drama in the classroom grades 4-6. Lincolnwood, IL: National Textbook Company.

Covell, K., & Howe, R. B. (2001). Moral education through the 3 Rs: Rights, respect and responsibility. *Journal of Moral Education*, *30*(1), 29–41. doi:10.1080/03057240120033794

Covell, K., McNeil, J. K., & Howe, R. B. (2009). Reducing teacher burnout by increasing student engagement: A childrens rights approach. *School Psychology International*, *30*(3), 282–290. doi:10.1177/0143034309106496

DeCoene, J., & DeCock, R. (1996). The children's rights project in the primary school 'De vrijdagmarkt' in Bruges. In E. Verhellen (Ed.), *Monitoring children's rights* (pp. 627–636). The Hague: Martinus Nijhoff.

Dockett, S. (1995). *You make me alive! Developing understanding through play*. Deakin, Australia: Australian Early Childhood Association.

Dorion, K. R. (2009). Science through drama: A multiple case exploration of the characteristics of drama activities used in secondary science lessons. *International Journal of Science Education*, *31*(16), 2247–2270. doi:10.1080/09500690802712699

Duveen, J., & Solomon, J. (1994). The great evolution trial: Use of role-play in the classroom. *Journal of Research in Science Teaching*, *31*(5), 575–582. doi:10.1002/tea.3660310510

Freire, P. (2000). *Pedagogy of the oppressed* (M. B. Ramos, Trans.). New York: Continuum Publishing. (Original work published 1970)

Gardner, H. (1991). *The unschooled mind: How children think and how schools should teach*. New York, NY: Basic Books.

Gazzaniga, M. S. (1989). Organization of the human brain. *Science*, *245*(4921), 947–952. doi:10.1126/science.2672334 PMID:2672334

Gomes, L. F. A. M. (2016). Teaching decision making analytical skills to engineers: A new paradigm. *Brazilian Institute of Capital Markets and Universidade Federal Fluminense*. Retrieved from http://citeseerx.ist.psu.edu/viewdoc/download?doi=10.1.1.201.3517&rep=rep1&type=pdf

Goodnough, K. (2001). Multiple intelligences theory: A framework for personalizing science curricula. *School Science and Mathematics, 101*(4), 180–193. doi:10.1111/j.1949-8594.2001.tb18021.x

Grady, S. (2000). *Drama and diversity: A pluralistic perspective for educational drama*. Portsmouth, NH: Heinemann Drama.

Heathcote, D. (1984). *Collected writings on education and drama*. Evanston, IL: Northwestern University Press.

Heaverlo, C. (2011). *STEM development: A study of 6th-12th grade girls' interest and confidence in mathematics and science* (Doctoral Dissertation). Retrieved from http://lib.dr.iastate.edu/cgi/viewcontent.cgi?article=1035&context=etd

HMI. (1989). Curriculum matters 17: Drama from 5-16. London: HMSO.

Hodson, D., & Reid, D. (1988). Changing priorities in science education: Part I. *The School Science Review, 70*, 101–108.

Howe, R. B., & Covell, K. (2005). *Empowering children: Children's rights education as a pathway to citizenship*. Toronto: University of Toronto Press. doi:10.3138/9781442674387

Jablon, P. (2013). Get real! Walking the walk to inform talking the talk: Full-time teaching in an urban high school. In M. Dias, C. Eick, & L. Brantley-Dias (Eds.), *Science teacher educators as K-12 science teachers: Practicing what we teach* (pp. 39–50). New York, NY: Springer.

Jablon, P. (2014). *The synergy of inquiry: Engaging students in deep learning across the content areas*. Huntington Beach, CA: Shell Education.

Jablon, P. (2010). *Analogous interrelationship role plays: Engaging the previously disengaged in science*. Paper presented at the Association of Science Teacher Educators International Conference, Sacramento, CA.

Jablon, P. (1989). *A Documentation and impact analysis of a project-oriented, interdisciplinary, team-taught, peer-influence program for secondary schools and its effects upon science learning* (Doctoral dissertation). (NYU No. 9016410)

Jablon, P., & Born, T. (1988). STS works in the city. *SSTS Reporter., 4*(2), 5–8.

Kamen, M. (1991). Use of creative drama to evaluate elementary school students' understanding of science concepts. In G. Kulm & S. M. Malcom (Eds.), *Science assessment in the service of reform*. Washington, DC: American Association for the Advancement of Science.

Kamen, M. (1991, August). *Creative drama and the enhancement of elementary school students' understanding of science concepts* (Doctoral dissertation). The University of Texas at Austin, Austin, TX.

Kase-Polisini, J., & Spector, B. (1992). Improvised drama: A tool for teaching science. *Youth Theatre Journal*, *7*(1), 15–19.

Kelly, G. J., Carlsen, W. S., & Cunningham, C. M. (1993). Science education in sociocultural context: Perspectives from the sociology of science. *Science Education*, *77*(2), 207–220. doi:10.1002/sce.3730770208

Kelner, L. B. (1993). *The creative classroom: A guide for using creative drama in the classroom, PreK-6*. Portsmouth, NH: Heinemann.

Kentish, B. (1995). Hypotheticals: Deepening the understanding of environmental issues through ownership of learning. *Australian Science Teachers Journal*, *41*(1), 21–25.

Kinsella, K. (1996). Designing group work that supports and enhances diverse classroom work styles. *TESOL Journal*, *6*(1), 24–30.

Kuenzi, J., Matthews, C., & Mangan, B. (2006). *Science, technology, engineering, and mathematics (STEM) education issues and legislative options*. Washington, DC: Congressional Research Service.

Llewellyn, D. (2004). *Teaching high school science through inquiry: A Case study approach*. Thousand Oaks, CA: Corwin Press.

Mardirosian, G. H., & Lewis, Y. P. (2009). How to use theater to teach at-risk students. *The Chronicle of Higher Education*, *55*(26), 34–37.

Maslow, A. H. (1968). *Toward a psychology of being*. New York, NY: Van Nostrand.

McCaslin, N. (1981). *Children and drama* (2nd ed.). New York: University Press.

McCaslin, N. (1986). *Children and drama*. New York: Longman.

McCaslin, N. (1996). *Creative drama in the classroom and beyond*. London: Longman Publishers.

McCaslin, N. (2006). *Creative drama in the classroom and beyond*. Boston, MA: Allyn and Bacon.

McGregor, D., & Precious, W. (2015). Dramatic science: Inspired ideas for teaching science using drama for ages 5–11. New York: Routledge.

Metcalfe, R. J. A., Abbott, S., Bray, P., Exley, J., & Wisnia, D. (1984). Teaching science through drama: An empirical investigation. *Research in Science & Technological Education*, *2*(1), 77–81. doi:10.1080/0263514840020109

Miller, C., & Saxton, J. (2004). *Into the story: Language in action through drama*. Portsmouth, NH: Heinemann.

NGSS Lead States. (2013). *Next generation science standards: For states, by states*. Washington, DC: The National Academies Press.

O'Neill, C. (1995). *Drama worlds: A Framework for process drama*. Portsmouth, NH: Heinemann.

Ødegaard, M. (2003). Dramatic science. A critical review of drama in science education. *Studies in Science Education*, *39*(1), 75–102. doi:10.1080/03057260308560196

Onwukwe, E., Akuezuilo, E., & Ijioma, B. (2014). Science drama: Clue to under achievement in secondary school chemistry? *STEMplanet*, 1. Retrieved from http://stemstates.org/journal/stemplanet-2014/2014-vol-11/science-drama-clue-to-under-achievement-in-secondary-school-chemistry.html

Osborne, J., Simon, S., & Collins, S. (2003). Attitudes towards science: A review of the literature and its implications. *International Journal of Science Education*, *25*(9), 1049–1079. doi:10.1080/0950069032000032199

Oxford, R. L., Hollaway, M. E., & Horton-Murillo, D. (1992). Language learning styles: Research and practical considerations for teaching the multicultural tertiary ESL/EFL classroom. *System*, *20*(4), 439–456. doi:10.1016/0346-251X(92)90057-A

Pieczura, P. (2013). Decidedly dramatic! The power of creative drama in social studies. *Social Studies and the Young Learner*, *25*(3), 9–12.

Pulkkinen, L. (1982). Self-control and continuity from childhood to adolescence. In P. B. Baltes & O. G. Brim Jr., (Eds.), *Life-span development and behavior* (Vol. 4, pp. 64–102). New York, NY: Academic Press.

Reiff, J. C. (1992). *Learning Styles. What research says to the teacher*. Washington, DC: National Education Association.

Roberts, D. (2011). *Fatal invention: How science, politics, and big business re-create race in the twenty-first century*. New York, NY: The New Press.

Salisbury, B. T. (1986). *Theatre arts in the elementary classroom*. New Orleans, LA: Anchorage Press, Inc.

Smilansky, S. (1968). *The effects of sociodramatic play on disadvantaged preschool children*. New York, NY: John Wiley & Sons.

Smilansky, S. (1990). Sociodramatic play: Its relevance to behavior and achievement in school. In E. Klugman & S. Smilansky (Eds.), *Children's play and learning: Perspectives and policy implications* (pp. 18–42). New York, NY: Teacher College Press.

Steinert, Y. (1993). Twelve tips for using role-plays in clinical teaching. *Medical Teacher*, *15*(4), 283–291. doi:10.3109/01421599309006651 PMID:8139402

Stencel, J., & Barkoff, A. (1993). Protein synthesis: Role playing in the classroom. *The American Biology Teacher*, *55*(2), 102–103. doi:10.2307/4449595

Strong, R. W., Silver, H. F., & Perini, M. J. (2001). *Teaching what matters most: Standards and strategies for raising student achievement*. Alexandria, VA: Association for Supervision and Curriculum Development.

Sturm, B. (2009). The drama of science. *Science in School*, *13*, 29-33. Retrieved from http://www.scienceinschool.org/sites/default/files/teaserPdf/issue13_drama.pdf

Thompson, J. (2003). *Applied theatre: Bewilderment and beyond*. Oxford, UK: Peter Lang. doi:10.3726/978-3-0353-0309-4

USDOE - United States Department of Education. (2006). *The Condition of education*. Washington, DC: National Center for Education Statistics, U.S. Government Printing Office.

Varelas, M., Pappas, C.C., Tucker-Raymond, E., Kane, J., Hankes, J., Ortiz, I., & Keblawe-Shamah, N. (2010). Drama activities as ideational resources for primary-grade children in urban science classes. *Journal of Research in Science Teaching*, *47*(3), 302–325.

Venkateswaran, T. V. (2004). Communicating science through plays, games and simulations. *Indian Journal of Science Communication*, *3*(1).

Wagner, B. J. (1979). *Dorothy Heathcote: Drama as a learning medium*. London: Hutchinson Education.

Webb, N. (1980). An analysis of group interaction and mathematical error in heterogeneous abilities groups. *The British Journal of Educational Psychology*, *50*(3), 266–276. doi:10.1111/j.2044-8279.1980.tb00810.x

Weld, J. D. (2000). Fly Guys. *Teacher Magazine*, *11*(8), 56–58.

Yager, R. E. (1993). The constructivist learning model. *Science Teacher (Normal, Ill.)*, *60*, 27–31.

KEY TERMS AND DEFINITIONS

Analogous Interrelationships: Fairly common interactions between persons that are familiar to a learner that are analogous to interactions between groups, societies, or nations that are not as yet understandable to learners, and can be used by educators to have their students better, and more easily, understand the larger, seemingly remote relationships.

Creative Dramatics: Creative drama is an improvisational, non-exhibitional, process-oriented form of drama, where participants are guided by a leader to imagine, enact, and reflect on experiences real and imagined. In education, used as a tool for better understanding by students of academic conceptual material.

Engaging Adolescents: The act of getting teens and pre-teens interested in some learning that they were previously not interested in.

Interdisciplinary Education: An educational approach in which two or more disciplines collaborate in the learning process with the goal of fostering interprofessional interactions that enhance the practice of each discipline.

Science Education: The field concerned with sharing science content and process with individuals not traditionally considered part of the scientific community.

STEAM: The use of the arts in engaging students and have them better understand the concepts and skills involved in the combination of Science, Technology, Engineering, and Mathematics classes.

STEM: The natural interaction and synergy of Science, Technology, Engineering, and Mathematics and the deeper understanding that occurs when they are not taught separately.

Theater: The activity of acting in, producing, directing, or writing plays.

Urban Education: The practice of having students learn who traditionally live and go to school in cities.

Section 3
Theme Based

Chapter 7
"Imagioneering" a New Mission Space

Kyle Seiverd
Toms River Regional School District, USA

EXECUTIVE SUMMARY

Incorporating STEAM into classroom practice can be both challenging and daunting. "Imagioneering" a new Mission: Space showcases a way to utilize STEAM into the environmental earth science classroom. Using the familiarity many students have with Disney, critical analysis is applied to the design of a famous park ride. Ride-layout is critiqued and improved upon via student collaboration. Students then use their ability to analyze design to build a structure that fits a particular purpose. This project is geared towards the high school level and students redesign the exterior and line-queue for Mission: Space.

LEARNER OUTCOMES

Next Generation Science Standards

- **HS-ESS1:** Earth's Place in the Universe.
- **HS-ETS1:** Engineering Design.

DOI: 10.4018/978-1-5225-2334-5.ch007

Copyright ©2017, IGI Global. Copying or distributing in print or electronic forms without written permission of IGI Global is prohibited.

National Visual Arts Standards

- **MA: Cr1.1.III:** Integrate aesthetic principles with a variety of generative methods to fluently form original ideas, solutions, and innovations in media arts creation processes.
- **MA: Cr2.1.III:** Integrate a sophisticated personal aesthetic and knowledge of systems processes in forming, testing, and proposing original artistic ideas, prototypes, and production frameworks, considering complex constraints of goals, time, resources, and personal limitations.

Common Core Mathematics Standards

- **N.Q.02:** Define appropriate quantities for the purpose of descriptive modeling.
- **S.IC.06:** Evaluate reports based on data.

LITERATURE REVIEW

Walt Disney World: The New Standard

Even though most think of Walt Disney World when they hear Walt Disney, the company got its start in animation. The most famous creation of the Walt Disney Company is Mickey Mouse. By using Mickey Mouse as the mascot, Walt Disney was able to convey messages and elicit emotions from audiences around the world. Walt's transition from animation to theme park design was inspired by a family trip to a carnival. While his daughters rode the carousel, Walt reflected on the grounds that surrounded him. He noted the poor conditions and the lack of family-friendly entertainment. These observations would be the seeds to what would become California's Disneyland.

As Walt Disney animators continued to pioneer music, film, and TV, Walt Disney was exploring other avenues. He wanted to provide safe, wholesome family entertainment to all ages. Upon completion of Disneyland, Walt continued to seek better a design for his 'land of make believe.' His restless pursuit of perfection gave rise to one of the most famous places on Earth:

In 1964, Walt Disney began secretly buying millions of dollars worth of Central Florida farmland. Some thought it was Howard Hughes; others, the space program. Speculation was rife almost to the very day, November 15, 1965, when Uncle Walt *arrived in town and announced his plans to build the world's most spectacular theme park. In a 2 year construction effort, Disney employed 9,000 people. The total cost of the project by its October 1971 opening was $400 million. Mickey Mouse escorted the first visitor into the Magic Kingdom, and numerous celebrities, from Bob Hope to Julie Andrews, took part in the opening ceremonies. In Walt Disney World's first 2 years, the attraction drew 20 million visitors and employed 13,000 people. The sleepy citrus-growing town of Orlando had become the 'Action Center of Florida,' and the fastest-growing city in the state. (Frommer's, 2016)*

IMAGINEERS: THE STEAM EXPERTS

When Walt Disney began planning Disneyland and later Disney World, he relied on his most creative staff to assist in novel project designs. He termed these members 'imagineers.' "Imagineers are the creative force behind the iconic Disney attractions and experiences" (Disney Careers, 2016). Walt Disney knew that greatness could not be achieved alone and insisted all major rides and attractions utilize a team of imagineers. Each imagineer brought ideas that contributed to the development of iconic rides such as the Haunted Mansion and Pirates of the Caribbean.

When developing an attraction for Disney Parks, imagineers had to know their limitations. For example when designing a space-themed ride, it is important that concepts presented are based on factual scientific data. Meaning, the idea of finding a singing elephant on the surface of the sun is such a far-fetched idea that guests lose the believability and 'wow' factor of the attraction.

Constructing a scale-model that is accurate in design requires both artistic and mathematical skills. If a real-life Moon was needed for a ride, a scaled down model would first be made. The size of the Moon's distinctive features, such as craters and valleys, are shrunk and designed using a medium that is believable and within engineering constraints.

Recently, imagineers have been using cutting-edge technology to virtually build their models before sending them for construction. Computer programs allow 3-D models to be designed and modified before being approved. This saves money and allows imagineers to catch flaws they may have missed. If the design is approved, 3-D printers make a scale-model of the object. In the 2015 expansion project to Fantasyland at the Magic Kingdom, the new area was designed virtually allowing creative and technological collaboration before construction.

EPCOT

The gemstone of Disney Parks in Orlando, Florida is the Walt Disney Magic Kingdom. When designing the Magic Kingdom, Walt reflected on design flaws from Disneyland. He had the imagineers build the entire park a story-higher, preventing park guests from seeing trash disposal and cast member movement. Although the Magic Kingdom attracts the greatest number of guests per year, Walt Disney conceived an idea to build another theme park showcasing cutting-edge research and cultures from around the world. "Epcot, the *E*xperimental *P*rototype *C*ommunity *O*f *T*omorrow, opened in 1982. EPCOT was conceived by Walt Disney to take its cue from the new ideas and new technologies that are now emerging from the creative centers of American industry" (Disney Information, n.d.a). EPCOT provides guests the opportunity to explore new frontiers in the fields of agriculture, sustainability, and more.

For an Image of Spaceship Earth at EPCOT go to: http://www.wdwlive.com/photos/epcot/future-world/mission-space/epcot-mission-space-exterior-1-9.jpg.

Ride Design and Scale

All attractions at Disney Parks began as simple ideas and blossomed into the cutting-edge rides they are today. When asked about how one should begin designing a ride, the response is to be a renaissance person in multiple areas. Brigante (2010) stated "… one should not be limited to just one or two specific tools. The ability to be creative across any medium is more important than expertise in one particular area. Universal designer Rich West emphasized the importance of the simplest design tool: a number 2 pencil." All designs must first be translated from mind to drawing. Often a flat 2-D or 3-D picture is drawn to illustrate the idea of a concept an imagineer has. The illustration will often be accompanied with the story-line the ride has. From there the idea is pitched and collaborated on, resulting in the building of a scale model.

For an image showing the concept showing the new Test Track overlay click on: http://i843.photobucket.com/albums/zz352/loaloauk/DLP%20that%20never%20was/Ride-X2.jpg.

Scale models provide one way to see ride designs in 3-D. Models are generally small-scale replicas of an actual object. Math skills are needed to construct an object to scale. By using models, imagineers are able to reflect and modify the design. Scale models can be made by hand using a variety of materials; however 3-D printing and computer software is becoming more common. Models come in a variety of scale, pending the stage of the design processes. Often preliminary models are done to a higher degree of scale. This provides a foundation as to how the model will appear and an overall sense of the ride. As the design progresses, models become lower in scale. Each succeeding model has greater detail and gives imagineers an opportunity to make refinements.

As the exterior of the ride is being constructed, ride vehicles and interactive guest experiences are 'play tested.' Reviews of the 'play tests' are taken into consideration and the ride is further modified. Once the 'play-testing' phase is complete, the ride goes into its final phase for construction. A ride is not solely the 'in seat' experience, but also the time waiting in the line-queue. Imagineers use technology, art, and science to entertain guests while they wait. A line-queue can have noise, interactive displays, and more. By providing guests an engaging experience while waiting, excitement builds and the chance of the ride being vandalized decreases.

NASA: Mars Exploration

When Walt Disney World was being constructed, space exploration was headline news. Families gathered to watch Neil Armstrong set foot on the Moon and many traveled far to watch NASA shuttle-launches from Fort Lauderdale, Florida. During the 1960s and 1970s much of NASA's attention was focused on the Moon. While the Moon is still of interest, NASA and much of society's attention is now on Mars. NASA's current goal is to send a manned mission to Mars by 2030. The idea of humans on Mars is exciting, but many challenges arise from a scientific standpoint. Current technology projects the trip to Mars will take over a year to complete. Research is being invested into how humans will age in space as well as how supplies will be provided to maintain a crew for over a year. To have a manned mission to Mars will require ingenuity and new technologies.

Mars Colony

Throughout Walt Disney World one can see the theme of 'Space' and the 'Future' interweaved into the park's design. The Magic Kingdom has 'Tomorrowland' which features one of the most famous rides, Space Mountain. EPCOT hosts various space-centered entertainments and in the early 2000s opened Mission: Space. "Mission: Space was created in collaboration with former NASA advisors, astronauts and scientists, along with the Walt Disney Imagineering team. The attraction uses new technology, which makes it the first ride system to take attraction visitors straight up in a simulated flight experience" (DI, n.d.b). In Mission: Space guests are divided into teams of four, each being assigned a role on a shuttle to Mars. A simulation-based ride takes guests on a shuttle to Mars with a few 'issues' along the way. The ride ends with the team on Mars and guests exit the ride into an interactive space-themed area.

For a picture showing the entrance of Mission Space at EPCOT click on: http://dlib.info/omeka/moblitas/archive/fullsize/a4f3edc6a5ce278b97dc15b2cd90b1d2.jpg.

All aspects of Mission: Space are space-centered. The exterior design was ultimately created using the model of planets and their rings as the focus. Imagineers used scale models of stellar-bodies in the design process and field-tested the attractiveness to park guests. The more intriguing and aesthetically pleasing the artistic exterior is, the more guests will be drawn to the ride.

One unique aspect of Mission: Space is that guests are given an option when they enter the ride. Imagineers created two experiences, one having greater G-forces than the other, but with the same plot. The option of choice gives guests prone to motion-sickness an opportunity not to miss out on the experience. A guest reviewed the ride as "(an) (i)ntense ride with a take on possible future missions to Mars [that] offers a great experience. The ride spins really fast to produce the G forces you experience in the ride" (Jacob F., 2016).

LESSON PLAN INFORMATION

The Art and Math Behind Ride Design

Mission: Space at Disney's EPCOT provides guests one perspective of our solar system. The entrance uses large-scale models of planets and rings to entice guests to the ride. For this activity, students will re-design the entrance to the ride. The ride entrance is not limited to the exterior, but also includes the interior line-queue. The line-queue has interactive displays and showcases technologies used by NASA. Due to the ever-changing pace of technology, space-themed rides can become outdated quickly. This activity gives students the opportunity to be imagineers and redesign one of the iconic attractions at Walt Disney World.

Materials

Markers, Colored Pencils, Rulers, Compass, Poster Board, Protractor, Computer Paper, Calculator, Pencils, Computers

Procedures

1. Divide the students into groups of no more than 3. Smaller groups enable members to contribute more.
2. **Engage:** Play the following video on the Haunted Mansion Line-Queue: https://www.youtube.com/watch?v=PUQxzdcL7ro

3. Lead a discussion as to what made the line-queue of the Haunted Mansion appealing to guests. Discussion Prompts could be:
    a. "How was art incorporated into the line-queue?"
    b. "During what parts were the guests most engaged and least engaged?"
    c. "How was the line engineered so no bottlenecks occurred?"
    d. "What technology was used to engage guests while waiting?"
    e. " Is safety an issue with this layout?"
4. **Explore:** Tell the groups they are going to become Imagineers. Their task is to draw a new entrance and line-queue for Mission: Space. Remind students to apply ideas presented in the 'Haunted Mansion Line-Queue' video to Mission: Space. Students can access ride basics at: https://disneyworld.disney.go.com/attractions/epcot/mission-space/
5. The design must meet the following criteria:
    a. Hold at least 100 people.
    b Expose guests to scientific data.
    c. Have an entrance and exit, and flow in a logical fashion.
    **Ride criteria is subject to revision by the instructor, however it is essential that a minimum requirement is placed.*
6. Give the students time to collaborate and sketch out their ideas on computer paper. A time limit is recommended to keep the groups progressing, but is it up to the teachers digression.
7. Teacher should rotate between groups and provide input into their designs.
8. **Explanation:** Have the groups select one representative to give a quick summary of their design to the class. Each group must ask the presenter one question.
9. **Elaboration:** After all groups present give students time to reflect, sketch, and label their Mission: Space entrance and line-queue. There should be two drawings, one showing the front of the ride and another of the overhead shot to show how the line will flow.
10. The drawings need be to scale. A scale must be included on both drawings.
11. A written explanation for how the groups plan to present Space information must be provided.
12. Have groups draw their displays on Space information with labels.
13. **Evaluation:** Using a rubric designed by the teacher, grade the projects.

CURRENT CHALLENGES FACING TEACHERS

Various hurdles are present, no matter the lesson taught. This lesson's main obstacle is time. Exposing students to engineering, discussing and planning in small groups, researching, and drawing all require valuable class time that an educator may not have. When working in groups, students can become side-tracked and drift into off-topic conversation. This creates time delays by hindering progress. One way to troubleshoot this problem is to give checkpoints the groups need to hit every 5 or 10 minutes.

Another hurdle is grading the projects. Each teacher has different expectations for their students so a uniform rubric does not apply to all classes. Rubistar.org is a free online resource that allows the teacher to design a rubric for various projects.

SOLUTIONS AND RECOMMENDATIONS

Disney's Imagineers stress that a budget and time-frame are always given when assigned a project. It is advisable that a time-limitation is placed to keep the project moving, however quality of design could be compromised. One way to limit classroom time is to have students only design the outside of Mission: Space and not the line-queue or vice-versa. Parts of the project and scientific research can be assigned as homework and students are responsible for each bringing in a different component. For example, if a group is planning on doing their line-queue display on Mars, each member could bring in a list of facts on Mars. Whichever student doesn't bring in the facts loses participation points on the project. However it is vital that students are given class time to discuss and collaborate on their project-design and layout.

Also if students are unfamiliar with Disney Parks, use a local amusement park or a common carnival ride to facilitate the idea of ride design. One could also change the ride selection from Mission: Space to a ride such as Pirates of the Caribbean as long as it aligns to the curriculum. If only one ride pertains to content presented in class, play the ride video from an online source. This will give students a perspective into the ride's theme and layout.

REFERENCES

Andrews, E. (2014). 6 Early amusement parks. *History Lists*. Retrieved from http://www.history.com/news/history-lists/6-early-amusement-parks

Brigante, R. (2010). Want to be a theme park designer? Disney Imagineer and Universal creative experts offer advice. *Inside the Magic*. Retrieved from http://www.insidethemagic.net/2010/04/want-to-be-a-theme-park-designer-disney-imagineer-and-universal-creative-experts-offer-advice/

Disney Careers. (2016). *Walt Disney Imagineering*. Retrieved from http://wdi.disneycareers.com/en/default/

Disney Information. (n.d.a). *EPCOT Overview*. Retrieved from http://www.wdwinfo.com/wdwinfo/guides/epcot/ep-overview.htm

Disney Information. (n.d.b). *Mission: Space overview*. Retrieved from http://www.wdwinfo.com/wdwinfo/guides/epcot/epfw-mission.htm

Frommer's. (2016). *Walt Disney World history*. Retrieved from http://www.frommers.com/destinations/walt-disney-world/683469

KEY TERMS AND DEFINITIONS

Aesthetic: Beauty to the designer.

Imagineering: Term coined by Walt Disney that combines imagination and engineering.

Line-Queue: Space for people waiting to board.

Mars: Planet in the Milky Way Galaxy that is fourth closest to the Earth's sun.

Modeling: A scaled representation of an actual object.

Play-Testing: When individuals interactive with a prototype for the purpose of error spotting.

Theme Park: Area designed to replicate a specific place and/or time.

Chapter 8
Bee Pollination

Kerry Carley Rizzuto
Monmouth University, USA

John Henning
Monmouth University, USA

Catherine Duckett
Monmouth University, USA

EXECUTIVE SUMMARY

The purpose of the chapter is to provide an exemplar of an inquiry-based unit on pollination for designing and implementing constructivist instructional practices while simultaneously providing outstanding teacher preparation. The unit on pollination was developed by preservice teachers through a partnership between the Monmouth Conservation Foundation and the Monmouth University School of Education. Through collective action, these institutions were able to enhance student learning on a vital part of the science curriculum, provide a rich clinical experience for pre-service teachers, and to familiarize teachers with a more constructivist approach to pre-school science instruction.

INTRODUCTION

Children's engagement with science begins with their first discovery (Bredekamp, 2014; Brooks & Grennon-Brooks, 1999; Tu, 2006). They are natural scientists, possessing an innate curiosity for investigating a world of things that are constantly

DOI: 10.4018/978-1-5225-2334-5.ch008

Copyright ©2017, IGI Global. Copying or distributing in print or electronic forms without written permission of IGI Global is prohibited.

beckoning: some living, some hard, some cold, some large, some metallic, and some dirty (Bredekamp, 2014; Piaget, 1955). Preschool children who engage in scientific exploration develop a better understanding of science concepts later in life. As a result, science education for preschool children is receiving considerable attention (e.g., Patrick, 2009; Conezio & French, 2002; Sharapan, 2012)

Despite its importance, children often do not engage in high-quality science experiences in early childhood classrooms (Gerde, Schacter, & Wasik, 2013; Fitzgerald & Schneider, 2013; Tu, 2006). Many early childhood educators are hesitant about introducing science in their classrooms because they did not enjoy studying science as students and feel ill prepared to teach it (Conezio & French, 2002). As a result, they teach science through a series of isolated experiments without any connection to a broader and more meaningful science curriculum. For example, one study showed that the preschool science curriculum consisted of children watering plants and displaying a random assortment of stones and shells in the back of their classrooms (Fitzgerald & Schneider, 2013).

Therefore, it is not surprising that a number of studies have discovered that preschool students are not prepared for science instruction when entering kindergarten (Conezio & French, 2012). Studies have found that children in kindergarten are already forming negative views of science. Student interest continues to fade through the elementary years, an occurrence that has been especially true for girls, who, even before kindergarten, are less likely to say they like science than boys (Sharapan, 2012). As a result, scarcely a third of kindergartners showed any knowledge of scientific processes and knowledge. The lack of interest in and knowledge of science has been attributed to "less investigative science practices" employed by their classroom teachers (Watters & Diezmann, 2007, p. 351).

Therefore, the purpose of the chapter is to describe how preservice teachers can learn to develop inquiry-based science units. This will be illustrated through a case study description of the design and implementation of an inquiry-based unit on pollination. The unit on pollination was developed through a partnership between the Monmouth Conservation Foundation and the Monmouth University School of Education. These institutions were able to enhance student learning on a vital part of the science curriculum, provide a rich clinical experience for preservice teachers, and to familiarize practicing teachers with a more constructivist approach to pre-school science instruction.

LITERATURE REVIEW

In this section, the authors provide a framework for understanding the purpose, goals and outcomes of the pollinator project. We begin by discussing the importance of pollination to the future of the environment and hence to the curriculum. To facilitate learning, instruction on pollination can begin as early as the preschool years while the brain is undergoing rapid development. This instruction is best accomplished through a constructivist approach that fosters a caring, play-centered approach to learning. This section concludes with an explanation of the importance of providing preservice teachers with an experience in creating inquiry-based instruction.

CONTENT INFORMATION

Importance of Pollination to the Curriculum

Pollination is a vital part of the curriculum, primarily because of its importance to food production. Approximately 75% of crop species rely on animal pollination (i.e. bees, butterflies, birds, etc.) to assure or maximize seed and fruit set (Klein et al., 2007). Biologists and agriculturalists agree that animal pollinators are responsible for significant services without which some crops would not be available, others would have significantly reduced yields, and additional land would be required to feed the world (Aizen, Garibaldi, Cunningham, & Klein, 2009; Klein et al., 2007; Losey & Vaughan 2006).

Due to their role in food production, the economic importance of animal pollinators is also highly significant to the curriculum. In the USA, the principal agricultural pollinator at this time is the honeybee (Aizen et al., 2009; Losey & Vaughan 2006). Honeybee pollination services in the USA were valued at approximately $240 million in 2006 and $15 billion in 2014. Native wild pollinators services have been valued at between approximately $3 and $9 billion (Losey & Vaughan 2006; The White House, 2014). Some crops, such as almonds and squash, are almost exclusively dependent on bee pollination for fruit set (Klein et al., 2007).

Unfortunately, honeybee colonies have declined precipitously and steadily in the US during the last 4 decades (Aizen et al., 2009; The White House, 2014). There is no clear cause of this decline; however, many forces including: mites, viruses, pesticide residues, loss of non-agricultural habitats clearly have negative impacts on honeybees and other pollinator populations (Long & Krupe, 2016). Whatever the precise causes, pollinator populations are declining sharply (Goulson, Nicholis, Botias, Rotheray, 2015; Potts et al., 2010). The decline is worldwide and will require societal attention if food security for most populations is to be achieved (Long & Krupke, 2016; The White House, 2014).

Apart from its vital role in the production of food, pollination is also a key part of the science curriculum. An understanding of pollination often makes related topics seem more relevant. A knowledge of the parts of the flower is essential for any significant study of botany and to the most basic identification of plants. The lessons developed for this project included botany and plant anatomy, life history, ecological interconnectedness, and biodiversity.

Because of its importance to the environment and the science curriculum, interested conservation organizations are creating potential teaching tools for teachers at all levels (For example, see the Xerxes Society, 2011, 2016). As part of the Pollinator Partnership, many organizations have promoted the establishment of pollinator gardens and teaching 'labs' for educating the community (See http://www.pollinator.org/). This project was dedicated by a collaborative effort to empower educators and students of all to make a positive difference in the world.

Constructivist, Inquiry-Based Learning

Science education that begins during the early years of schooling can capitalize on the sensitive time for brain development (Brooks & Grennon-Brooks, 1999). In early childhood classrooms, teachers have the opportunities to design and implement curricula that will enhance the child's development by providing opportunities for the brain to recognize and classify naturally occurring patterns (Gerde, Schacter, & Wasik, 2013). Constructivist theorists argue that those connections are best effected through direct interactions with the natural world that are supported by developmentally appropriate information and the opportunity to construct new understandings through interactions with peers (American Association for the Advancement of Science, 1993; Bianchini and Colburn 2000; Bredekamp, 2014; Llewellyn, 2002). Inquiry approaches are especially important in preschool classrooms, where children are developing foundational knowledge on which all future science understanding will be based (Conezio & French, 2002).

Challenges

Although early childhood teachers recognize the need to engage their students in constructivist-based science lessons, they often struggle to enact school science experiences that are truly constructivist in theory and practice. Their resistance may also be due to the mistaken idea that science education is incongruous with a constructivist approach, perhaps because science is "perceived and presented as too formal, too abstract, and too theoretical – in short, too difficult for preschool children and their teachers" (Gerde, Schacter, & Wasik, 2013; Johnson, 2007).

These challenges are compounded for preservice teachers who are learning how to teach (Schwarz & White, 2005). The self-efficacy of preservice teachers has a direct impact on how much and how well they teach science content (Sharapan, 2012). To foster their self-efficacy, preservice teachers should be given the opportunity to design and implement inquiry-based units during clinical experiences. A well designed clinical experience can help preservice teachers develop their professional self-efficacy, increase their professional knowledge regarding science content, and learn how to integrate science content into an early childhood constructivist curriculum framework. The purpose of this project was to provide such an experience for preservice teachers through a partnership between the Monmouth Conservation Foundation and Monmouth University.

Design Features

The design for the constructivist approach in this case study is based on five essential features that are at the core of inquiry-based approaches to science teaching and learning (National Research Council, 2000): the learner: (1) engages in scientific questioning, (2) searches for evidence to support ideas, (3) hypothesizes possible explanations based on evidence, (4) connects those explanations to science understanding, and (5) shares findings and explanations with larger classroom community. These features support students in becoming science learners who are comfortable in formulating answers to scientific questions. Teachers use these features to support students in their science learning as well as their development. Teachers serve as guides through children's wonderings, play centered activities, social learning, as well as serving as a caretaker in a sense.

Wondering

The first step in creating scientific thinking preschoolers, is to create environments in which young children feel inspired and motivated to wonder like scientists. An environment that fosters both wondering and scientific thinking is one that gives them the freedom to engage in child-centered explorations, experimentations, and time to engage in social play situations (Brooks & Grennon-Brooks, 1999). They need places where they can explore the outdoors, play in mud, and handle tools of the trade, such as magnifying glasses, pails, and shovels.

Play-Centered

Moreover, teachers need to be aware of the many ways science can be naturally integrated into a play-centered curriculum. For example, as a child experiments with a mixture of water and sand, she is making observations as well as predictions (Chaille & Britain, 2003). The child is also testing those predictions, building theories, and perhaps sharing the theory with a classmate. When preschool children are engaged and using all three of the early childhood domains (social and emotional, cognitive, and physical), they are engaged in a well-rounded scientific experience.

Social Learning

Children should be encouraged to share their observations and ideas with each other when "building theories, testing those theories, and then evaluating what worked, what didn't, and why" (Conezio & French, 2002, p. 13). Shared inquiry can be especially beneficial in fostering curiosity and stimulating new ideas (Chaille & Britain, 2003). Shared inquiry generates a number of diverse ideas about the problem, while simultaneously challenging children to communicate the reasoning behind their ideas.

Caring

The first task a teacher needs to accomplish in the classroom is making the students feel cared for and happy. When the teacher attends to the expressed needs of her students, they don't develop the aggressive impulses resulting from an unsuccessful struggle to satisfy deep emotional needs (Noddings, 2005). Meeting students' needs requires that teachers be careful and caring listeners as children wonder aloud. Attending to what children say provides teachers with a large, rich repertoire of the ideas that can catch the hearts and minds of the students.

CASE DESCRIPTION

Designing an Inquiry-Based Unit

In November of 2015, Monmouth University School of Education was contacted by board members of the Monmouth Conservation Foundation (MCF). The MCF is Monmouth County's only county-wide land trust. The MCF is a not-for-profit organization that has been preserving land and protecting the natural habitat of Monmouth County for over 38 years. In an effort to engage young children, the MCF founded the Kids for Conservation project, primarily as an academic enrichment program whose purpose is to teach young children about the environment. The curriculum is based upon age appropriate themes that teach children about land conservation, wildlife protection, and the preservation of farmland and open space.

The goal of the partnership was to develop a new curriculum that would focus on teaching preschoolers about the importance of pollination. The authors initiated the project by convening several meetings with ten junior and senior preservice teachers who were pursuing their early childhood teaching endorsements. Several meetings were held during December of 2015 in order to form teams of preservice teachers; once the teams were established, a meeting was held with members of the MCF in order to form common goals and ensure mutual expectations. A meeting schedule for writing lessons plans and forming the curriculum was scheduled for the winter break.

During the first team meeting, the preservice teachers reviewed the scientific method, by engaging in a bubble making experiment. The group started with the following research question. Which of the following home products would produce the best bubbles: dish soap and water, dish soap and glycerin, or dish soap and corn syrup? Next, they designed an experiment to test their hypothesis. The group recorded observations as they mixed the ingredients and then came to a conclusion about their initial hypothesis. At the completion of the lesson, the group outlined the scientific method together. The net result was that all eight of the participants reported feeling more confident working with the scientific method. At the next meeting, preservice students who had seemed less interested in taking writing and teaching roles in the past few meetings were now more comfortable volunteering for more responsibilities.

Before the curriculum unit or lesson plan writing began, key terms were reviewed with the group such as constructivism, developmentally appropriate practice, and inquiry-based learning. These terms were looked at carefully in the context of teaching early childhood science. The group discussed subtle but powerful shifts in their own experiences and learning of science. They drew on their previous experiences to discuss the importance of developing and teaching the pollination curriculum from a place of active exploration. As the group connected the teaching of science with familiar learning theories, the teaching of science became more accessible. Moreover, as they spoke about the constructivist teaching methods that they used in other courses, they began to make connections to their science instruction.

Next, the group had to decide exactly what information was developmentally appropriate for their three and four-year-old student audience. Groups of the preservice teachers regularly met with the course instructor in order to discuss the progress of the lessons and unpack their experiences. Finally, consensus was reached and the following learning goals for the preschool students were developed:

The Goals for the Preschool Students

1. Students will understand the importance of the pollination process.
2. Students will demonstrate knowledge of healthy pollinators by speaking, writing or drawing, with an emphasis on honeybees.
3. Students will also be able to demonstrate knowledge of the pollination process by speaking, writing or drawing.

4. Students will also demonstrate understanding of what they can do to help pollinators.
5. Finally students will be able to identify the steps of the scientific process by writing, speaking or drawing.

The entire set of lesson plans for the pollinator project is listed in Appendix A. This plan includes a complete listing of the relevant STEAM standards for the content. All of the materials for each lesson and learning center were purchased and paid for by a grant from the Monmouth Conservation Fund. Materials included art supplies and manipulatives such as: seeds, soil, bucket, tangible and realistic food play items, age appropriate nonfiction books about bees, paper plates, Popsicle sticks, Kool-Aid, crayons, markers, and glue.

IMPLEMENTING AN INQUIRY-BASED LESSON

The preservice teachers decided that one preservice teacher would lead the lesson, e.g. introduce the learning goals in a whole group context and then describe the goals for each learning center. The learning center activities were based on a series of constructivist activities that started with the life cycle of a plant and moved towards the process of pollination. A typical lesson began with a whole-class meeting in the rug area of the classroom and then moved to the learning center activities as described below.

- **Reading *a Bee and Me*:** During the whole class lesson, the lead preservice teacher read a book called *A Bee and Me*. The lead preservice teacher paused every few pages and checked for understanding, asking questions such as, "Why is the boy afraid of the bee? Have you ever been afraid of a bee? What does the bee tell the boy that makes him unafraid? What important information did this book teach us?" After reading, talking, listening, and speaking about the book, the preschoolers had a chance to visit each of the four learning centers. One preservice teacher supervised each center, so that all children were able to participate to the best of their abilities.
- **The Honeybee Waggle Dance:** Helped students learn how honeybees communicate with one another. It also teaches students important facts about honeybees, for example, bees create one third of the United States' food supply. The lead preservice teacher marked an area in the middle of the room to represent the hive. Then, the students practiced the Round Dance by waggling (shake their abdomens) while moving around. The preservice teachers reminded the students that this dance is used when a food source is near to

the hive. Then the students practiced dancing towards entrance, or dancing upwards in a hive, which means "fly towards the sun." Dancing away from entrance or dancing downwards in hive is bee-speak for "I will fly away from the sun."

- **Plant a Garden:** The students planted a class garden consisting of the flowers, Nasturtium, Sweet Pea or an *Aquilegia.* Students used a variety of materials to create replicas of the different parts of flowers: roots, stems, leaves, seeds, stamens, and stigma. Each class created a list of edible plants that led to a discussion that all of the foods listed depend on bee pollinators. Questions that could be asked to further the discussion with the student include: "Are these foods part of your favorite meal or snack? Think back on what you ate for breakfast or lunch… Could you have eaten all those foods if bees were no longer around to pollinate our crops?" Then children sorted real food items into categories of the parts of the plant it represents. For example, flax seeds, celery, spinach leaves, apples, and carrots are all sorted into different categories corresponding to the different parts of the plant.
- **The Kool-Aid Pollinator Activity:** Intended to help the students recreate the act of pollination, using moist cotton balls and Kool Aid powder to represent pollen. The children recreate the pollination process by dipping moist cotton balls in different colors of Kool-Aid, which signifies different types of pollen from different flowers. This hands-on learning experience enables students to visually see the process of pollination.
- **Pollination Relay Race:** Each class is divided into two and each team holds a honeybee. The students then race to buckets filled with pollen coins to collect the coins as if they were the pollinating bees. Each student (holding the bee) must visit the three flower buckets and deposit pollen coins into each flower, then come back to the line.

Reflections

During each of the learning center activities, the students were encouraged to reflect on what they learned through speaking, writing or drawing. In the following paragraphs, the Kool Aid pollinator activity will be used as an example to demonstrate how student thinking was evoked in the various activities.

Learning Goal 2: Asking Open-Ended Questions About the Role of the Pollinator

The students used opened ended questions to reflect on the role of the honeybee in pollination and what their job entails. For example, during the Kool-Aid pollination activity the students discussed the "what-ifs" such as if the honeybee was not able to live up to its duties and what the repercussions would be if they could not perform their tasks due to the decline in healthy honeybees. They also used the "Busy Bumble-Bee Song" as a visual and auditory medium for the students to understand about bees and the role they have in pollination. By constructing their own open-ended questions and identifying the different parts of the bee, the students demonstrate understanding of what makes a healthy pollinator.

Learning Goal 3: Developing a Hypothesis About the Process of Pollination

Students created their own hypothesis of the process of pollination for the Kool-Aid pollination activity. Their hypothesis is centered around whether or not the Kool-Aid, representing the pollen, would stick to the cotton ball, representing the bee, when we dragged it across the Kool-Aid, representing the actual process of pollination. Students used their understanding of pollination to draw conclusions and make educated guesses as to what was going to occur during the experiment. During the honeycomb transfer activity, students will see the other tasks that the bees haven involving pollen transfer such using the pollen they collect to feed the baby larvae in the hive.

Learning Goal 4: Brainstorming Ways to Help Pollinators

The pollination Kool-Aid experiment opened a line of inquiry regarding the effects of pollinations for all parties involved, the bees, the plants, and us as well. Students identified their favorite flowers as well as food that bees are responsible for making to create a personal connection with the students to reinforce what was taught during the lesson. The students then brainstormed ideas as to what they can do to help the pollinators do their jobs. Some responses from the students included: "Plant flowers that the bees will like to visit,"" If I see a bee in my house I will let it go outside, I won't hurt it," "Tell the bee 'thank you" for all of is hard work," "Stand really still if I see a bee flying around me instead of trying to hit it," and "Don't knock down their hives if I see one in a tree."

Learning Goal 5: Identifying the Steps of the Scientific Process

Students will work through the seven steps of the scientific process through the Kool-Aid Pollination experiment. Students will start the experiment by asking the questions they have about the experiment after looking at all of the materials. The questions focused on what they wanted to know about the experiment. These questions assessed the student's prior knowledge as well as making the activity child centered and have the students feel as if they are in control of their education.

After the students asked their questions they conducted background research for their experiment. A discussion was conducted regarding the different jobs of a bee and what happens during pollination. Students drew upon their observations to predict what how the materials related to the functions of a bee. The students then constructed their own hypotheses and made an educated guess as to whether or not the powder would stick to the cotton ball.

The students then tested the hypothesis by conducting the experiment. The students each took turns dragging the cotton ball over the powder. Then they made their observations and stated what they saw during the experiment as well as what they noticed about the materials used. Each student concluded what the "bee" did in the experiment. Students concluded what each part of the experiment represented and what role each component of the experiment played in the process of pollination. After the experiment, the students reflected on the experiment, drew conclusions, and discussed what would happen without bees and pollination.

Assessment

Assessments included opened questions, competition of crafts and different activities, participation in discussion as well as completion of the KWL chart at the conclusion of the lesson. The lesson leader also recorded anecdotal notes on the different lessons at each school as lesson reflection and refinement for future implementation. Three other preservice teachers also facilitated learning and assessment at each learning center.

KWL Charts

In order to do an assessment for each class, the unit began with a KWL chart. A KWL chart has three large columns; the first column is labeled *K* for What I *K*now, the *W* for What I *W*onder, and the *L* for What I *Learned.* KWL charts are useful before beginning an activity to activate children's background knowledge and to learn their questions and interests. KWL charts are also useful at the end of an activity to help children remember what they learned. KWL charts are one of the best ways to assess what preschool children have learned (Bredekamp, 2014). Additionally, they help children learn to use nonfictions texts to find answers to their questions, learn that nonfiction texts are read differently than works of fictions, and learn about text features such as table of contents and captions (NAEYC, 2010).

Science Journals

In addition to using the KWL charts, the students kept science journals. Journals provide a motivation for children to observe carefully. When the children record their drawings, they choose among many possible features to represent. The actual features included in the drawing provide a window onto children's conceptual understandings. For example, even when children are observing and documenting the outside of a flower, a number of them will include the stem or stamen after being engaged in lessons regarding the pollination process. Although the variance in artistic abilities may limit the details some children can symbolize clearly, teachers can engage in follow-up discussions so that children can describe their observations.

Verbal Assessment

Throughout the curriculum students were assessed using open-ended questions, group discussions, identification and recall questions. The overarching form of formal assessment used was the KWL chart, which was used to assess the knowledge gained by the students throughout the curriculum. The children were also able to answer all of the questions with an 80% degree of accuracy through questioning and answering.

Observations

Students were also encouraged to be careful observers and take anecdotal notes while in the classrooms with the preschool children and they brought those qualitative findings back to the group once all of the teaching sessions had been completed. Students were assessed informally using open ended how and why questions throughout each activity. Students were asked to identify different components of the experiment and the activities as well as recall question such as "what was the powder on my flower called?" and "what did I call my educated guess in the beginning of our experiment?" The final products the students created such as the worksheet from the pollination experiment, the bee craft with the paper plate as well as the finished sorting activities were all formal assessments as well.

IMPACT ON PRESERVICE TEACHERS

While the preservice teachers followed the schedule for implementing the pollination curriculum, the early childhood faculty member met with the group twice to unpack their experiences. At the beginning of the project typical responses included:

I am afraid to teach science
I don't feel like I will be good at it [teaching science]
I am not sure what the scientific process is, I would have to look it up
I think that a big part of me volunteered to be a part of this project in order to help me feel more confident about teaching science.

At the conclusion of the pollination unit, the responses to the preservice teacher comments were more positive. Each of the preservice teachers said they were feeling stronger about the teaching of science in early childhood. Three reported that they thought that their confidence might be connected in some way to being familiar with the curriculum. Examples to some prompts were:

I actually enjoy the thought of planning for science – I look at it so differently now.
I can see how the science was there all along, the children are the ones doing all of the work.
I just have to help them figure some of the stuff out and they can do the rest. I am so impressed with the level of their thinking, they literally blew me away),
I really do believe that some of these kids are going to save the world.
I think that even we, as early childhood student teachers even gave them [the preschool students] enough credit.
I have a whole different feeling about teaching science
Now I can see how it connects to everything and how it was all already there and how important it is.

In addition, the preservice teachers growing confidence was reflected in the preservice teachers' reflective journal. An example of one reflection:

I can't believe how much success I had with this science unit. I never even understood the value of teaching science in early childhood until we began this unit. Once I knew and understood how important the content [pollination] is and that I had some of my friends from class with me, it made me so much more into it...I didn't feel stupid for saying that I was always hoping that I would never have to teach science by itself. It is like night and day now. Katlyn really stepped up and showed us how to approach it from a P through 3 way.

The cooperating teachers, who hosted the preservice teachers and their teaching of the curriculum, were also enthusiastic about the impact of the lessons on their students. For example, one of the teachers said:

I am quite surprised by the depth and wonder that so many of my students brought to the science, especially learning about so many details about pollination. I would not have guessed that they would have been able to sustain so much interest in this subject. I am really shocked.

Another at teacher another school site said:

Originally, I thought that the topic of pollination would be way over the kids' heads. However, I was very impressed with how these young teachers taught the curriculum. The real thing that captured my interest was how much my own students ate up the all the science that the student teachers presented. They just could not get enough of the pollination. We had to continue studying the subject after the student teachers left. I was more surprised with how much science my preschoolers were able to take in and how much interest they had in the subject. I think that science is an area that I have to place more consideration in in the future in my own lesson planning.

SOLUTIONS AND RECOMMENDATIONS

Preschool children want and need to learn about their worlds, and as developmental research asserts, young children become engaged when they are given chances to discover the environments and the natural world around them (Bredekamp, 2014; Piaget, 1955; Vygotsky, 1978). The goal of science education in early childhood should be to encourage and refine children's instinctive passion for discovering their natural worlds and to assist them in channeling their joyful curiosity for exploring the world around them by assisting them in shaping their energies and helping them grow enthusiastic true scientific literacy (Editors, Scientific American, 2012). Children are biologically prepared to learn about the world around them, just as they are biologically prepared to learn to walk and talk and interact with other people.

1. This case showed how children can be engaged in STEM learning in early childhood.
2. It also showed how teacher preparation can be enhanced through experiential learning.
3. Most importantly is showed how accomplishing those goals together can create s synergy for learning at both levels.

In a constructivist preschool classrooms children are viewed as powerful and capable of seeking information that builds knowledge (Willis, 1993). Learning, even when it involves mistakes, is considered an important process of growing and living. Thus, it was important to develop and teach the pollination curriculum from a place of active exploration, with the preservice teachers as facilitators, connecting the content to the children's experiences as well as ensuring that the lessons and the content were flexible and integrated with other curricular areas. When the group connected the teaching of science with learning theories that they were familiar with such as constructivism, it made the teaching of science more accessible. Moreover, as they spoke about the teaching methods that they used in other courses such as a balanced approach in early literacy and mathematics, they began to make connections to the constructivist teaching methods that they could use in their science instruction.

However, developing appropriate constructivist practice that combines the social, cognitive, and content related aspects that meet the needs of all students is an exceedingly difficult task (Bredekamp, 2014). More research is needed in how to facilitate teachers as they struggle to enact the inquiry-based classroom approaches called for in reforms (Sharapan, 2012). The data from the interviews at the beginning of the project demonstrated that the preservice teachers' low level of self-efficacy surrounding their ability to teach science developed as a result of their own science-related experiences in elementary and high schools. These negative beliefs and attitudes held by the preservice teachers in this study are reflected throughout the research (Abell & Roth, 1992; Stevens & Werner, 1996). In addition, the preservice teachers' low sense of self-efficacy initially prevented them from effectively learning about science content and curriculum. Many of the students in the study were slow to take charge and assume responsibility for becoming a team or lesson leader.

The positive shift in the preservice teachers' beliefs and attitudes about teaching science, including their self-efficacy, can be attributed to a few factors. First, they were able to learn the specific information that they needed to create the curriculum in a way that held direct meaning and purpose. Secondly, the social nature of the collaborative work assisted the preservice teachers in sustaining in their efforts to persevere. Next, they were given ample time to teach the pollination curriculum in authentic preschool classrooms, which were highly supportive of their efforts. As students, they were encouraged to explore their fears and misconceptions about teaching science in a risk-free environment and they came to recognize that others also held similar self-doubts about teaching science. Last, their prior knowledge from other coursework about the tenets of constructivism such as the recognition of the power of activating prior knowledge, active engagement with challenging and developmentally appropriate tasks scaffolded by more knowledgeable peers and teachers, was critical in the shift in the dispositions of the preservice teachers regarding their self-efficacy regarding the teaching of science.

Just as discovery learning is created for preschool students, the preservice teachers took part in their own discovery learning as they balanced crafting instruction for each preschooler in the zone of proximal development (Vygotsky, 1978). For example, the preservice teachers invested time in getting to know the children individually as well as in a preschool class. They prepared many questions for the cooperating teachers, such as "What are the children's greatest interests?" and "How will we be able to teach this particular group of children to learn this unit at a developmentally appropriate level for them at this time in their growth?"

This group of preservice teachers accepted the challenge of teaching science, overcame their fears, and found joy in learning and teaching science content and methodology. Most importantly, they learned to believe in themselves and their own abilities to learn. To do so, they needed to know that they had a support system in their new learning venture, just as they were learning to instill care into their science curriculum for their preschool students. When learners of any age feel cared about, they can be inspired to take great leaps of faith. By participating in a caring teaching and learning community, the preservice teachers were able to grow significantly in their self-efficacy for teaching science. Their growth was also greatly aided by the clinical field work that allowed them the time to unpack their experiences with their peers and a course instructor (Conzeio & French, 2002).

REFERENCES

Abell, S. K., & Roth, M. (1992). Coping with constraints of teaching elementary science: A case study of a science enthusiast student teacher. *Science Teacher Education*, *76*(6), 581–595.

Aizen, M. A., Garibaldi, L. A., Cunningham, S. A., & Klein, A. M. (2009). How much does agriculture depend on pollinators? Lessons from long-term trends in crop production. *Annals of Botany*, *103*(9), 1579–1588. doi:10.1093/aob/mcp076 PMID:19339297

American Association for the Advancement of Science. (1993). *Benchmarks for science literacy*. New York, NY: Oxford University Press.

Bianchini, J. A., & Colburn, A. (2000). Teaching the nature of science through inquiry to prospective elementary teachers: A tale of two researchers. *Journal of Research in Science Teaching*, *37*(2), 177–209. doi:10.1002/(SICI)1098-2736(200002)37:2<177::AID-TEA6>3.0.CO;2-Y

Bohart, H., Benson-Collick, H., & Charner, K. (Eds.). (2016). *Spotlight on young children: Teaching and learning in the primary grades*. The National Association of Young Children.

Bredekamp, S. (2014). *Effective practices in early childhood education: Building a foundation*. Upper Saddle River, NJ: Pearson.

Brooks, M. G., & Grennon-Brooks, J. (1999). The courage to be constructivist. *Educational Leadership*, *57*(3), 18–24.

Chaille, C. M., & Britain, L. (2003. The young child as scientist: A constructive approach to early childhood science education (3rd ed.). Boston, MA: Allyn & Bacon.

Conezio, K., & French, L. (2002). Science in the preschool classroom: Capitalizing on children's fascination with the everyday world to foster language and literacy development. *Young Children*, *57*(5), 12–18.

De Palma, A., Kuhlmann, M., Roberts, S. P., Potts, S. G., Börger, L., Hudson, L. N., & Purvis, A. et al. (2015). Ecological traits affect the sensitivity of bees to land-use pressures in European agricultural landscapes. *Journal of Applied Ecology*, *52*(6), 1567–1577. doi:10.1111/1365-2664.12524 PMID:27546902

Fitzgerald, A., & Schneider, K. (2013). What teachers want: Supporting primary school teachers in teaching science. *Teaching Science*, *59*(2), 7–14.

Gerde, H., Schachter, R., & Wasik, B. (2013). Using the scientific method to guide learning: An integrated approach to early childhood education. *Early Childhood Education Journal*, *41*(5), 315–323. doi:10.1007/s10643-013-0579-4

Goulson, D., Nicholls, E., Botias, C., & Rotheray, E. L. (2015, February26). Bee declines driven by combined stress from parasites, pesticides, and lack of flowers. *Science*, *347*(6229), 1–16. doi:10.1126/science.1255957 PMID:25721506

Johnson, C. C. (2007). Whole school collaborative sustained professional development and science teacher change: Signs of success. *Journal of Science Teacher Education*, *18*(4), 629–662. doi:10.1007/s10972-007-9043-x

Klein, A.-M., Vaissiere, B. E., Cane, J. H., Steffan-Dewenter, I., Cunningham, S. A., Kremen, C., & Tscharntke, T. (2007). Importance of pollinators in changing landscapes for world crops. *Proceedings. Biological Sciences*, *274*(1608), 303–313. doi:10.1098/rspb.2006.3721 PMID:17164193

Llewellyn, D. J. (2002). Inquire within: Implementing inquiry-based science standards in grades 3-8. Thousand Oaks, CA: Corwin Press.

Long, E. Y., & Krupke, C. H. (2016). Non-cultivated plants present a season-long route of pesticide exposure for honey bees. *Nature Communications Nat Comms*, *7*, 11629. doi:10.1038/ncomms11629 PMID:27240870

Losey, J. E., & Vaughan, M. (2006). The economic value of ecological services provided by insects. *Bioscience*, *56*(4), 311–323. doi:10.1641/0006-3568(2006)56[311:TEVOES]2.0.CO;2

Noddings, N. (2005). *The challenge to care in schools: An alternative approach to education* (2nd ed.). New York, NY: Teachers College Press.

Patrick, H. (2009). *Engaging young children with informational books.* Thousand Oaks, CA: Corwin.

Piaget, J. (1955). *The construction of the reality in the child.* Basic Books.

Potts, S. G., Biesmeijer, J. C., Kremen, C., Neumann, P., Schweiger, O., & Kunin, W. E. (2010, January/February). Global pollinator declines: Trends, impacts and drivers. *Trends in Ecology & Evolution*, *25*(6), 345–353. doi:10.1016/j.tree.2010.01.007 PMID:20188434

Schwarz, C., & White, B. (2005). Meta-modeling knowledge: Developing students understanding of scientific modeling. *Cognition and Instruction*, *23*(2), 165–205. doi:10.1207/s1532690xci2302_1

Scientific American Editors. (2012). *Science education: Back to school.* Washington, DC: National Academy Press.

Sharapan, H. (2012). From STEM to STEAM: How early childhood educators can apply Fred Rogers' approach. *Young Children*, *67*(2), 36–40.

Stevens, M., & Werner, D. (1996). New elementary teachers' knowledge and beliefs about instructional representations: A longitudinal study. *Journal of Elementary Education*, *22*(3), 327–347.

The National Research Council. (2000). Retrieved from http://www.nationalacademies.org/nrc/

The White House. (2014). *Fact Sheet: The economic challenge posed by declining pollinator populations.* Retrieved from https://www.whitehouse.gov/the-press-office/2014/06/20/fact-sheet-economic-challenge-posed-declining-pollinator-populations

Tu, T. (2006). Preschool science environments: What is available in an early childhood classroom. *Early Childhood Research Journal*, *33*(4), 245–251. doi:10.1007/s10643-005-0049-8

Vygotsky, L. (1978). *Mind in society: The development of higher psychological processes*. Boston, MA: Harvard University Press.

Watters, J., & Diezmann, C. (2007). Multimedia resources to bridge the praxis gap: Modeling practice in elementary science education. *Journal of Science Teacher Education, 18*(3), 349–376. doi:10.1007/s10972-007-9051-x

Willis, S. (1993). Educators seek 'developmental appropriateness'. *Teaching Young Children., 45*, 35–42.

Xerces Society. (2011). *Attracting Native Pollinators*. Portland, OR: Timber Press.

Xerces Society. (2016). *Gardening for Butterflies*. Portland, OR: Timber Press.

KEY TERMS AND DEFINITIONS

Assessment: The process of determining the level of student dispositions, skills, and understanding.

Discovery Learning: A constructivist-based approach to learning that engages children in an inquiry of their environment.

Environmental Education: Engages young children with the natural world to promote their understanding of conservationism.

Inquiry-Based Learning: Challenges children to learn through investigating and reasoning about a problem.

Project-Based Learning: An approach to teaching that engages children in exploring their environment.

Science Education: Engages young children in learning opportunities that capitalize on their natural curiosity for observing, exploring, and discovering the world around them.

Scientific Method: Involves teaching young children how to identify a problem, collect data, and test their hypotheses.

Chapter 9
Finding and Using the ART in Science Lessons

Kevin D. Finson
Bradley University, USA

EXECUTIVE SUMMARY

Teachers can find art in their science lessons if they know what to look for. The art in many science lessons is in the form of visual representations, which can be seen to exist on a continuum ranging from very concrete and iconic to very abstract and symbolic. There should be a progression through a lesson that guides students in identifying, interpreting, and analyzing different visual representations, and ultimately help them know how to create their own. The teacher should be very deliberate in his/her selection, generation, and/or use of visual representations. Five critical things the teacher should be mindful of include: (1) how the learner's mind processes visual information, (2) what visual literacy is, (3) what is meant by cognitive loading, (4) different types of visual representations, and (5) augmenting student learning by using a variety of visual encryptions.

LITERATURE REVIEW

Incorporating art within science lessons may at first seem daunting to teachers, but many elements of art can be easily utilized and are, in fact, often already present in lessons if one knows what to look for. For example, it is a very common practice in middle grade life science classes to have students draw what they see when observing cells through a microscope. The inclusion of that act of drawing is the

DOI: 10.4018/978-1-5225-2334-5.ch009

Copyright ©2017, IGI Global. Copying or distributing in print or electronic forms without written permission of IGI Global is prohibited.

incorporation of art within the science lesson. A key consideration by teachers at this juncture is what to do with those drawings. Most of the time, the drawings are relegated to being just a part of a laboratory report, and little more is said about or done with them. This is unfortunate. With some further consideration, the teacher could make use of those drawings to more accurately assess what their students are conceptually gleaning from their microscopy work, and could use those drawings as springboards to help students develop their concepts about cells more efficiently and more thoroughly. The drawings can become tools useful in the teaching and learning of science conceptual information. In short, the art can facilitate science learning.

The importance of art in the form of visual data in science education is underscored by an increasing number of references to it in standards such as the Next Generation Science Standards (National Research Council, 2012). Among the practices elucidated in the standards are some directly calling for conceptualization through visualization of data, interpreting and analyzing it, generating it, and communicating with it. In these ways, the NGSS framework provides direction to science educators about what should be included in instruction. Yet educators must know why those art-related things should be included and how their use affects students' cognitive development, including why a certain visual representation should be utilized in one context but not another or why visual representations should be sequenced in a particular order.

For teachers to capitalize on the power of art in their science lessons in ways that address NGSS standards, they should understand how some essential and critical aspects of art fit together. These aspects include . . .

- How the learner's mind processes visual information,
- What visual literacy is,
- What is meant by cognitive loading,
- Different types of visual representations, and
- Augmenting student learning by using a variety of visual representations.

Mental Processing of Visual Information

A good beginning point is to understand a basic tenet derived from a cognitive psychological perspective on human brain function. Without going into too much depth, the essential point is that the human brain is hardwired to process images such as drawings, pictures, and photographs much more readily and efficiently than it is to process more abstract and symbolic images such as alphanumeric text. For example, it is much easier for learners to quickly grasp the idea of how to set up a laboratory apparatus (such as a stream table in an earth science lab setting) by looking at a diagram or picture of it as compared to reading a paragraph of text describing it and explaining how to set it up. When given a choice, students tend to prefer working with images rather than reading text. Brumberger (2010) studied students' communication abilities based upon the time they spent involved with video media, such as watching television, playing video games, using digital cameras, and using computers. Results showed students preferred using visual media that provided guidance with graphics. Much earlier, when explaining his theory of cognition, Jerome Bruner noted that visual representations play significant roles in student learning (Cognition and Development, n.d.; Simply Psychology, n.d.).

The Visual Literacy Component

Making sense of the art embedded in visual images used in science requires us to understand something about visual literacy. Visual literacy is one of three kinds of literacy: audio, print, and visual. In this context "literacy" is someone's knowledge and understanding about a set of knowledge, and the ability to articulate meanings from it. The three kinds of literacy mentioned above together comprise a literacy trio from which other types of literacy might arise. Together, the trio is important to effective learning, and a well-balanced use of them must be considered in instructional design and delivery. The use of visual representations and artforms is a significant component of visual literacy. Simply exposing a learner to visual images and art does not mean he/she will learn from them. The learner must have a certain degree of visual literacy, and this is something the teacher can help them learn.

Sibanda and Sibanda (2013) noted that, "Generally, in human development, visual literacy precedes verbal literacy and most children are capable of interpreting or reading visuals long before they can read the same information in words and sentences" (p. 41). That being said, it is also notable that attention to the preparation and selection of visual representations receives less focus than do materials from print media. Consequently, students are directed to focus more on print than on visual information presented to them. Unfortunately, this can lead to lower comprehension and decreased understanding on the part of students. When teachers approach the use of visual representations appropriately, they include in their instruction opportunities for students to practice using them. That practice includes how to make sense of them (e.g. identifying them, analyzing them, etc.) and how to generate some of their own that demonstrate their level of learning and understanding of concepts and abstract principles. In sum, children need to learn how to represent their own ideas in depictive and descriptive ways (Lesh & Doerr, 2003; Valanides, Efthymiou, & Angeli, 2013). It is vitally important that children be given opportunities to express what they think, know, and feel; and allowing them to do so through the construction of representations can clearly be a way to make learning more meaningful, more personally relevant, and more valuable to them.

Cognitive Load

Cognitive load refers to the amount of demand that is placed on the brain to process information. The human brain can only hold in its "working" and manipulative memory a certain amount of information. Once that limit is reached, something that is in that working memory must be let go before something new can be added. This is an obvious problem when students must deal with huge quantities of information. Once a learner reaches cognitive overload, his/her response seems to shift to memorizing with a decrease in ability to draw conclusions from information and data being viewed (Coleman, 2010). Typically, visual representations incorporate very large quantities of information. Even so, the cognitive load they present is less than that required to process text, symbols, formulae, and other more abstract forms of information. Remember from an earlier section of this chapter that we know the human brain is more efficient in handling and processing artful visual representations than alpha-numerics. Consequently, visual representations are often seen as being in an artform that is mentally manageable and flexible, and does not significantly contribute to cognitive overloading. One way to ameliorate the cognitive load issue is to improve the quality of the visual representations that are used, decrease the quantity of text, and insuring visual representations are clearly addressed in the text in a complementary fashion rather than being ancillary additions.

However, simply putting visual representations in front of students is not an effective teaching practice in and of itself. At the heart of visual literacy is one's ability to decode and encode meanings from visual images; to select and create visual representations; in being able to analyze, interpret, and evaluate visual images; and being able to communicate meanings of visuals and through various visual representations (Yeh & Lohr, 2010). So, teachers need to be aware that they need to teach their students how to manage visual imagery (Gilbert, 2005; Vavra et al., 2011). One management approach that can be effective is walking through construction stages of visual representations with students, which helps them learn how to deconstruct other representations they encounter and consequently come to more efficiently interpret what information is being conveyed by them – and, in turn, serves as a model guiding students in how to construct their own visual representations. Consider the following example. Many years ago when print encyclopedias were popular, one included a section on a frog. That section contained multiple pages of a frog diagram, but each page was different. The pages were printed on a transparent film so that when stacked on top of each other (or unstacked), different complexities of the frog could be seen. The first page showed the frog's skeleton. The next page showed the frog's organs. When overlapped with each other, the viewer could clearly decipher where the organs were with respect to parts of the skeleton. The next page showed the frog's veins and arteries, and the next showed muscles, with the last page showing the skin. By placing one page atop the other, one could visually build or construct an increasingly-complex image of the frog. This approach helps with the cognitive load issue. A more recent example was provided by Bodzin and Cirucci (2009) in their work with helping students learn about land use planning using geospatial layering.

Many science-related visual representations are complex, and showing them to students can contribute to cognitive overload and confusion. In short, this is not the most humane way to teach children. If the teacher takes time to deconstruct the complex image and introduce it to students in parts that build upon each other, like in the encyclopedia frog, students are more likely to understand the final image and be able to relate understandings of the conceptual aspects of it. Similarly, students can learn from this how to build up their own conceptually-oriented visual representations to demonstrate their levels of understanding. Such an approach would reduce cognitive load, and be efficient and effective in terms of visual representation utilization, and give students more ownership in what they are learning. This point is important. As Brumberger (2010) found, students are often skilled at finding and using visual representations, such as those available in software built within computers or smart phones, but they do not typically understand the complexities of those images nor do they consider themselves skilled in creating their own illustrations. So students may be able to put together something like a PowerPoint presentation that contains images while really not understanding much about the quality of the images or whether those images really convey what they want conveyed. When pressed, students may find it difficult to explain the depth of detail and understanding the teacher is looking for with respect to a particular science concept. And without a sense of ownership, what was learned with the PowerPoint presentation may not be retained much beyond the termination of the class.

Some Types of Visual Representations

In science, there are multiple ways of finding out about something, whether it be a phenomenon, an object, or something else. One might think about different types of approaches to teaching or learning science. For example, one might find a closed-ended lesson to be the most appropriate way to teach something. An example of this might be teaching middle school students about the parts of a microscope and how to handle it before turning them loose with one to do an investigation. Another type of approach might be more open-ended, such as exploration or perhaps discovery/inquiry, such as that necessary to investigate what happens to the population of microorganisms in a jar of pond water over a period of time. A teacher selects a particular type of approach to help students arrive at a targeted point of learning and understanding. In a similar way, there are different types of visual representations

Figure 1. The visual representation continuum

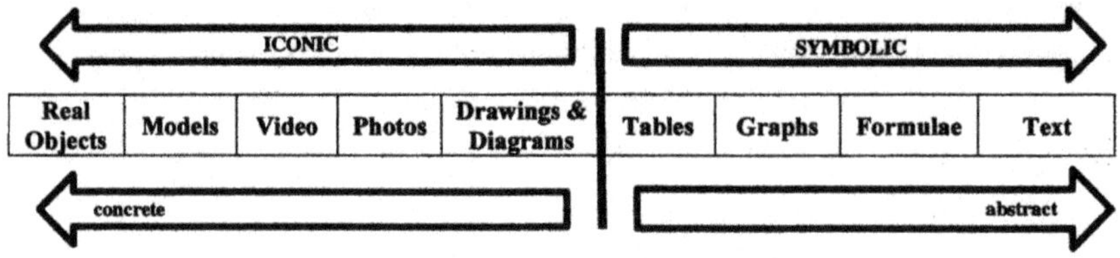

that can be used in science instruction. For each, there are both positive and negative aspects, depending upon the context in which they are employed and exactly how they are used. For example, a single picture of an object might convey information about the physical attributes of the object, but little about how it changes as variables impacting its state are changed. Conversely, a visual representation might convey something about how something relates to other objects, but little about the makeup of the object itself. Hence, it is important for teachers to understand something about the different types of visual representations that exist, and what that means in terms of cognitive loading and learning.

A number of researchers have identified categories of visual representations. Typically, these categories include items that are concrete (e.g. photos, models, drawings) and those that are abstract (e.g. graphs, symbolic text, etc.). Those that are more concrete are called iconic. Those that are more abstract are called symbolic. As noted earlier, the human brain is wired to accommodate iconic images much more readily than it is symbolic images. Briefly, this occurs because symbolic images require much more interpretive skill and have much higher cognitive load requirements. Saunders (1994) identified nine categories of visual representations, and Pozzer, Ardengi & Roth (2005) showed how those representations actually fit along a continuum or spectrum from most iconic to most symbolic. Olson (2013) modified that continuum to include other types of visual representations, and it is that continuum that one might find the most usable and applicable to today's science classrooms (see Figure 1). Beginning with the most iconic representation, the continuum starts with actual physical objects. The next would be models, followed by videos, then photographs, and finally drawings and diagrams.

Note that each of those iconic representations is, in some way, a form of art. Moving beyond drawings and diagrams, the continuum shifts to more symbolic representations, moving to tables, then graphs, then formulae, and finally text. The reader might find it odd that text is at the far end of the symbolic continuum. However, research in reading cognition clearly shows that each part of each letter must be processed by the brain, and then those parts are combined for us to recognize letters, and then the letters are combined to form recognizable words, and so forth.

In short, dealing with text is extremely complex and has very high cognitive load demands. Most of us in education are so familiar with reading that we don't realize how much demand is placed on us when we read. As a contrast, consider for a moment the difficulties encountered when teaching ESL/ELL students. Not only must they learn to process (i.e. interpret) a different language -- in and of itself a large cognitive load issue, but they must do so at the same time they are processing the text. The cognitive demands on them are vastly increased over that imposed on native English-speakers. This can be significantly ameliorated by the teacher's effective utilization of iconic representations with ESL/ELL students, resulting in improved classroom success and learning.

CONTENT INFORMATION

There are, of course, a number of possible content-specific standards that can be addressed in a lesson that integrates STEAM disciplinary areas. For the case that will be discussed later, the focus begins with a look at physical science, with some specific emphasis on speed and energy transfer of balls rolling on an incline. Obviously, physical science content standards dealing with force and motion and mathematics standards dealing with scale, dimension, and graphing would be clear targets at the beginning. From a purely science standpoint, applicable standards for the targeted grade level (middle elementary in this case, approximately grades 4-5) might be using evidence to construct an explanation relating to the speed of an object to the energy of that object (NGSS 4-PS3-1) and perhaps asking questions and predicting outcomes about changes in energy that occur when objects collide (NGSS 4-PS3-3). The mathematics connected standards could include modeling with Common Core mathematics (MP.4) and perhaps summarizing numerical data sets in relation to their context (6.SP.B.5). Consideration of other standards that may be applicable would necessarily expand as one integrates other areas cogent to the lesson, such as the Common Core Reading/Language Arts standard prescribing students write informative or explanatory texts to convey ideas and information (W.4.2), or with standards that more closely align with visual representations. Some visual representation standards that have utility include NGSS MS-PS3-1 in which students are asked to construct and interpret graphical displays of data, and Common Core Reading/Language Arts SL.8.5 about integrating multimedia and visual displays into presentations to clarify information, and strengthen claims and evidence. There are even applicable standards from the National Core Arts Standards (NCAS). Among the NCAS standards are "anchor standards" about students conceiving, developing, and organizing new artistic ideas and work to convey meaning through the presentation of artwork. All of the above standards can clearly be integrated into the science lesson. Indeed, if one were to create a matrix including the science, mathematics, reading, and art standards noted above, it would be evident there is considerable cross-discipline re-iteration, overlap, and support between the sets of standards.

Enterprising teachers who wish to make other curricular connections with the science lesson and the standards mentioned in the previous paragraph might even consider integrating some from social studies. For example, from a history perspective, the early Egyptians probably employed ramps in their construction of the pyramids, and the Romans certainly employed ramps in their roads, buildings, and even in warfare. From a societal and geographic perspective, one can almost imagine the energy at work on cars or trucks attempting to manage their traverse up or down steep hills such as famous Filbert Street in San Francisco (an amazing slope of 31.5% -- compared to an intended maximum slope on interstate highways of 4%). Perhaps more common are boat ramps at lakes and rivers across the country, upon which people move their watercraft into or out of the water, or even the spiraling ramps inside parking garages. And lest we forget something closer, there are the slipper slides on playgrounds at many schools and parks throughout the country, down which children gleefully descend into patches of sand or water (and sometimes into the waiting arms of other children or adults).

The connections of the science lesson (described later) to many facets of learning and life – both past and present – are many, varied, and can be easily scaffolded to students' learning. An important key in all this is that the instruction connect in substantial ways with the students' prior and common experiences, and is scaffolded to match and support what is being experienced in the classroom. As students build their understandings with these scaffolds, they develop mental models and images in their minds that they can hopefully illustrate in artful ways that not only convey they understand the concepts, but also in ways that have deep personal meanings to them – and as a consequence, learn better, learn more efficiently and effectively, retain information much longer, and come to see how it all connects to so many facets of their everyday world.

CASE DESCRIPTION

The case presented below is one that shows how, with a little additional attention, the teacher can identify art-related components that can readily be part of a science lesson, and integrate them in a way that provides a nearly seamless progression of learning experiences for the student. Carefully planned, such an integration can help students tap into what many of them find as more natural ways of understanding their world (e.g. iconic representations), thereby increasing their ability to understand the science being presented to them while doing so in a manner that is less intimidating than traditional (and more sterile) science instructional approaches. Note this does not mean the science is less demanding or less substantial, but that it is capitalizing on the way learners' brains are pre-designed to receive and process information.

Lesson Plan Concerns

Harkening back to the beginning of the chapter, one of the goals of the NGSS is to have increased and improved conceptualization through visualization of data, interpreting and analyzing it, generating it, and communicating with it – both for students and teachers. Hence, a clear goal for science teachers is to help their students arrive at a point in their learning where they can convey their understandings of science concepts through visual representations. In his theory of cognition, Jerome Bruner (Cognition and Development, n.d.; Simply Psychology, n.d.) claimed that visual representations play significant roles in student learning. According to Bruner, a key to cognition is that learners can recover from their mental schema iconic and symbolic representations when the initial stimulus is absent. At that point, learners begin to effectively create their own representations. Later, those learners can then manipulate those mental images, and derive more sophisticated concepts and engage in more effective problem-solving.

The conundrum science teachers face is exactly how to accomplish getting their students to that desired end point. The solution is multifaceted. It includes exposing students to different types of visual representations in carefully planned sequences, helping students deconstruct representations to more effectively understand what they contain and what information is being conveyed by them, and helping students construct their own visual representations that demonstrate their conceptual understandings and learnings. As mentioned, this requires careful and deliberate planning on the part of the teacher. Consider, for example, the kind of planning necessary for a chemistry activity in which students mix calcium chloride with baking soda, water, and phenol red. The ensuing reaction produces a gas, a yellow liquid, and a yellowish-white precipitate. It also produces heat. In most published activities about this, students investigate what ingredient was responsible for the heat, and whether doubling the amount of that ingredient would double the heat produced. This activity could be extended by having students do some inquiry to determine what those reaction products are. There are, of course, specific tests that one can employ to find out. For example, students may conjecture that the precipitate might be sugar, or flour, or salt, or chalk. There are chemical tests for each of those. But before doing the tests, one must know how to separate the precipitate from the other products. This is a process the teacher must teach well in advance of doing the chemical reaction activity. In addition, the teacher must be deliberate in teaching about each of the chemical tests in advance. Those tests might be the addition of iodine solution to test for starch, the use of Benedict's qualitative solution to test for sugars, the use of dilute hydrochloric acid to test for chalk, and so forth. In a similar way, the teacher would have to plan ahead to teach about how to test for certain gases, such as oxygen or carbon dioxide. Also in the mix would be teaching about pH and indicators. From a visual representation perspective, students can draw and color things they observe reaction-wise or chemical test-wise. They can draw their experimental apparatus or set-ups for doing their extension testing.

Perhaps to simplify and focus more exclusively on visual representations as identified in the Olson (2013) continuum, it might be helpful to look at the kind of deliberate planning the teacher must use, and the visual representations that he/she should utilize throughout a particular lesson. In 1999, Gabel investigated high school chemistry teachers' teaching, and found that many move from iconic to symbolic visual representations as they progress through a lesson, yet do little to guide their students to see how those representations are interconnected. In 2003, Kozma observed that there are key differences between how students and scientists view and utilize visual representations. Scientists typically look at attributes or properties of multiple visual representations and arrange them as they do their reasoning and when sharing their work with others. In contrast, students seemed to be limited in their abilities to examine inscriptions across types of representations, or to examine single representations beyond superficial attributes. Brante et al. (2013) reported that most learners come to comprehend a topic by shifting attention back and forth between iconic (picture) and symbolic (text) information, and that the more representations employed in the learning, the more effectively the topic is learned. They also found that poorer readers rely heavily on iconic representations, whereas more efficient readers strike more of a balance between reliance on iconic and symbolic representations, shifting back and forth between them. Similar results were found by Valanides et al. (2013) in their study of elementary students' thinking as they made use of drawings and textual explanations about shadows. One important finding from such studies is that overt and intentional instruction on interpreting visual representations, making use of them, and having students generate their own is critical to more engaged and effective student learning (Carlsen, 2009; Wu & Krajcik, 2006). In contrast, overreliance on one primary type of visual representation, such as mathematical equations alone, leads students to potentially miss connecting the more abstract with the more concrete ways of understanding a concept or phenomenon.

In a teacher's planning, consideration should also be given to use of visual representations for normative and for pedagogical means. Normative visual representations are those used in scientific practice. These include such representations as graphs of data or the use of mathematical formulae. This is a way scientists make sense of what they are finding from data they collect. Pedagogical visual representations are those used to help convey ideas that make understanding the representations simpler.

This could include creating a diagram or a drawing of an idea. For example, drawing the water cycle is a pedagogical visual representation. Effective science teachers guide their students through both normative and pedagogical visual representation usage. Both used together, shifting back and forth from one to the other in carefully planned sequences, can lead to more effective and efficient student learning (Ozcelik & McDonald, 2013).

STEAM Content: The Physical Science Lesson

Let us now turn to walking through a specific science lesson to see how all this might actually be implemented by a science teacher. The lesson is taken from the Science Curriculum Improvement Study (SCIS) developed in the 1960s at the University of California-Berkeley (Karplus, 1964). The particular activity is called *Ramps and Sliders*. As an overview, the activity is a middle-elementary activity that uses an adjustable ramp that can be set at several degrees of incline or steepness. The ramp has a groove on the center of its top surface running along its length that allows for steel balls to roll down it in a controlled manner. The mass of the steel balls can be measured, as can the horizontal, vertical, and inclined length of the ramp. The slider is similar to a half-cone shaped device into which the steel ball can roll as it comes off the ramp. The slider will slide along the floor as the steel ball transfers its momentum to the slider, and the distance it moves can be measured. The slider can accommodate masses on its top surface. The attributes of these materials allows for tests to be conducted when variables of ramp height/incline are changed, mass of the steel ball is changed, and slider mass is changed. Students are tasked with determining the relationship between the ball's mass and the distance the slider moves, and between the mass of the slider and the distance it moves. Modified versions of the activity exist, usually involving different kinds of ramps and sliders. It is important to note that at each step of the lesson, the teacher actively engages students and helps them interpret each of the types of visual representations they encounter. This includes helping students see how one type of representation connects to the others, and how all the representations are related to the same object/event. Also note that the teacher guides the students in learning how to generate their own visual representations.

The following steps might illustrate how the lesson can be taught incorporating visual representations (Table 1 summarizes the steps).

1. A good place to begin the lesson is to start with something concrete with which the students are familiar. (Table 1 provides a summary of this and what follows.) For example, the teacher might ask students if they have ridden in a wagon down a hill or sidewalk. (Alternative scenarios might be sledding or

Table 1. Lesson components related to visual representations

	Lesson Component	Content	Relationship to Visual Representation
1	Introduction	Rolling in a wagon downhill	Concrete/Real Objects/Experiences
2	Trade Book Story	Character in a moving wagon	Drawings
3	Showing Ramp & Slider	Connecting the ramp and ball to the hill and student in the wagon, using the ramp and ball as a model of the wagon on the hill	Models
4	Watching a wagon on a hill	Watching a wagon roll on a hill or slope, and connecting that to the ramp and ball	Photos and/or Videos
5	Rolling the ball down the ramp	Determining the speed of the ball on the ramp	Formulae
6	Rolling the ball down the ramp	Testing speed of balls on ramps of differing inclines	Models
7	Making an illustration of the ball moving down different ramps	Students make a drawing of what occurred with their ball on each of the inclined ramps	Drawings/Diagrams
8	Using the sliders	Students measure their ramps and ball speeds as well as travel distance of their sliders	Tables
9	Plotting ramp and slider data	Students create a graph of the ball speeds versus slider travel distance	Graphs
10	Comparing graphs to drawings	Students make comparisons between their graphs and the drawings they made earlier	Drawings
11	Writing a story	Students write a story to explain how the ramps and sliders activity illustrates what might happen to them if they sat in a wagon and rolled down a slope	Text

skiing down a slope, rolling a rock or a can down a slope, etc.) If the teacher is willing and able to arrange it, he/she might have students go outside and actually experience sitting in a wagon that rolls down an incline. These actions begin at the most iconic end of the visual representation continuum: concrete object/event.

2. After listening to student responses about their experiences and what they might know about this, he/she might follow up with a story, perhaps from a trade book, that illustrates a character riding in a wagon that moves downhill. This action moves students along the continuum from the concrete object/experience to drawings/diagrams, a place on the continuum that is markedly less iconic. The teacher should be clear and careful to help students see how the drawings represent and/or connect to their concrete experience.

3. The teacher then introduces the ramp and slider. The items are set up as if ready to be used. The teacher explains what the ramp is and guides students to connect the sloping ramp to the sloping hill in their concrete experience. The teacher explains that the small steel ball at the top of the ramp represents the student sitting in the wagon.
4. The teacher may augment this by comparing the ramp set-up with photographs of a wagon on a hill or by showing a short video segment of a wagon rolling down a slope.
5. Next, the teacher demonstrates to students how to use the ramp and ball. This would involve placing the ball at the top of the ramp, releasing it, and allowing it to roll down the ramp. When the ball is released, an assisting student should begin timing with a stopwatch. When the ball reaches the bottom of the ramp, the assisting student should stop the stopwatch and report the number of seconds it took for the ball to roll down the ramp. Another student can measure the height of the upper end of the ramp, and the horizontal length of the ramp (see Figure 2). Then, the teacher guides the students in determining the speed of the ball on the ramp. This would involve learning that the speed is the horizontal distance the ball travels divided by the time for the ball to travel that distance. The teacher should again be explicit in helping students connect the ramp to the hill and the ball to the student sitting in the wagon. With respect to the visual representation continuum, the teacher has now moved the students into the symbolic portion of the continuum: formulae.
6. The teacher asks students what they think will happen to the time it takes for the ball to roll down the ramp if the upper end of the ramp is raised higher. After listening to student responses, the teacher then instructs students to set up their own ramps and balls, and do some testing. Students should find the speed of their steel balls on their ramps that are set at the same height as the teacher's, and then change the height of their ramps and retest to determine the speed of their steel balls on the steeper slope. The teacher should be sure

Figure 2. Ramp and Slider set-up

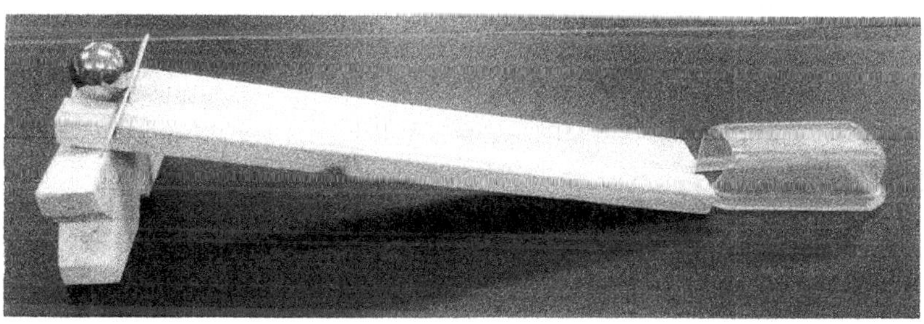

to guide students in relating the steeper ramp to a steeper hill. On the visual representation continuum, this is moving the students more toward the iconic portion of the continuum to models.

7. Now, the teacher has students make a drawing to show what they found out about the balls rolling down the ramp when at its two heights. Students should label their drawings, and – if possible – make them to scale (perhaps life-size). The students may use other things in their drawings to illustrate what is occurring in their drawings, such as arrows pointing in the direction of the moving balls. On the visual representation continuum, the teacher has now shifted the students to drawings/diagrams.
8. Next, the teacher poses the question of what would happen if the ball rolls down the ramp and hits the slider. The students make their responses, and the teacher then instructs them to test it to see what happens. The teacher is clear with students that they must make measurements of the ramp, the ramp heights, the ball speed on each ramp incline, and the distance the slider moves. Students are instructed to record their data in an organized manner in a data table. By doing this portion of the lesson, the teacher has had students move to the symbolic portion of the visual representation continuum to tables.
9. Once students have completed step 8, the teacher has them graph their data with slider distance on one axis and ball speed on the other axis. There will be two graphed lines, one for each of the ramp heights. This shifts students into the graph portion of the visual representation continuum.
10. After graphing of data is completed, the teacher has students compare their graphs to the drawings they made earlier. The teacher guides students in connecting the graph plots with what was illustrated in student drawings. This moves students back toward the iconic portion of the visual representation continuum to drawings.
11. Finally, the teacher has students write a story to explain what they have found out from using the ramps and sliders, putting it in terms of students sitting in wagons and rolling down slopes. This causes students to move toward the far end of the visual representation continuum to symbolic text.

In analyzing what transpired in the *Ramps and Sliders* lesson, the teacher might begin with using a visual representation continuum grid. Such a grid would look something like that in Figure 3. Each of the visual representation types are shown in the top row, and the number of lesson components are enumerated in the left column.

Figure 3. Visual representation grid

Step	Real Objects	Models	Video	Photos	Drawings & Diagrams	Tables	Graphs	Formulae	Text
1									
2									
3									
4									
5									
6									
7									
8									
9									
10									
11									

Once the visual representation continuum grid is constructed, the teacher then plots what occurred in each step of the lesson. Directional arrows can be inserted to show the directionality of the ebb and flow of the lesson components along the continuum. From the above example, that plot would look like that shown in Figure 4.

Examining the visual representation continuum plot in Figure 4, the back-and-forth movement along the continuum through the lesson progression is readily apparent. The overall progression is from the most concrete/iconic portion of the continuum to the most abstract/symbolic portion of the continuum. Between those endpoints, there is a specific pattern that emerges. More iconic representations are followed by representations that are somewhat more symbolic. Symbolic representations are typically followed by iconic representations. When there is a relatively large shift toward the symbolic portion of the continuum, there is a nearly equal

Figure 4. Visual representation progression plot

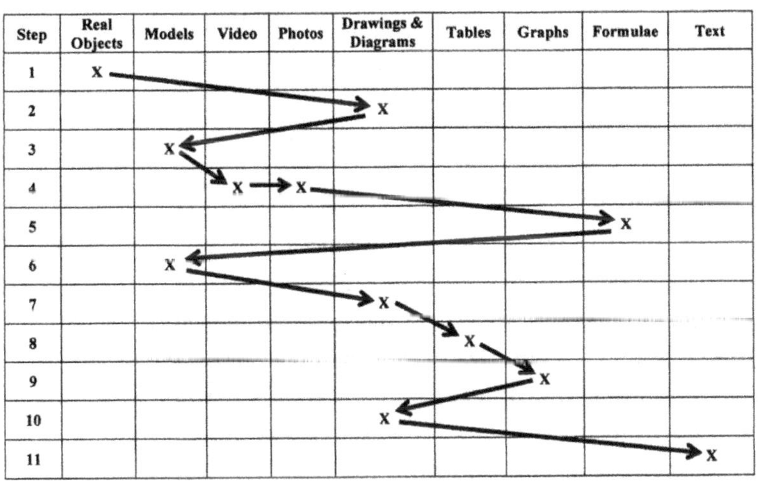

sized (but slightly smaller) shift back toward the iconic portion of the continuum in the next step of the lesson. The pedagogical rationale for this is that as one pushes students into more abstract/symbolic thinking and activity, that must be anchored by more iconic imagery with which students are likely to be either more familiar or can more easily process within their cognitive loads. The three critical aspects of this are that at each visual representation point the teacher insures students . . .

1. Understand how that representation was created, what it means, and what information it conveys,
2. How each representation relates to and connects with each of the other representations, and
3. Have opportunities to generate their own visual representations to convey their conceptual understandings of what occurred during the lesson.

CURRENT CHALLENGES FACING TEACHERS

With respect to challenges facing teachers with integrating components from the disparate content areas that comprise STEAM, there are perhaps five that might rise to the top of one's list. Fortunately, none of them are insurmountable, and the integration of them can successfully be accomplished with a mental reorientation or redirection of the habit of mind a teacher may have. Mostly, it is a way of thinking that can help the teacher get to the integrated endpoint. It requires a breaking away from the old traditional mode of thinking, and involves working to keep an open mind to what before might have seemed as unrelated things.

The top five conundrums a teacher may face as challenges are (1) learning about cognitive loading and what is appropriate for a specific group of students, (2) understanding what visual literacy is and how that is both similar and different from reading literacy, (3) familiarization with the visual representations continuum, (4) integrating seemingly disparate subjects, and perhaps the biggest one: (5) the curriculum and testing conundrum.

1. **Learning About Appropriate Cognitive Loading:** As already mentioned earlier in this chapter, each human being is capable of mentally handling only so much information at one time. That amount of information can vary depending upon the complexity of the information, the experiences a student has had in dealing with information, the developmental level of the student, and the age of the student. There is no one-size-fits-all solution to this issue. Every student a teacher has in class will vary in these characteristics to some

extent. As a general rule, most people can handle between 5 and 7 different pieces of information in their minds and still manage to be effective dealing with them. In a sense, it is like the human mind is a bucket that has a limited amount of space in it, and can hold only so much material. If something new gets dumped into the bucket, then something has to either be removed or spill out. Once that occurs, that spilled information is no longer available to be utilized in the working memory. In science, teachers often feel confronted with the need to convey huge amounts of information to students – and usually in relatively short amounts of time. Researchers in neuroscience and cognitive psychology would be quick to tell us this is not only inefficient, but is not good for the learner. Fortunately, there is an answer to this problem. Neuroscience has shown us that the human brain can handle huge amounts of information if it is in a form that the brain is already hard-wired to process. That form is iconic visual images. The brain can process increased volumes of information with more efficiency and faster speeds if that information is provided or created in iconic forms. The more abstract the information becomes, the more mental processing that needs to occur and, therefore, the slower the processing – and consequently the more limited the available space in the the working memory. Hence, as teachers learn to find, select, create, and/or use more iconic science illustrations and representations with their students, the better the learning will tend to be for them.

2. **Understanding Visual Literacy:** Earlier in this chapter, the issue of visual literacy was discussed. Notably, one's visual literacy typically precedes one's verbal and connected reading literacy. Visual literacy is one's ability to make sense of what he/she sees in visual images. This requires one to observe, analyze, and interpret those visual inputs. Once that is successfully accomplished, then one can progress to articulating what he/she understands from that imagery. In too many instances, teachers do not think very deeply about the visual images to which they expose students. Textbooks are often rife with complex images that confront students as they attempt to read and make sense of the text. Teachers cannot assume their students have an inherent ability to automatically decipher and understand the images they encounter any more than they have the ability to write a narrative before they have learned anything about the alphabet and sentence structure. The problem, then, is how to help students learn to be more visually literate. Again, the answer to this is relatively straightforward, although it might require some additional time. The teacher needs to deconstruct complex images (complex to the student, not necessarily to the teacher) so students can first come to understand the simplified components and structures within the images, and then progressively add onto those images more "layers" of imaging until the entire representation is taught. The

teacher needs to be deliberate in this approach, and provide students opportunities to create their own visual representations in a similar building-up process to eventually reveal their understanding of concepts and information. As with many things in teaching, some extra time invested in the beginning can pay dividends later. The issue with visual literacy is no exception. As students become more familiar with the skills and tools for deconstructing images and analyzing and thereby interpreting them, they will become more efficient in the process.

3. **Becoming Experienced with the Visual Representations Continuum:** A problem teachers may encounter with science illustrations encountered in curriculum and other materials is that those illustrations are often isolated from others that would appear on either side of them within the visual representations continuum. This situation can exacerbate problems with visual literacy. Consequently, the teacher needs to be able to identify where on the continuum a particular type of representation fits, and have a clear sense of the kinds of representations that are both more iconic and symbolic. As shown in the ramps and slider lesson example, it is important to provide adequate scaffolding to guide students through connecting the more abstract and symbolic representations with the concrete iconic representations with which they are likely more familiar. This is not unlike a review of the prior day's lesson before embarking into today's lesson material. The number and extent of movements within a lesson to the more iconic visual representations will vary depending upon how well the students are grasping the conceptual parts of the learning, but should certainly not be overlooked. The teacher needs to know the different types of representations along the continuum, be able to assess with which ones his/her students function well, and determine how far the student can be pushed in dealing with more symbolic (abstract) ones. This is very similar to assessing a student's zone of proximal development (as per Lev Vygotsky), and is good pedagogy.
4. **Integrating Seemingly Disparate Subjects:** Teachers, particularly at the upper elementary level, are often confronted with too little time to teach too many things. The school day is segregated into specific times for reading/language arts, for math, for social studies, and – if time permits – science and art. Specialists from each of those disciplines are quick to point out the importance of their respective disciplines for student learning. But it is the teacher who must ultimately decide how much time is devoted to each subject. As will be touched on in the next paragraph, when the testing is high-stakes, most attention and time will be devoted to the subjects tested. Although one can argue this is not necessarily pedagogically sound practice, it is certainly seen by teachers as a necessary survival strategy. So the problem becomes one

of how to actually cover so many disciplinary subjects within available time structures. A quick answer is via integration of subjects. As shown in the ramps and sliders lesson example, literacy (writing), mathematics, science, and art can all coexist in a smoothly choreographed lesson that flows seamlessly for students. If the mathematics topic needing to be covered in a certain week is graphing, then the teacher can select a science activity in which graphing is included. If reading or writing brief narratives about experiences or observations is the topic of the week, then the teacher can select a science activity in which students write their interpretive stories of what has occurred through the science lesson. Along with their writing, students can be taught and asked to include drawings of what they observe and/or predict, to construct data tables, and more. The overall result is a lesson that clearly integrates STEAM components, that has more meaning and realism to students (i.e. is more like what they would encounter in the real world), and successfully addresses multiple subject content within available time parameters.

5. **Curriculum and Testing Conundrum:** Perhaps the biggest problem perceived by teachers is how to cover the content that will be on the test – specifically state (and national) mandated tests. All too often, it is math and reading first and only. Science and art as afterthoughts. In more recent versions of some states' tests, science has re-emerged as a testable subject. As a whole, the problem confronting teachers is how to match the curriculum they must teach to the mandated tests. Integrating the STEAM content areas within lessons can help address this conundrum. The essential key is to have students actively engaged in each of the disciplinary components during a lesson. It would, for example, be unacceptable to have students simply read about science (to fulfill reading requirements) rather than actually engage in doing science. Likewise, it should not be acceptable to ignore the art components that complement and strengthen mathematics and science. The more connectedness students can see between the disciplines, the more deeply rooted the learning becomes, and the better the learning. When confronted with problems to solve (such as those on the tests – not to mention in real life), students will be able to draw from a larger framework of understanding to address and solve the problems. Such multiplicity of learning can only improve students' chances for success.

SOLUTIONS AND RECOMMENDATIONS

In looking more broadly at the application of art in science instruction, one must first agree that most of the visual representations in the continuum are forms of art. With respect to models, teachers may have their students make clay sculptures of mountains or of tectonic faults, or models of cells or of the solar system. The sophistication of the models can vary a great deal, but they nevertheless remain models. With respect to videos, teachers can have their students create videos using simple digital cameras or even smart phones. In the *Ramps and Sliders* lesson, the teacher could have students make a video of other students riding downhill in a wagon. Similarly, the teacher could have students take photographs of such an event. Alternatively, the teacher might locate existing videos or photos on the internet to incorporate into the lesson. Videography (including photography) is a form of art, and is something students can learn to do relatively well with limited instruction. A classic art form is drawing. As with the other visual representations, drawings show a great variety in quality, whether done by students or teachers or professional artists. Drawings can be clear, detailed, or colorful; or they can be fuzzy, only show specific elements of an object, or be in black and grays on white. One might even argue that graphs and even text require artistic touches.

Whatever visual representations the teacher chooses to use, he/she should be cognizant that as one moves along the continuum, there are things to be gained as well as things to be lost. For example, models convey many of the characteristics of the actual object, including shape, proportion (if done well), coloration, etc., but may be awkward to transport. What elementary student doesn't get excited to see that volcano model "erupt" and spew out lava made from foaming baking soda and vinegar? Videos or photographs of actual objects clearly show much of the detail of the actual object or of the model, but lack the three-dimensionality and heft (weight, mass, feel) of the object. Again, students can become mesmerized by videos of volcanoes erupting. But photographs are often much more portable than the actual objects. This is particularly true today with digital photos that can be stored on smartphones or tablets. Drawings and diagrams allow us to highlight specific aspects of an object while decreasing focus on less relevant aspects, and can also be used to bring clarity to what students might see in the actual object. Those teachers who have had their students dissect frogs know that having a diagram of the internal anatomy of the frog is much more helpful to students than just relying on the mass of dissected tissue and flesh most students see in the dissecting pan. Diagrams and drawings can also be useful in showing flow and movement or directionality that we would otherwise find difficult to observe, such as what we see in diagrams of the water cycle that include directional flow arrows. Diagrams and drawings can be very portable, but they lack much of the detail and attributional quality of the real objects or models. Continuing further, we later learn that symbolic representations such as formulae also tell us much about the real objects and models, but they are more difficult to interpret and understand. Yet they can be manipulated in ways that actual objects and models cannot. Hence, the teacher needs to carefully consider what visual representations he/she needs to include in instruction, clearly having a particular purpose in mind for each. In other words, the teacher must be very deliberate in his/her selection, creation, and/or use of visual representations with students – and be deliberate in planning opportunities for students to learn how to use them and even generate their own.

Indeed, art has a significant role to play in science. The art does not necessarily have to be professional grade or perfect. In fact, much art associated with science classrooms is anything but. Even so, those artworks have deep meanings for students and help them connect their learning in ways difficult or impossible without them. Used properly, art can be the key component in making science come alive and move for students. It puts the movement in STEAM instruction.

REFERENCES

Bodzin, A. M., & Cirucci, L. (2009). A land-use-planning simulation using Google Earth. *Science Scope*, *32*(7), 30–38.

Brantc, E. W., Olander, M. H., & Nystrom, M. (2013). Exploring the impact of contrasting cases in text and picture processing. *Journal of Visual Literacy*, *32*(2), 15–38. doi:10.1080/23796529.2013.11674708

Brumberger, E. (2010). Visual literacy and the digital native: An examination of the millennial learner. *Journal of Visual Literacy*, *39*(1), 19–47.

Carlsen, M. (2009). Reasoning with paper and pencil: The role of inscriptions in student learning of geometric series. *Mathematics Education Research Journal*, *21*(1), 54–84. doi:10.1007/BF03217538

Cognition and Development. (n.d.). *Jerome Bruner*. Retrieved from http://psychology4a.com/develop8.htm

Coleman, J. (2010). Elementary teachers instructional practices involving graphical representations. *Journal of Visual Literacy*, *29*(2), 198–222. doi:10.1080/23796529.2010.11674681

Coleman, J.M., McTigue, E.M., & Smolkin, L.B. (2011). Elementary teachers' use of graphical representations in science teaching. *Journal of Science Teacher Education, 22*(7), 613-644. doi:.10.1007/s10972-010-9204-1

Gabel, D. (1999). Improving teaching and learning through chemistry education research: A look to the future. *Journal of Chemical Education*, *76*(4), 548. doi:10.1021/ed076p548

Gilbert, K. J. (2005). Visualization: A metacognitive skill in science and science education. *Visualization in Science Education, 1*, 9–27. doi:10.1007/1-4020-3613-2_2

Karplus, R. (1964). The science curriculum improvement study. *Journal of Research in Science Teaching, 2*(4), 293–303. doi:10.1002/tea.3660020406

Kozma, R. B. (2003). Technology and classroom practices: An international study. *Journal of Research on Technology in Education, 36*(1), 1–5. doi:10.1080/15391523.2003.10782399

Lesh, R. A., & Doerr, H. M. (Eds.). (2003). *Beyond constructivism: Models and modeling perspectives on mathematics problem solving, learning, and teaching*. Mahwah, NJ: Lawrence Erlbaum Associates.

National Research Council. (2012). *A framework for k-12 science education: Practices, crosscutting concepts, and core ideas*. Washington, DC: National Academy of Sciences. Retrieved from http://www.nap.edu/catalog.php?record_id=13165

Olson, J. K. (2013). Education rather than training: Cognitive pluralism and preservice science teacher education. In K. D. Finson & J. E. Pedersen (Eds.), *Visual data and their use in science education* (pp. 95–124). Charlotte, NC: Information Age Publishers.

Ozcelik, A. T., & McDonald, S. P. (2013). Preservice science teachers uses of inscriptions in science teaching. *Journal of Science Teacher Education, 24*(7), 1103–1132. doi:10.1007/s10972-013-9352-1

Pozzer-Ardenghi, L., & Roth, W.-M. (2005). Making sense of photographs. *Science Education, 89*(2), 219–241. doi:10.1002/sce.20045

Saunders, A. C. (1994). Graphics and how they communicate. In D. M. Moore & F. M. Dwyer (Eds.), *Visual literacy: A spectrum of visual learning* (pp. 183–192). Englewood Cliffs, NJ: Educational Technology Publications, Inc.

Sibanda, J., & Sibanda, L. (2013). Visual literacy development through the mediation of grade 4 English textbooks. *Journal of Visual Literacy, 32*(2), 39–66. doi:10.1080/23796529.2013.11674709

Simply Psychology. (n.d.). *Jerome Bruner – Learning Theory in Education*. Retrieved fromhttp://www.simplypsychology.org/bruner.html

Valanides, N., Efthymiou, I., & Angeli, C. (2013). Interplay of internal and external representations: Students drawings and textual explanations about shadow phenomena. *Journal of Visual Literacy, 32*(2), 67–84. doi:10.1080/23796529.2013.11674710

Vavra, K. L., Janjic-Watrick, V., Loerke, K., Phillips, L. M., Norris, S. P., & Macnab, J. (2011). Visualization in science education. *Alberta Science International Journal*, *4*(1), 22–30.

Wu, H.-K., & Krajcik, J. S. (2006). Exploring middle school students use of inscriptions in project-based science classrooms. *Science Education*, *90*(5), 852–873. doi:10.1002/sce.20154

Yeh, H.-T., & Lohr, L. (2010). Towards evidence of visual literacy: Assessing preservice teachers perceptions of instructional visuals. *Journal of Visual Literacy*, *29*(2), 183–197. doi:10.1080/23796529.2010.11674680

KEY TERMS AND DEFINITIONS

Cognitive Load: The amount of demand that is placed on the brain to process information.

Conceptual Understanding: The extent to which the learner can incorporate new learning at the conceptual level in his/her mental schema and then can draw from and utilize that information to solve problems.

Iconic: Visual representations that are primarily pictorial in form, such as actual objects, photographs, videos, and drawings.

Manipulative Memory or Working Load: The amount of information the human brain can manage to hold and manipulate at a single time, usually limited to 5-7 pieces of information.

Symbolic: Visual representations that involve abstract mental processing to understand them, such as data tables, graphs, formulae, and alpha-numeric text.

Visual Continuum: A spectrum of types of visual representations that begins with the most concrete at one end and ends with the most abstract at the other end.

Visual Literacy: One's knowledge and understanding about a set of visual imagery and the ability to make sense of it and articulate meanings from it.

Visual Representations: A variety of ways in which visual information is exhibited, ranging from very concrete to abstract constructions. Note: Some recent developments in the field of visual data bring researchers to differentiate between visual representations and visual inscriptions. Representations are now being seen as the mental ideas one holds, whereas inscriptions are the external ways those representations can be exhibited (such as through drawings or graphs). In this chapter, the term "visual representation" has been used so as not to require an in-depth discussion on the psychology regarding the differentiation between representations and inscriptions.

Chapter 10
Tower Design as a STEAM Project

Judith A. Bazler
Monmouth University, USA

EXECUTIVE SUMMARY

The next generation science standards promote the teaching of engineering skills including the designing, testing and building of models. Tower building can provide a real world experience that not only provides the students with physics and mathematics through motion and stability but the explanation of the use of models and the engineering practice of design, redesign and testing of these models. Tia Pilskow (2014) used the project of building a tower with her middle school students in order to provide a cooperative team long term project. She focused first on the design using background information on existing towers. She required each team to design their tower first using graph paper and scale. This process stressed the need for Art and Mathematics in the STEAM project. The science, technology and engineering also played a major part in the design. The case included in this article expands her process by including a cost analysis attempting to promote real world engineering.

INTRODUCTION

Designing and building a tower project provides students with all aspects of the STEAM process. The physics of force and motion combine with the affects of earth movement, wind, and other geological factors; plus, the mathematical calculations affect the design and structural integrity of the tower. The history of towers,

DOI: 10.4018/978-1-5225-2334-5.ch010

Copyright ©2017, IGI Global. Copying or distributing in print or electronic forms without written permission of IGI Global is prohibited.

including the newest and largest structures, may motivate students to proceed on a fascinating journey that perhaps trends back to the pyramids then fast forwards to the Burj Khalifa. This journey also provides the student with historical and present engineering role models and information concerning their interests and their passion for this art form.

The approach of researching past and present structures complete with the research of the scientific and mathematical foundations before the design and before the actual activity is the basis for this lesson's STEAM approach to learning. All disciplines combine to develop the problem and to formalize a new solution. The building and testing and redesigning not only strengthen the process but stress the importance of the research process.

This approach varies from past practice by embracing the research process and the STEAM philosophy. This chapter provides the practitioner with a sampling of the history of a number of towers with suggestions for further research. The history intertwined with science, mathematics, and engineering provides the foundation needed to proceed to design and redesign. The actual activity leads the student into the building and testing which automatically forces the student back to the design. Learner Outcomes complete with possible assessments for this project are:

The students will:

- Practice structural design, model development and engineering, assessed by their designs, their models, and their redesigns.
- Focus on motion and stability, assessed by observation questions and applied design.
- Use mathematics and computational thinking, assessed by observation questions and applied design.
- Explain the use of models in writing, assessed by observation questions.
- Research towers globally, assessed by observation questions.
- Interact in teams, assessed by teacher informally and by observation questions.

Standards

Next Generation Science Standards

Dimension I: Scientific and Engineering Practices (1,2,3,4,5,6,7,8) including asking questions and defining problems, developing and using models, planning and carrying out investigations, analyzing and interpreting data, using mathematics and computational thinking, constructing explanations and designing solutions, engaging in argument from evidence, and obtaining, evaluating, and communicating information.

Dimension II: Crosscutting Concepts (2, 3, 4, 6, 7) including cause and effect: mechanism and explanation, scale, proportion, and quantity, systems and system models, energy and matter: flows, cycles and conservation, structure and function, and stability and change.

Dimension III: Disciplinary Core Ideas/Physical Science (PS1, PS2, PS3) Matter and its interactions, motion and stability: forces and interactions, Energy / Earth and Space Science (ESS2, ESS3) the Earth's systems, Earth and human activity/Engineering, Technology and Application (ETS1, ETS2) Engineering design, links among engineering, technology, science, and society.

Content Information

The Physics of Building a Tower

The foundation of a tower has a number of forces acting upon it. There are the compressive/download forces caused by the weight of the tower and there are the forces of the wind causing an uplift effect. Three major items also need to be addressed including the load, the wind speed, and seismic forces.

From a mathematical standpoint, the equation for the tower is shown in Figure 1 where *f(x)* is the half-width of the tower at height *x*, f_0 is the half-width of the tower at the ground and *w(x)* is the maximum wind pressure the tower can withstand at a height *x*.

(This is a root of the quadratic equation describing a parabola resting against the right side of the tower, so the curve of the tower's right side is described by the negative arm of the parabola).

For example, the fastest winds recorded at the tower reached a speed of 214 km/h in 1999 and would have produced pressures of just 2.28 kN/m^2.

Figure 2 provides a graphic representation of the mathematics involved in towers.

The force of the wind (dF) produces a torque around the bottom left corner of the tower which is countered by the force of the tower's weight (dW).

See examples of towers in Appendix.

Figure 1. Mathematical equation for the tower

$$f(x) = xw(x) - \sqrt{x^2 w^2(x) + f_0^2}$$

Figure 2. Graphic representation of mathematics involved in towers

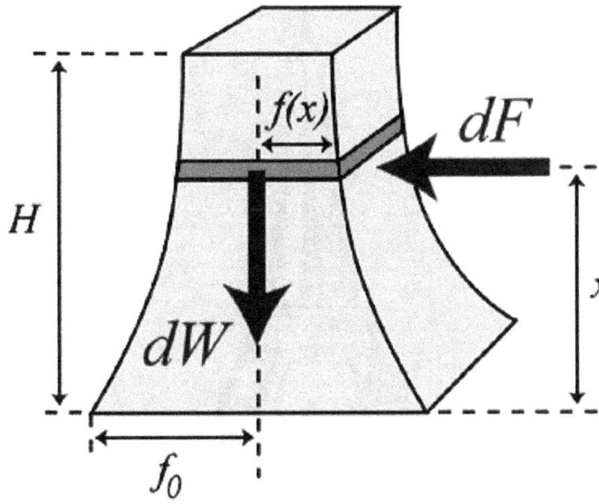

CASE DESCRIPTION

Lesson Plan Concerns

This is a STEAM project; therefore, this is not a one day activity. The time for part A and part B is necessary for the students to reflect on how much time is given to the planning and designing components of building a tower. This project may take months to complete. The materials are easy to find and not costly.

The most difficult portion of this project is NOT jumping to the building but instead spending time on the design. A great show on LIFEtime television is Project Runway and Project Runway Kids. This show creates a challenge, perhaps to create an outfit for women/girls out of material from a hardware store. The participants first sketch an example of what they might create. Then most of the participants get their materials but first create a pattern/model. Then and only then do they construct the garment. The viewer senses that both the pattern and the actual finished garment are a model for future distribution because in both phases the designer faces many challenges that they must resolve. In the tower project, building the design may present additional challenges that the designer did not focus on during the design process but that need to be resolved in order to proceed with the design. The research prior to the design should provide the incentive but the teacher needs to elaborate on the structural science and mathematics, including, but not limited to, force and motion. Even though there is not a section for redesign, those challenges that were not initially identified will arise and redesign becomes implicit in the building process.

Figure 3. Eiffel Tower

Project Plan

Tower Design and Building Challenge

How do engineers design and build towers? Engineers need to develop many solutions by asking questions and defining constraints. Designs are then reviewed for cost, improvements, time, and function. Models are then developed. The engineering design process includes: defining the problem, gathering pertinent information, generating multiple solutions, analyzing and selecting a solution, and testing and implementing the solution.

Students will first research other existing towers and then design a tower and generate a test to determine its stability and strength. Students will then build the designed towers out of straws, paper clips, and tape. Students will be assessed on the height of the tower, its stability, and its cost.

Materials

- 50 straight plastic drinking straws.
- Scissors.
- Two feet of masking tape.
- 30 paper clips.
- One sheet of graph paper.

- Two sheets of lined paper.
- One sharpened pencil.
- Metric ruler.
- Can of soda.
- Cost of Materials Data Sheet.

Safety

Be careful using scissors and the sharpened pencils, both can cause damage if used improperly. Do not drink or open the soda.

Procedures

Designing-Part A:

1. Keep track of time spent on Parts A and B!
2. Read all materials and explore on the internet the tallest buildings in the world and how they were designed and constructed. Look up the Burj Khalifa, the Eiffel Tower, and La Torre di Pisa.
3. Graph the design that you will try to make in class, make certain that your graph uses metric measurements.
4. Determine the height of the tower that you will build.
5. Design a test for the tower that is stable enough to hold a soda can.

Building-Part B:

1. Gather all the materials that you will use.
2. Build the tower using your design graph.
3. Measure the height of your tower, record the height on the data sheet.
4. Test the tower using your original test design.
5. Determine the cost of building your tower using the Cost of Materials Data Sheet.
6. Answer the Questions on Observations list below on a separate piece of paper.
7. Clean up the area.

Observations

1. What was the height of your tower?
2. What was the cost of building your tower? How did design compare to cost?
3. How much time was spent on both the design and the building of the tower?

4. If you had a chance to do this project again, what would your team have done differently?
5. Do you think that this activity was more rewarding to do as a team, or would you have preferred to work alone on it? Why?
6. If you could have used one additional material (glue, wooden sticks –as examples) which would you choose and why?
7. Do you think that once a building is designed and approved for construction that many aspects are changed during the building process? Why or why not? How does the design consider any imposed state or federal regulations, as well as public safety, health, and/or intellectual property?
8. Describe any constraints the design may pose relative to health or safety hazards (noise, radiation, allergies, etc.).
9. How long do you think it will take before a building is constructed that surpasses the height of the Burj Khalifa? Where do you think it will be built? Why?
10. What environmental constraints exist for engineers to take into account when designing a building?
11. Pick one of the towers you researched in Designing-Part A and write a short one page report on the tower. Include in your report the location, the height and dimensions, any constraints in building the tower, the cost of building the tower, and important information about the tower.

Table 1. Cost of materials data sheet

Materials/Labor	Cost/Unit/Item	How Many Used	Total Cost
50 straws	$1.50/box 100		
Scissors	$1.89 each		
2 feet masking tape	$.06/meter (1m=3.28ft)		
30 paper clips	$0.01/each		
One sheet of graph paper	$1.34/ream of 500		
Two sheets of lined paper	$1.34/ream of 500		
One sharpened pencil	$2.41/pack of 12		
Metric ruler	$19.34/24 rulers		
One 12 ounce soda	$3.56/6 cans		
Civil engineer each one	$40.00/hour/engineer		
Total			

CURRENT CHALLENGES FACING TEACHERS

Integrating the design with a hands on component, including the building of a tower with design, enables students to focus first on the design/art of a tower and on the science involved in the design, then on the practicality of their design. Without the design, students rarely look at the serious problems with force on any structure and on the aesthetics of their product. Included in the activity is an economic component, again looking at the economic feasibility of the design. Student responses to this case included both the positive motivation factor and also the understanding of both the science, mathematics, engineering and art design involved in the process of towers. The time factor for any project is greater than in the traditional forty-five minute or double class activity. Usually any STEAM project will be spread over a longer time schedule.

SOLUTIONS AND RECOMMENDATIONS

This project can be completed as a competition as evidenced in a number of Spaghetti Tower challenges on YouTube. One could substitute spaghetti if desired but the project building is more efficient using straws due to the spaghetti fragility. Some of the challenges use both spaghetti and marshmallows for structural integrity instead of tape or glue. Both the marshmallows and glue seem to be more messy. One could choose to change the goal of the project from being a tower that is tall and structurally sound to just tall. Examples of tall towers that were built for this project are in Appendix.

Finally, this project lends itself to group teaming. There seems to be no need to assign tasks to each member due to the internal motivating factor implicit in the process itself. Overtime, the motivation of achieving success with the structural strength and sturdiness seems to be incentive enough. Perhaps that alone is the challenge.

REFERENCES

Ashley, S. (2001). When the twin towers fell. *Scientific American*. Retrieved from https://www.scientificamerican.com/article/when-the-twin-towers-fell/

Burj Khalifa. (2011). *Civil Engg. Seminar*. Retrieved from http://civilenggseminar.blogspot.com/2011/09/burj-khalifa.html

Case study: The Leaning Tower of Pisa. (n.d.). *Madrid Engineering*. Retrieved from http://madridengineering.com/case-study-the-leaning-tower-of-pisa/

Craven, J. (n.d.). The World Trade Center twin towers, 1973-2001. *About.com*. Retrieved from http://architecture.about.com/od/worldtradecenter/ss/worldtrade.htm#step1

Eiffel Tower. (2006). *Engineering.com*. Retrieved from http://www.engineering.com/Library/ArticlesPage/tabid/85/ArticleID/62/Eiffel-Tower.aspx

Gallant, J. (2002). The shape of the Eiffel Tower. *American Journal of Physics*, *70*(2), 160–162. doi:10.1119/1.1417530

Leaning Tower of Pisa. (n.d.) In *Wikipedia*. Retrieved from https://en.wikipedia.org/wiki/Leaning_Tower_of_Pisa

Pliskow, T. (2008). *Reaching new heights in middle school science: Straw tower competition builds*. Academic Press.

Pratap, T. (2015, October 29). *What were the engineering challenges in building the Burj Kahalifa?* [Msg 3]. Message posted to https://www.quora.com/What-were-the-engineering-challenges-in-building-the-Burj-Khalifa

Willis Tower. (n.d.). Retrieved from http://www.willistower.com

KEY TERMS AND DEFINITIONS

Engineering Design: The process of devising a system, component, or process to meet desired needs. It is a decision-making process in which the basic science and mathematics and engineering sciences are applied to convert resources optimally to meet a stated objective. Among the fundamental elements of the design process are the establishment of objectives and criteria, synthesis, analysis, construction, testing and evaluation. The engineering design component of a curriculum must include most of the following features: development of student creativity, use of open-ended problems, development and use of modern design theory and methodology, formulation of design problem statements and specification, consideration of alternative solutions, feasibility considerations, production processes, concurrent engineering design, and detailed system description. Further it is essential to include a variety of realistic constraints, such as economic factors, safety, reliability, aesthetics, ethics and social impact.

Force: A quantitative description of the interaction between two physical bodies. Force is proportional to acceleration.

Load: The forces to which a structure is subjected due to superposed weight or to wind pressure on the vertical surfaces; *broadly*: the forces to which a given object is subjected.

Torque: A force that produces or tends to produce rotation or torsion <*an automobile engine delivers torque to the drive shaft*>; *also*: a measure of the effectiveness of such a force that consists of the product of the force and the perpendicular distance from the line of action of the force to the axis of rotation.

Tower: A building or structure typically higher than its diameter and high relative to its surroundings that may stand apart or be attached to a larger structure.

APPENDIX

The Eiffel Tower

For an image of the Eiffel Tower go to HTTP://.wpclipart.com/phpimageeditor / editimagesworkwith/eiffel_tower_picture_1482095012_71_71_221_206.jpg.

Eiffel's tower was completed in a matter of months with a small labor force. His results started a revolution in civil engineering and architectural design. With the completion of the tower, Eiffel earned the nickname, 'magician of iron.' Over time the tower faced demolition but then became an important radio and television transmission tower.

For an image of the Eiffel Tower go to HTTP://.wpclipart.com/phpimageeditor /editimagesworkwith/eiffel_tower_detailed_1482095133_71_71_221_206.jpg.

The tower's dimensions are a height of 1069 feet (320.75 meters), which is about the equivalent of a 105-story building. The public has access to heights of 189 feet (57.63 meters), or 19 stories; 380 feet (115.73 meters), or 38 stories; and 896 feet (273 meters), or 89 stories.

Other statistics include:

- 2.5 million rivets.
- 300 steel workers, and 2 years (1887-1889) to construct it.
- Sway of at most 12 cm in high winds.
- Height varies up to 15 cm depending on temperature.
- 15,000 iron pieces (excluding rivets).
- 40 tons of paint.
- 1652 steps to the top.

Eiffel, Alexandre Gustave 1832-1923, French engineer: A noted constructor of bridges and viaducts, he also designed the Eiffel Tower and the internal structure of the Statue of Liberty. He was initially charged with corruption in the 1888 scandal of Ferdinand de Lesseps's failed Panama Canal project. Lesseps (1805-1894), the successful promoter and construction supervisor for the Suez Canal (1859-69), later assumed presidency of the French company building the Panama Canal starting in 1881. Eiffel was hired to provide engineering services for the project. Seven year later the project was forced into bankruptcy amid charges of corruption. Lesseps was sentenced to prison for misappropriation of funds, but the sentence was not carried out. Eiffel was cleared of all wrongdoing by a French appeals court in 1893. Nonetheless, he withdrew from commercial life and spent the rest of his years studying aerodynamics" (Eiffel Tower, 2006).

The Burj Khalifa

Burj Khalifa is a skyscraper in Dubai, United Arab Emirates, and is currently the tallest structure in the world, at 828 m (2,717 ft). Construction began on 21 September 2004, with the exterior of the structure completed on 1 October 2009. The building officially opened on 4 January 2010.

For an image of the Burg Khalifa go to HTTPS://en.m.wikipedia.org/wiki/Burj_Khalifa

The tower's architecture and engineering were performed by Skidmore, Owings, and Merrill of Chicago, with Adrian Smith as chief architect and Bill Baker as chief structural engineer. The primary contractor was Samsung C&T of South Korea. The total cost for the project was about US $1.5 billion.

Burj Khalifa has more than 160 stories. It is the tallest building in the world. It has a spiraling "Y" shape (see Figure 4).

The top section of the tower consists of a structural steel spire utilizing a diagonally braced lateral system. The structural steel spire was designed for gravity, wind, seismic, and fatigue in accordance with the requirements of AISC Load and Resistance Factor Design Specification for Structural Steel buildings. The exterior exposed steel is protected with a flame-applied aluminum finish.

Willis (Sears) Tower in Chicago

"One of the tallest buildings in the world and the tallest building in America. In 1969, Sears Roebuck and Company was the largest retailer in the world, with about 350,000 employees. Deciding it needed a central office space for its many

Figure 4. Typical floor plan

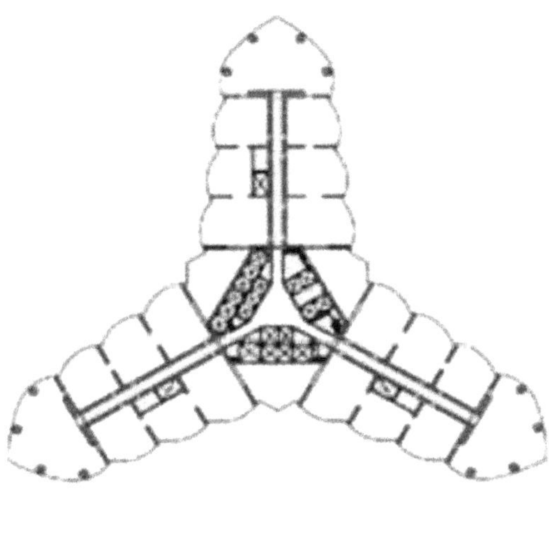

employees, the company hired architects Skidmore, Owings, and Merrill to design what would become one of the largest office buildings in the world. After breaking ground in 1970, it took three years to complete and used enough concrete to make an eight-lane, five-mile-long highway. In 1988 Sears moved out and it was renamed Willis Tower in 2009.

Designed by architectural firm Skidmore, Owings, & Merrill.

1,450 feet high (443 meters); 1,730 feet high (520 meters) including twin antennae.

Eighth-tallest building in the world; tallest in Western Hemisphere.

Cost more than $175 million to build" (Willis Tower, n.d.).

For an image of the Willis Tower go to HTTP://.wpclipart.com/phpimageeditor/editimagesworkwith/Wilis_Tower_formerly_Sears_1482094735_71_71_221_206.jpg.

Pisa in Italy

For an image of the Leaning Tower of Pisa go to HTTP://.wpclipart.com/phpimage-editor/editimagesworkwith/Leaning_Tower_of_Pisa_1482094532_71_71_221_206.jpg.

"The *Leaning Tower of Pisa* is a freestanding tower of the city of Pisa. It is known for its unintended tilt. The tower's tilt began during construction, caused by an inadequate foundation on ground too soft on one side to properly support the structure's weight. The tilt increased in the decades before the structure was completed, and gradually increased until the structure was stabilized (and the tilt partially corrected) by efforts in the late 20th and early 21st centuries. The height of the tower is 55.86 metres (183.27 feet) from the ground on the low side and 56.67 metres (185.93 feet) on the high side.

There has been controversy about the real identity of the architect of the Leaning Tower of Pisa. For many years, the design was attributed to Guglielmo and Bonanno Pisano. However, recent studies seem to indicate Diotisalvi as the original architect due to the time of construction and affinity with other Diotisalvi work.

Galileo Galilei is said to have dropped two cannonballs of different masses from the tower to demonstrate that their speed of descent was independent of their mass. During WWII, the Germans were using the tower as an observation post.

Numerous efforts have been made to restore the tower to a vertical orientation or at least keep it from falling over. Most of these efforts failed; some worsened the tilt. On February 27, 1964, the government of Italy requested aid in preventing the tower from toppling.

On January 7, 1990, the tower was closed to the public After a decade of corrective reconstruction and stabilization efforts, the tower was reopened to the public on December 15, 2001, and was declared stable for at least another 300 years. In May

2008, engineers announced that the Tower had been stabilized" (Leaning Tower of Pisa, n.d.).

Twin Towers in NYC

"Designed by American architect Minoru Yamasaki (1912-1986), the original World Trade Center consisted of two 110-story buildings (known as the "Twin Towers") and five smaller buildings. The North Tower (1 WTC) was finished in 1970 and the South Tower (2 WTC) was completed in 1972. Tower One was 1,368 feet (414 meters) tall, 110 stories high. Tower Two was 1,362 feet (412 meters) tall, 110 stories high. Both Towers weighed about 1,500,000 tons together. On August 7, 1974 Philippe Petit used a bow and arrow to assemble a steel cable between the two towers and then he walked across the tightrope. Other stunts included parachuting from the top and climbing from the bottom of the Towers" (Craven, n.d.).

For an image of the Twin Towers in NYC. HTTP://www.wpclipart.com/phpimageeditor/editimagesworkwith/Twin_Towers_c_2000_1482094065_71_71_221_206.jpg

"After first describing the highly redundant structural system that kept the 110-story twin towers standing for decades despite hurricane-force winds and a terrorist truck bomb, the engineers then delineated how that system was breached and finally overcome on that fateful day when America was attacked. The main culprits in bringing the famously lofty buildings down, they concluded, were the two intensely hot infernos that erupted when tens of thousands of gallons of aviation fuel spilled from the doomed airliners. Once high temperatures weakened the towers' supporting steel structures, it was only a matter of time until the mass of the stories above initiated a rapid-sequence "pancaking" phenomena in which floor after floor was instantly crushed and then sent into near free fall to the ground below. Significantly, the panel stated that any mitigating reinforcements and redundancies added to these buildings could have only delayed the inevitable failure, though they would have bought more time for the evacuation of the occupants. No existing or foreseeable economically viable skyscraper structure, they agreed, could have withstood this kind of cruel onslaught. Clearly, prevention is the best defense against this kind of assault." (Ashley, 2001)

Chapter 11
Using Air to Move Paper Airplanes and Balloon Rockets:
The Great Race

Letitia Graybill
Monmouth University, USA

EXECUTIVE SUMMARY

This is a case in which students build paper airplanes from templates provided by the instructor as well as those which they design themselves. They extend their ideas on flight by using the principles developed in the paper airplane race to power simulated rockets made out of balloons. They consider such variables as materials, mass and design to see which combination of material design and mass are most effective in constructing an airplane or a rocket that flies the fastest and the furthest in a competition. Contestants are rated on team consideration of variables needing to be controlled in order to have a fair assessment of the designs. When the designs are agreed upon and constructed, a race is conducted. The ideas developed in the paper airplane competition are then used to design a rocket carrying a paper airplane capable of flying across the classroom in the fastest time with the most direct route. This is a simulation of the space shuttle flights. The parameters of the races are developed by the participants.

DOI: 10.4018/978-1-5225-2334-5.ch011

Copyright ©2017, IGI Global. Copying or distributing in print or electronic forms without written permission of IGI Global is prohibited.

LITERATURE REVIEW

Humans have yearned to fly like birds for eons. Many designs were created through the ages that had some disastrous results. But because of the desire, inventive attitudes, and almost fanatic devotion to the creation of new knowledge, the search went on and on and on. It took until 1903 in Kitty Hawk, North Carolina (Wikipedia) for Wilbur and Orville Wright to produce the first sustained flight. Engineering took over after that and resulted in a revolution in travel by air. The history of the saga has been well documented in the literature. We have learned through experiment and engineering principles to develop crafts that can fly supersonically and transport people and material around the earth in increasingly shorter times. We have revolutionized travel to the point where travel beyond the Earth is within possibility and may even be accessible for the average human. Those who were alive in 1969 to see Neil Armstrong walk on the Moon were enthralled by the vision. What an exciting moment that was! This initiated a series of flights in which more and more powerful rockets were used to launch these spacecraft beyond the Earth's gravitational field. More extensive interplanetary flight has already been accomplished by non-manned space flights. Two Voyager crafts were launched in September of 1977. These Voyagers spent 11 years exploring Jupiter, Saturn, Uranus, and Neptune before escaping the gravitational pull of the sun in 2012 to begin interstellar explorations (NASA, n.d.). The Voyager space crafts are the third and fourth human creations to fly beyond all the planets in our solar system. They were preceded by Pioneers 10 and 11 in escaping the gravitational pull of the sun. This was accomplished in 1998. Voyager 1 passed Pioneer 10 to become, as far as we know, the most distant human-made object in space. As of September 2013, Voyager 1 was at a distance of 18.7 billion kilometers from the Sun while Voyager 2 was at a distance of 15.3 billion kilometers. All of this has been achieved in a mere one hundred and eleven years. Imagine what is possible in the future. This review of the literature is sufficient for a basic understanding of a historical representation of the mechanics of flight. There are many other books that can give a deeper understanding. *The Completer History* by R. G. Grant is a well-received source that gives additional, technical detail. This could serve as a reference for teachers.

What a wonderful history of flight we can discuss in our classrooms! Students can become excited by the prospect of future developments. How many of them might become space travelers of the future? How many of them might become part of interplanetary missions that are now being planned? If teachers can "touch the future," as the first teacher in space, Christa McAuliffe, said, (McAuliffe, n.d.) how many science teachers can actually affect the future of spaceflight through the students that they teach? It is an awesome responsibility that most teachers, especially

Figure 1. Titan Rocket with Voyager spacecraft mounted on it

science teachers, should gladly accept. It could be inspiring to have students develop their own ideas of designing for flight by using simple materials that are inexpensive and yet effective in considering questions of design, speed, and flight.

CONTENT INFORMATION

The lessons included in this case are interdisciplinary. Cross curriculum concepts in mathematics, science, social studies, and language arts are integrated with standards in technology and design. The Core Curriculum Mathematics Standards, the Next Generation Science Standards, and the latest standards in technology and social studies education are supported by the activities described within this paper. These standards include:

1. Next Generation Science Standards http://www.state.nj.us/education/aps/cccs/science/
    a. Standard 1 MS PS 2-2: Plan an investigation to provide evidence that the change in an object's motion depends on the sum of the forces acting on the object and the mass of the object. Designers will take into account Newton's Laws e.g. balanced and unbalanced forces and consequent changes in motion. They must also consider specification of units and changes in variables.
    b. Standard 2 MS-PS2-4: Construct and present arguments to support the claims that gravitational forces are attractive and depend on the masses of the interacting object.

2. New Jersey Student Learning Standards Mathematics http://www.state.nj.us/education/aps/cccs/math/
    a. Organize, represent, and interpret data with up to three categories; ask and answer questions about the total number of data points, how many in each category, and how many more or less are in one category than in another.
    b. This standard is found in all grades from Grade 1 through high school.
3. New Jersey English Language Arts Standards http://www.state.nj.us/education/aps/cccs/lal/
    a. Reading, writing, and speaking grounded in evidence from texts, both literary and informational.
4. Social Studies Standards http://www.state.nj.us/education/aps/cccs/ss/
    a. Social studies education provides learners with the knowledge, skills and perspectives needed to become active citizens in local national and global communities in the digital age.
5. Technology Education http://www.state.nj.us/education/aps/cccs/tech/
    a. Students must be given the opportunity to study technology that provides an opportunity for them to learn about the processes and knowledge related to technology that are needed to solve problems and extend human capabilities (ITEA). They must have an opportunity to:
        i. Design, develop, and utilize technological systems such as communication, transportation, manufacturing, and construction technologies.
        ii. Perform open-ended, problem-based design activities.
        iii. Learn cognitive, manipulative, and affective learning strategies.
        iv. Apply technological knowledge and processes to real world experiences using up-to-date resources.
        v. Work individually as well as in a team to solve problems (ITEA).
6. National Core Arts Standards http://nccas.wikispaces.com/Overarching+Anchor+Standards
    a. This site gives an overview of the current thinking of arts education in the nation. They were adopted by New Jersey in 2014.

Air is what makes airplanes fly. Aerodynamic forces operating on the airplane must be considered in designing crafts. Four forces, drag thrust, lift, and gravity (mass) must be balanced in order to have the plane fly any appreciable distance (Ross-Nazzal & Blackburn, 2009). Drag is the force that opposes an aircraft's motion through the air. Drag is caused by the resistance of the air to the plane flying through it. This can easily be felt by running a hand through the air. Students can feel the resistance of the air as their hands pass through it. We try to minimize drag by designing a streamlined plane. They can see a reduction in drag on their hands by streamlining the shape of their hand as it passes through the air. Thrust is the force that opposes drag. This is the force that moves the plane forward. Lift must oppose gravity. Gravity is the force due to the Earth's gravitational field and is well recognized by students. Lift is related to the mass of the plane and is accomplished when the pressure below the wings is greater than the pressure above them. Design of the wing is an important factor in minimizing mass and maximizing lift. These opposing forces keep the plane in the air. The plane's mass must be kept to a minimum in order to maximize lift. The effect of air on a balloon can be demonstrated by filling a balloon with air and then letting go. What a topsy - turvy flight is initiated and what child has not experienced this for him or herself! If only that random flight can be controlled, then what? So with this background which has hopefully engaged the interest of the students, the teacher then can begin with the students to design and build an aircraft that will fly far and true and perhaps increase its flight by taking into account the jet effects of air escaping from a balloon and increasing thrust and by considering streamlining design. These ideas will incorporate mathematics, engineering principles, social studies topics, science concepts, and aesthetics.

The world record for a paper airplane flight is held by Joe Ayoob and aircraft designer John M. Collins (both USA), at McClellan Air Force Base on February 26, 2012, in North Highlands, California. (Guinness World Records, n.d.). The plane was constructed from a single sheet of uncut A4 paper. A4 paper is a 20 lb. copy paper that is readily available. The aircraft was designed by John M. Collins and flown by his colleague, Ayoob. The record stands since 2012. The plane flew 69.14 meters (226 feet 10 inches) (Guinness World Records, n.d.). The highest altitude paper plane launch was 35,043 meters (37,963 yards) and was achieved by David Green (UK) in Elsworth, Cambridgeshire, UK, on 24 June 2015 (Guinness World Records, n.d.).

Aerodynamic forces operating on the airplane must be considered in designing the airplanes. The four forces, drag, thrust, lift, and gravity (mass) must be balanced in order to have the plane fly any appreciable distance. Lift, opposing gravity, is accomplished when the pressure below the wings is greater than the pressure above them. You can demonstrate this by blowing air across the top of a strip of paper. The paper rises from the difference in air pressure above it and below it. Design of the wing is an important factor in minimizing mass and maximizing lift. These opposing forces keep the plane in the air. The plane's mass must be kept to a minimum in order to maximize lift (NASA, n.d.).

Another force, thrust, is responsible for pushing the plane through the air and enabling it to fly some distance. Modern airplanes use jet engines powered by expensive fuels. Paper airplanes are powered by the thrust from a child's arm launching it. How interesting to consider the effect of using a balloon to increase thrust! You can demonstrate air rapidly escaping from a balloon. How can we harness the power caused by this air? The wrestling with ideas generated by the design can bring students to a basic understanding of Newton's Laws of Motion.

Learners will be required to research current information on flight. They will present written information from informational texts (Common Core English Language Arts), recognize the importance of space exploration (Social Studies Standards), and design an airplane with respect to aesthetics and function (Core Arts). They will use mathematics in collecting and analyzing data (Common Core Mathematics Standards). Principles in science will form the basis of instruction on the possibilities of flight (Next Generation Science Standards).

It is evident from classroom experience that such integration works. Knowledge should not be compartmentalized. All facets of an integrated curriculum need to be utilized in order to produce an individual with multiple varied skills of investigation, collaboration, and scholarship.

CASE DESCRIPTION

Before planning simulated flights with paper and balloons, many questions could arise in the minds of students and maybe even the minds of teachers. How can a heavy airplane carrying hundreds of people and tons of baggage and freight actually fly? How can these massive rockets and spacecraft escape from the Earth? Many questions such as these can be elicited from students while discussing the history of travel above and beyond the Earth. The interest and discussion that results from the research into the history of flight can set the stage for research into the basic principles of flight by simulation of the flights of paper airplanes and balloon rockets. Students can build airplanes, consider their designs, and test their designs by holding competitions to see which is most successful in flight. Then they must consider the development of jet engines to increase the speed and accuracy of the paper aircraft that simulate space crafts that could eventually escape the Earth's gravitational field.

Students of all ages enjoy playing with paper airplanes that they make themselves. This is a motivating activity that teachers can capitalize on to generate ideas on variables, construction design, aesthetics, and even practical results with their students. Teachers can use this intrinsically stimulating experience to practice skills of design, testing, collection, and interpretation of data with students with investigations that support the Next Generation Science Standards and which provide opportunity for integration of mathematical and engineering concepts within the science curriculum. Artistic principles must also be incorporated in producing a paper aircraft with some degree of beauty. Students involved in designing and flying paper airplanes can perhaps replicate some of the excitement generated by the Wright brothers, and all who followed in perfecting the flying machines that we have today.

The purpose of this experience is to use the basic principles of flight to design and test a paper airplane and to participate in a contest to determine which design and which materials build an airplane that flies the longest distance in the shortest time. This activity proved quite popular with pre-service teacher candidates in a methods of teaching science course. It can be used from kindergarten through col-

Figure 2. Possible airplane designs from templates

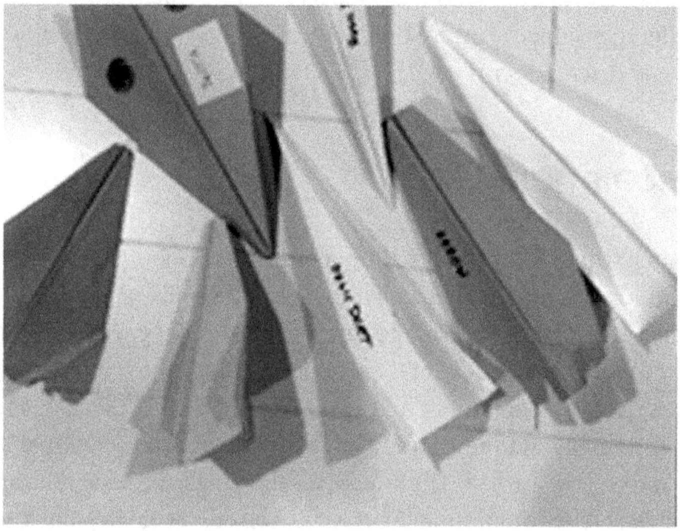

Figure 3. Dart design constructed by grade 6 student

lege depending upon the concepts and mathematics used and the complexity of the design chosen. It can be challenging at any grade level. This activity is cost effective since it only uses materials that are almost always available in schools.

The students discuss the variables that can affect the speed and distance traveled by a paper airplane in relation to these forces and the materials available for construction. Discussion centers on what really matters. Students discuss the effects of mass, the sleekness of the design, the force behind the thrust, decorative effects, and the weather. They begin to demonstrate understanding of the effects of these variables and come to realize the importance of varying one condition while keeping all others equal. The degree of understanding will be related to the ages and developmental levels of the students. Differentiated outcomes must be considered here. The idea of planning a fair race can be related to the concept of what really matters for most students or what is *fair*. Students at a higher developmental level will be able to demonstrate a deeper understanding of the importance of variables. A competition to determine which group builds the best airplane can be planned including all students while taking into account developmental levels and specific talents of those involved. That will be the ultimate test of the various designs considered.

All students working on this activity can become excited about the race and can be actively engaged in discussing conditions. They observe the designs of the airplanes, classify according to the variables discussed, make predictions based on size, mass, and streamlining to hypothesize, collect and analyze data, communicate results, and collaborate. Much discussion should take place concerning what conditions are important in ensuring a *fair* race. Some things that students have identified as necessary to control are:

1. Mass,
2. Type of paper,
3. Initial thrust,
4. Friction,
5. Effect of any identifying or decorative items (mass), and
6. Streamlined design.

These items are mostly controllable and can result in the design of a plane that will fly fast and relatively far. Thrust will probably be difficult to control since it will be accomplished by a student initially launching the plane with a push from the power of arm muscles. Students will see that the same student should be selected to give the initial thrust as the plane takes flight. Some students are stronger than others. This, of course, will vary from group to group and will be considered in part 2 of this lesson plan.

Students also volunteer additional items that they cannot control on the day of the race. These things include weather and wind speed. The race may be done in the gymnasium on days that the weather is not favorable. In fact, it might be better to use the gymnasium for the race if it is available.

The discussion of the importance of considering variables in an experiment and identifying what is important and what can be controlled is vital to the scientific process. Younger students begin to show understanding of the concept that, all things being equal, which elements really produce an airplane that will fly the farthest in the shortest time. Older students will recognize the relationship between the independent and dependent variables.

LESSON 1 DESIGNING AND BUILDING THE AIRPLANE

Essential questions to be answered:

1. What are the forces that enable an airplane to fly?
2. What design principles must be considered in designing an airplane that will fly?
3. How will flight time and distance be measured?

Enduring understandings:

1. Science is a creative process that results in the development of new knowledge through experiment, data collection, and analysis.
2. The importance of evidence in making generalizations about the natural world.
3. The value of aesthetics in design of a flying craft.

Students must:

1. Ask questions that can be investigated within the scope of the classroom or outdoor environments.
2. Be encouraged to develop hypotheses based on observations.
3. Plan and carry out experiments that involve multiple variables.
4. Work collaboratively to identify independent and dependent variables as well as controls.
5. Identify necessary controls.
6. Identify variables, independent as well as dependent.
7. Identify critical measurements.
8. Determine tools to use.

9. Recognize what data to record.
10. Discern how much data is needed.
11. Establish how to communicate results.
12. Identify important variables that need to be measured.
13. Analyze and interpret data that is collected during an experiment.

This experience ideally integrates the five disciplines; science, social studies, art, technology, and mathematics. It can be used as an interdisciplinary activity in each of these content areas. Hopefully, through this activity, participants will come to realize that scientific progress is based on connections between evidence and explanations and that solutions to difficult problems require content from other curriculum areas. They must practice engaging in discussions based on evidence and data collected.

After a discussion of the research that they have done on the history of flight, students are given a packet of information about paper airplane design and construction. Groups of younger students are required to construct a basic airplane for each of the models selected based on the information in the packet. This is done before the time the race is run. Groups should be designed for maximum scaffolding and for differentiating instruction. The designs listed below were chosen to use with an elementary school class. They are simple to construct and give good results. More complex designs can be used with older students. There is much information on the Internet for airplane design and construction. Students research these designs and build those that they are developmentally ready for.

1. Delta,
2. Arrow, and
3. Classic Dart.

For older designers, research on models must be done before actual construction of the airplanes. Students present their ideas based on research to the class. This might take several classes to accomplish. Construction can begin at home for all students, thus making the family part of the project. Different designs can be chosen according to their complexity and the developmental levels of students.

The discussion of the importance of considering variables in an experiment and identifying what is important and what can be controlled is vital to the scientific process. All things being equal, students begin to show understanding of the concept that design, including the forces of drag, thrust, lift, and gravity (mass), is what is really important in constructing an airplane that will fly the farthest in the shortest time. Again, what will produce a fair race?

Figure 4. Balloon rocket

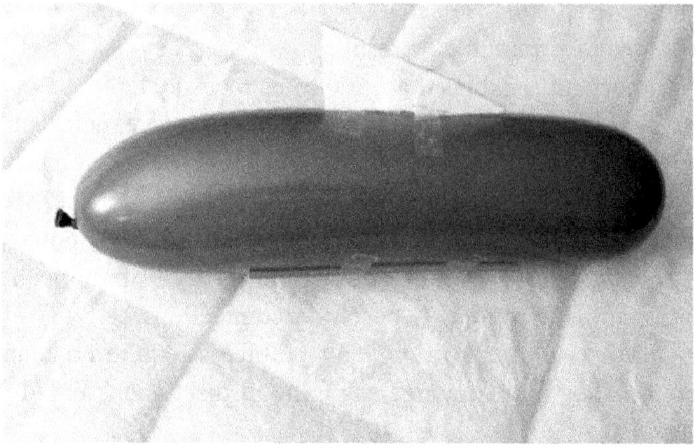

Figure 5. Balloon rocket

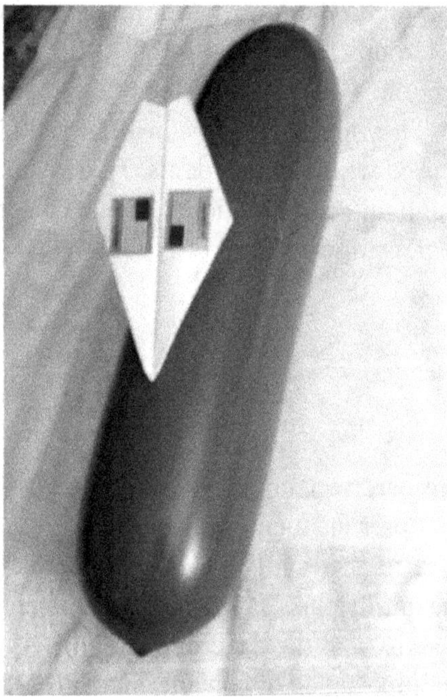

Materials Needed Parts 1 and 2

- Standard copy paper,
- Card stock paper,
- Scotch tape,
- Balance,
- Clipboards,
- Video cameras (iPads),
- Balloon pump,
- Paper Clips,
- Crayons and stickers to decorate airplanes,
- Stop watches,
- Staple guns and staple,
- Measuring wheel or meter sticks,
- Airplane templates, and
- Computer.

Safety Precautions

Students should be advised to be careful with scissors. Be sure to obtain permission of the principal to take the class outdoors to test the airplanes. Parents may also be enlisted in this activity to help keep track of the students as they spread around school grounds to set up their tests. Students should not fly the airplanes in the classroom nor should they fly them towards other students. The same rules will apply if conducting the race indoors in the school gymnasium. Safety concepts must also be established for the construction phase.

Procedure Lesson 1

Pre - Construction Phase: Research

1. Many students will have prior knowledge of paper airplanes but will need more detailed information on their construction. Begin with a search of the literature on paper airplanes. This can be done using search engines on a computer (Dewar). For younger students, the teacher can provide some appropriate reading material on paper (Collins).
2. The teacher can also provide templates for the children to use in construction.
3. Conduct a discussion on the results of the research.
    a. What have others done?
    b. How can previous work add to understanding of the process of flight?

c. What is an operational definition for fastest and furthest traveled by the planes?

4. After discussion the students can begin construction of their planes

Sample Note Taking Form for Literature Search

Names ______________________________

What is the research question? ______________________________________

____

What have you found in your research that will help you construct an airplane that will fly?

1. ______________________________________
2. ______________________________________
3. ______________________________________
4. ______________________________________
5. ______________________________________

References

At least 5 resources should be listed.

1. ______________________________
2. ______________________________
3. ______________________________
4. ______________________________
5. ______________________________

Construction Phase

- Distribute the material and the templates before the actual race takes place. The students are then given time to construct the airplanes according to the templates of the designs that they choose.
- This is a collaborative effort. Teams of three to four work well. Scaffolding of teams is recommended to insure individual strengths of all students is considered.
- Students are directed to build a model of each airplane using the two weights of paper (copy machine paper and card stock) available.
- Students are given free rein to design their planes and to decorate them as they desire (always considering the variables and the purpose of the design).

- Airplanes can be decorated to personalize them for each team.
- Work with students to construct data tables that represent the variable data.
- Consider the value of graphical analysis. Remember one picture is worth one thousand words.
- On the day of the race, each group discusses the variables to be considered and the construction of the airplanes.
- Conduct the race under the conditions established by the class with consideration of all of the variables identified previously in class discussion.
- Video the race so that students can evaluate results and look for ways to improve flight. This can be done in a post flight reflection.

Lesson Outcomes

Lesson outcomes can be differentiated to ensure that all students become involved in the endeavor and so that all students become engaged and feel a valuable part of the project.

At the conclusion of this activity most learners will be able to:

1. Research the literature on the construction of paper airplanes.
2. Identify relevant information.
3. Design and build an airplane using the three basic patterns available.
4. Choose an alternate design based on Internet search.

Figure 6. Grade 6 student planning to launch balloon rocket

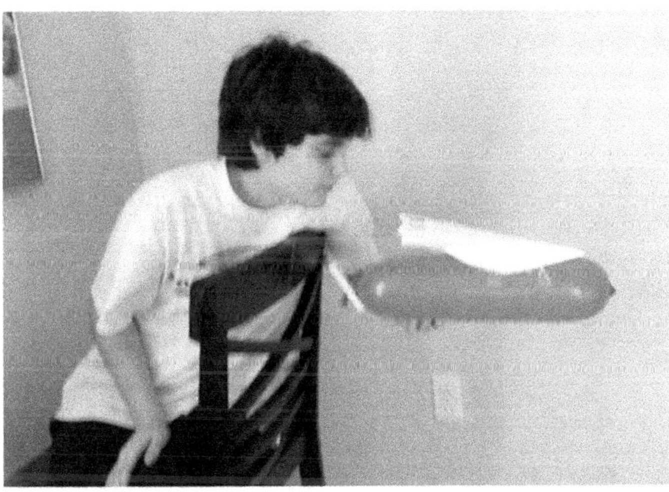

5. Demonstrate understanding of the forces that interact on the airplane that must be considered in its design through oral discussion of their designs.
6. Use templates to construct a device that is attractive and that will possibly overcome the opposing forces.
7. Consider all controllable and uncontrollable variables affecting the flight of the airplane.
8. Extend the idea of the importance of variables in conducting an investigation.
9. Conduct a perceived *fair* race of the airplanes constructed.
10. Design a data table to present data.
11. Demonstrate an understanding of the value of graphical analysis.
12. Demonstrate a basic understanding of Newton's Laws of Motion.
13. Reflect on the procedures used in order to improve the designs in another race.

Students must be involved in developing the data table necessary for comparing the speed and distance that the constructed airplanes flew. This will represent averages of all the trials by the different collaborators involved.

Table 2 can then be used to summarize all data collected by the individual groups allowing for comparison of multiple trials of the airplanes constructed.

The students will be able to identify the design of the airplanes according to the definition for fastest and shortest time developed before the race. A simple histogram can be used to present the data graphically. Students should also write a reflective narrative description of the data in Table 2 and create a histogram that presents the data visually.

Table 1. Student Record keeping Form. Students keep procedural details using clipboards.

Experimenters ______________ Date and Time ________________
Design Used ________________
Materials ________________
Weather ________________
Wind speed ________________
Distance flown
Trial 1 ______________
Trial 2 ______________
Trial 3 ______________
Average time in seconds ___________
Time in the air
Trial 1 ______________
Trial 2 ______________
Trial 3______________
Average Time in seconds______________

Table 2. Data collection form for individual group results

Design	Material	Average Weight (g)	Average distance flown (m) (3 trials)	Average time in the air (Sec) (3 trials)
Design 1	Card Stock			
Design 2	Card Stock			
Design 3	Card Stock			
Design 1	Copy paper			
Design 2	Copy paper			
Design 3	Copy paper			

Table 3. Evaluation rubric: Team Members ________________________________

Item	Not Met (0 points)	Met (2 points)	Target (3 points)
Preparation	The group did not prepare properly for the construction of the paper airplanes.	The group did prepare but were not careful in following the templates provided.	The group worked effectively and collaboratively to construct the planes and followed the templates as closely as possible.
Construction	Construction was sloppy and did not meet the requirement of streamlining.	Construction was adequate but the airplanes produced were not attractive and streamlined enough.	Airplane produced was streamlined, well-constructed, and showed collaboration and good planning on team members.
Planning	Planning was not thorough and did not consider all of the variables mentioned and discussed prior to the race.	Planning was adequate but did not take into account all of the details	Planning was thorough with consideration of most of the variables. Team members worked cooperatively in conducting the race.
Data	Did not complete all phases of the race. Data incomplete and not explained in a narrative or histogram.	Completed all of the race but data table missing some measurements.	Completed all phases of the race and produced acceptable data that was capable of analysis.
Explanation	Was unable to explain the results. Data incomplete and not explained in a narrative or histogram.	Produced an acceptable data table but was unable to write an explanatory narrative.	Produced an exemplar data table. Constructed a relevant histogram and explained the results in a narrative.
Reflection on results	Not able to produce or present a coherent reflection on the racing process.	Produced and presented an adequate analysis of what went well and what needed improvement in their procedures.	Produced and presented to the class a well thought out analysis of the race procedures and made valuable suggestions for improvement in procedures.

LESSON 2 INCREASING AND CONTROLLING THRUST USING A BALLOON ROCKET

After conducting the race and seeing which design won out over the others, older students can then consider the variables that were difficult to control, especially the amount of thrust by the individual students who were selected to launch the planes. Using the demonstration of the balloon that flew irregularly when the air was let out of it, the students begin to think about how to standardize thrust by controlling the air escaping from the balloon. The teacher can introduce the design of the space shuttles that were mounted on the Space Shuttle Solid Rocket Boosters (NASA, n.d.) that took the astronauts within the shuttle beyond the Earth's gravitational field and into orbit. These were the most powerful rockets ever built.

Using the best airplane design identified in Lesson 1 as a result of the competition, the students now plan another race using balloons filled with air. Newton's Laws can be applied in planning the design of the paper airplane shuttles and balloon rockets. The Third Law, stating that for every action there is an equal and opposite reaction, provides the basis for this experiment. The SRB rockets burned fuel and the hot gases that were produced by the burning fuel within the body of the rocket provided thrust by the hot gases pushing against the walls of the rocket and fast exiting gases pushing in the opposite direction. Pressure builds up within the walls of the rocked and the only place that pressure can be released is at the relatively small opening at the base of the rocket. In our case, air will be used to exert pressure on the walls of the balloon and the small opening at the base of the balloon will provide rapid escape of air and consequent reduction in pressure within the balloon.

Procedure Lesson 2

This portion should possibly be used with older students. It could be handled by middle level and above. Examination of the photo of the Voyager flight should generate discussion about the thrust needed to have the Voyager spacecraft escape the Earth's gravitational field. Ask the students if it might be possible to simulate such a flight by using our paper airplanes attached to a balloon. Air would escape from the balloon and serve as the power to launch our simulated spacecraft. This then should begin construction phase 2. How is the airplane attached to the balloon?

Problems to be solved through class discussion:

1. What kind of balloon should be used?
2. How can we be sure that the balloons are inflated equally?
3. How does Newton's Third Law predict the movement of the balloon rocket?
4. How can we control the flight of the balloon rocket?

Figure 7. Teacher candidates working on airplane templates

Discussion:

1. Streamlined balloons should be used.
2. An air pump should be used to inflate the balloons as equally as possible.
3. Measurements of the lengths and circumferences of the largest part of the balloons should be taken to ensure that all start out containing the same amount of air.
4. Balloons should be mounted on a device that will control its flight and give it direction.

Safety Precautions

Be cautious in assembling and launching the balloons. Do not overinflate the balloons. Avoid loud noises from the popping of overinflated balloons. Do not point them at other students. Conduct this competition in a large room- perhaps the school gymnasium. If spectators are present keep them out of the paths of the balloons. Have at least one adult present to help insure the safety of the participants and spectators.

Additional Materials Lesson 2:

- Long balloons.
- Clothespin to keep air inside of the inflated balloons until they are released.
- Masking tape.

- Fishing line used for rocket guidance.
- Scissors.
- Drinking straw to attach balloon rocket to fishing line.

Preparing for the competition:

1. Choose a design for the paper airplanes from Lesson 1
2. Use these design(s) for Lesson 2
3. Identify the variables in using balloons for thrust (recorded above)
   a. Size of balloon
   b. Degree of inflation of the balloons
   c. Total weight of the assembled balloon and shuttle including ballast

Conduct a similar competition as was done in Part 1 to determine the distance a plane attached to the "rocket" could fly when attached to a balloon. Students work in groups to construct the rockets and airplanes using different designs. The balloons should be fairly constant in size and pressure.

Further questions during construction:

1. What shape of balloon should be used?
2. Does the color, texture, or shape of the balloon matter?
3. How should the airplane be attached to the balloon?
4. How can balance be achieved in order for the model to fly straight with the airplane in the proper position?
5. How can damage to the airplane and the hypothetical occupants within the simulated shuttle be avoided?

Table 4. Sizes of balloons. All balloons should be inflated to the same degree. Length and circumferences as well as number of pumps to fill the balloons should be recorded in the following sample table.

Balloon Trials	Circumference-Widest Part (cm)	Length (cm)	Number of Pumps Used with Manual Pump.	Weight of the Assembled "Rocket "Balloon and Paper Airplane "Shuttle".	Distance Flown	Problems Encountered
1						
2						
3						
4						

6. How can the air be released from the balloon in a controlled manner?
7. How should the “rocket” be designed to maintain balance?

Once the questions above have been discussed as thoroughly as possible, students then construct the rocket launchers and test fly them. Data is collected by individual groups as in Table 2 above. Again, aggregate data tables should be constructed using individual group data. Graphical as well as narrative analysis must be used. A narrative reflection should be required explaining positive as well as negative aspects of the tests. Modifications in design are then discussed. These modifications may include:

- The use of a smooth string to guide the rocket balloons
- Method of holding the string straight and taut
- Length of the string needed
- Support of the string
    - Ring stand supports
    - Student support at launching point

Outcomes

At the conclusion of this lesson, learners will be able to:

1. List questions relevant to constructing a simulated rocket and shuttle launch.
2. Attempt, through researching and discussing design principles, to construct a simulated rocket/spacecraft launch.

Figure 8. Apparatus that can be used in balloon launch. Balloon on plastic straw with fine gauge string pulled through it. Tape is used to provide ballast and to keep the balloon attached to the straw. A clothespin is used to keep the air inside the balloon until the “rocket” is launched. The paper airplane is secured to the rocket by glue or scotch tape.

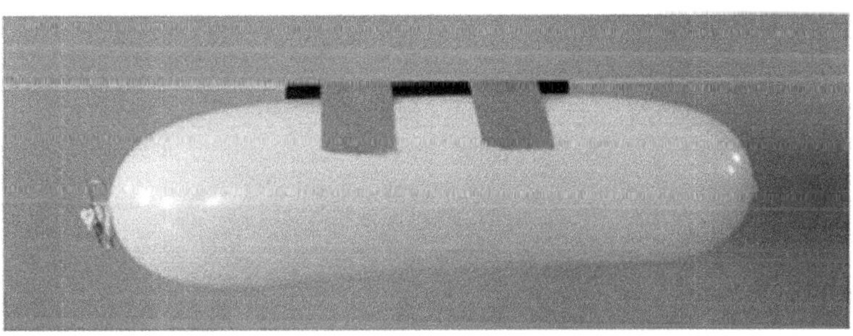

3. Conduct a test fly with the designs that they produce.
4. Demonstrate understanding of the constraints of the test by outlining the variables involved.
5. Reach a conclusion based on this experience; which “spacecraft” and which “rocket launcher” was the best one based on the test fly results.
6. Suggest improvement in design that would provide a smoother and speedier flight.

Table 5. Evaluation rubric

Item	Not Met	Met	Target
Discussion of Variables	Does not recognize the importance of control of variables.	Does account for some variables in the design.	Accounts for as many variables in design as possible and recognizes these variables that are not possible to control.
Construction of Rocket	Poorly constructed.	Adequately constructed but unable to assemble parts effectively.	Well-constructed, spacecraft attached securely. Ready for launching.
Selection of shuttle	Selects a design.	Selects a better design.	Selects best design.
Experimental setup	Does not demonstrate the importance of design in the launching of spacecraft and rocket launchers.	Adequately demonstrates the process of design but is unable to produce an acceptable design.	Completely demonstrates understanding of the necessities for construction. Actually constructs a rocket launcher that successfully launches.
Data collection	Data collection missing critical portion or no data collection design.	Has data collection plan but misses one or more critical measurements.	Complete data table and collection of critical data.
Graphic Analysis	No plans for graphic analysis.	Plans incompletes.	Identifies independent and dependent variables and where to place them on a histogram. Has enough data for an adequate analysis.
Explanation of results (reflection)	No explanation possible.	Explains what data has been collected but since plans were incomplete unable to explain results.	Complete plans allow experimenters to initiate an explanation based on evidence.
Future improvements	None	limited	Recognizes problems with plans and able to initiate suggestions for improvement in design.

Culminating Assessment

Students will be required to present the results of their investigations orally as well as in written form. A symposium ion flight could be held in the school where children could discuss all facets of their investigations to the other students. This might also be a means of attracting and motivating other students to STEAM careers.

Comments on These Two Lessons

Lessons 1 and 2 will help students to recognize that designs need adjustments and improvements in order to increase chances for success. Experiences such as these can be used to engage students and to create intrinsically motivational activities to develop engineering and design principles based on scientific thinking. They can be used to help students understand some of the principles of flight, especially space flight. The plans that the National Space and Aeronautics Agency has for the future are exciting and present inherent problems that must be solved before going further. These young citizens of the future should be introduced to the complications of design, problem solving skills, action and reaction of force, and testing procedures that scientists and engineers go through in planning such endeavors. Hopefully they may be encouraged to pursue careers in science, mathematics, and engineering that will be so much a part of tomorrow's needs.

CURRENT CHALLENGES FACING TEACHERS

The challenges facing teachers are based mostly on the lack of understanding of what science process skills really are and how important these processes can be in developing critical thinking skills in students. This is something that takes interest, training, and commitment on the part of teachers and students alike. Some teachers may feel that physical science projects are not of interest to their students. Perhaps they have little interest in such projects themselves. Older teachers might look on these projects as something too difficult to take on and to manage. They may lack training in overcoming these attitudes. They do not see the value in the outcomes possible in such investigations. They also perhaps do not realize the need for scientists and engineers in the future. It is often said that we are living in a technological world. But what does this really mean? What are the implications for the future? There is a need for more scientists and engineers in the future to help solve such problems as global warming, pollution, use of fossil fuels, alternative forms of energy, and ethics in biomedical advances- to cite just a few. We are in the midst of technological change and the implications of these changes are currently largely unforeseen. This era can be compared to the industrial revolution of the 19th century and its effects on daily life. Many of those living through the revolution of the 19th century did not realize the profound changes that were to occur as a result of the change from cottage based manufacturing to machine age manufacturing. Those living now in the 21st century are in a similar predicament. They cannot really understand or predict the changes in society caused by advances in science and engineering. Just imagine the robot revolution and its effects in the work force.

Another challenge could also be found in the attitudes of school administrators who see little value in the possible increase in class time needed to work on physical science projects. They may not be capable of relating these projects to the development of future scientists and technicians who will need these skills to tackle the problems that future generations will face with the environment, food supply, genetics, energy shortages, population growth, climate change, and land and water supply - to mention just a few. Perhaps more importantly they do not realize the ethical concerns that we will face in the future. School administrators must accept the fact that a firm foundation in science and technology will help future citizens to deal with problems that will most certainly arise. Curriculum designers must consider integrating science topics within the entire curriculum. Science knowledge and process depend upon skills important in reading mathematics and social studies. In fact, science knowledge impacts all facets of the curriculum. In addition, we need a scientifically literate populace of voting citizens to understand the impact that science development can have on lifestyles and living standards. We need scientifically literate citizens and scientifically literate politicians to work together to improve conditions on the Earth that we all share.

The current emphasis on standardized testing in mathematics and literacy results in schools assigning large blocks of time for these subjects while minimizing the amount of time during the school day for science content and activity. They do not realize, but must be made to realize, the importance of cross curriculum projects in a real-world learning environment. It is understandable that the emphasis on test scores in literacy and mathematics is important in developing an educated populace capable of following a variety of career paths. But scientific literacy is just as important and must be integrated within the entire school curriculum. This is not as difficult to do as it might seem.

Another challenge could be the perception that projects are too expensive and would take up too much of the school's supply budget. There are many projects that can be done using ordinary materials that are commonly found in schools. Of course, there are projects that can be expensive but a good start on developing concepts of design and consequent problems can be made at all instructional levels using ordinary materials as suggested in the lessons described.

Challenging also is the current realization that all children, despite disabilities, are entitled to an education in an inclusive classroom. Science is for *all* children. All children are equally valuable in classes but not all children are equally competent. This is especially apparent in science classes where some ideas can be too abstract for disabled students to master. In fact, they can often be too abstract for average students to master. We need to work together with special education teachers to develop differentiated methods so that all children become scientifically literate to some degree.

SOLUTIONS AND RECOMMENDATIONS

Perhaps all of the challenges and faulty perceptions in the previous section could be overcome by teacher training programs that emphasize the importance of incorporating such topics within interdisciplinary curricula. Techniques of incorporating science concepts throughout the curriculum need to be further explored. Science can be integrated seamlessly within the mathematics, language arts, social studies, and even art and music curricula. Preservice teacher candidates as well as experienced teachers need to be offered programs that can show how science concepts can be built into all curricula. Candidates in Schools of Education must be trained in science methods that consider the processes as well as the factual science content.

Preservice teachers in science methods classes should be practicing interdisciplinary techniques that do not isolate science concepts and processes from other curricula. This must be done in science methods classrooms at the college and university level. In supervision of student teachers from Monmouth University, integration of science concepts within mathematics, reading selections, and social studies is encouraged. In this way, perhaps some of the challenges facing teachers will be lessened in future members of the profession.

Experiences in college level methods courses have been successful in generating ideas and interest in preservice teachers. These lessons cited in this paper were done towards the end of the course after the students conducted a series of experiments designed to instruct them in a science process approach to elementary and middle level science instruction. Hopefully they will be inspired to take this approach in their future science classrooms.

Teachers currently in service do a valiant job of educating students under increasing pressures of testing and accountability. It is understandable the they, as well as administrators, feel that the value of their institutions is being tied to standardized test scores and are therefore reluctant to change curricula and time slots designed to maximize scores. In-service programs conducted in universities by science trained professionals could be made available to in service teachers and administrators. With the adoption of the Next Generation Science Standards, this becomes imperative. If these standards are to have an impact on science curricula throughout the nation we must assist teachers and administrators in comprehending the implications of these standards and to develop methods of incorporating them within current curricula. This will take individual as well as statewide efforts. Those of us currently involved in science education need to take leadership in this initiative.

The challenge presented by the idea that science is for all children needs to be faced by science educators and special educators. Special needs children will be just as much a part of the future as their peers in general education classrooms. Methods of making science concepts comprehensible by all children must be explored by partnerships between special educators, general educators, and science educators. Incorporating the strategies, techniques, and art of co-teaching needs to expand. Co-teaching is being implemented widely in New Jersey and its benefits must be recognized in all venues. Teachers must be given the opportunity to collaborate in using tried and true strategies already used with special needs children. The concept of Universal Design suggests that we use multiple methods of presenting knowledge, engaging students, and assessing students to have effective, inclusive classrooms for all despite learning problems that might exist.

The solutions to the challenges are not easy to implement. There can be resistance on the part of preservice as well as in-service teachers to an emphasis on science in the curriculum. Asking teacher preparation students in my science methods classes whether they enjoyed science in previous schooling resulted in the not so surprising response that they did not. They speak of science courses with reliance on mountains of facts, and that memorization of these facts as well as a whole new vocabulary was daunting and boring. Students mostly say that loss of interest in science began in middle and high school courses. Science in the secondary schools is often lacking in inquiry methods. This is understandable since teachers at the secondary level often have classes that are college preparatory and are concerned with covering information that might be asked on achievement tests. Perhaps a whole new curriculum needs to be developed that stresses scientific literacy for the citizenry of the future. Hopefully the emphasis on problem solving and engineering design in elementary schools will continue to motivate those planning on careers in science and engineering to take the more challenging courses that still should remain in the secondary curriculum.

In summary, some solutions to the challenges include:

1. In-service training for in-service teachers and administrators.
2. More cooperation among special educators and science educators.
3. Emphasis on interdisciplinary curricula.

4. Methods of integrating the Next Generation Science Standards within current science curricula.
5. Exploration of the secondary curricula to develop more science-society courses for the students who might have other talents besides science.

These solutions suggest even further inquiries. Teachers must prepare themselves to meet these challenges.

REFERENCES

Collins, J. M. (2013). *The new world champion paper airplane book.* New York, NY: Crown Publishing.

Dewar, A. (2014). *One minute paper airplanes kit.* Rutland, VT: Tuttle Publishing Company.

Grant, R. G. (2007). *Flight: The complete history.* New York, NY: Dorling Kindersley.

Guinness World Records. (n.d.). Retrieved from http://www.guinnessworldrecords.com

McAuliffe, C. (n.d.). *BrainyQuote.com.* Retrieved fromhttp://www.brainyquote.com/quotes/quotes/c/christamca134582.html

NASA. (n.d.). *Fast facts.* Retrieved from http://voyager.jpl.nasa.gov/mission/fast-facts.html

National Aeronautics and Space Administration. (2015). *Welcome to the beginner's guide to aerodynamics.* Retrieved from https://www.grc.nasa.gov/www/k-12/airplane/bga.html

Ross-Nazzal, J. (Interviewer), & Blackburn, G.C. (Interviewee). (2009). *NASA Johnson Space Center oral history project edited oral history transcript.* Retrieved from Johnson Space Center Oral Histories Project Website: http://www.jsc.nasa.gov/history/oral_histories/BlackburnGC/BlackburnGC_5-4-09.htm

Wright brothers. (n.d.). In *Wikipedia.* Retrieved August 2016, from https://en.wikipedia.org/wiki/Wright_brothers

KEY TERMS AND DEFINITIONS

Aerodynamics: A science that studies the movement of air and the way that objects (such as airplanes or cars) move through air: the qualities of an object that affect how easily it is able to move through the air.

Drag: The force of wind or air resistance pushing in the opposite direction to the motion of the object.

Gravity: The force that attracts a body toward the center of the earth, or toward any other physical body having mass. For most purposes Newton's laws of gravity apply, with minor modifications to take the general theory of relativity into account.

Lift: The force that directly opposes the weight of an airplane and holds the airplane in the air. Lift is generated by every part of the airplane, but most of the lift on a normal airliner is generated by the wings. Lift is a mechanical aerodynamic force produced by the motion of the airplane through the air.

Mass: The property of matter that measures its resistance to acceleration. Roughly, the *mass* of an object is a measure of the number of atoms in it. The basic unit of measurement for *mass* is the kilogram.

Template: A form, mold, or pattern used as a guide to making something. Here are some examples: A ruler is a *template* when used to draw a straight line. A document in which the standard opening and closing parts are already filled in is a *template* that you can copy and then fill in.

Thrust: The force used to overcome the drag of an airplane, and to overcome the weight of a rocket.

Chapter 12
A Mathematical Approach to Designing Insulators

Kathryn E. Pedings-Behling
SC Governor's School for Science and Mathematics, USA

EXECUTIVE SUMMARY

How do we keep hot drinks hot and cold drinks cold? Companies such as Tervis, YETI, and Thermos spend their time researching and designing products around that very question. In this lesson, students will discover, through mathematical modeling, which materials provide the best insulation and be tasked with designing their own insulator. This lesson has been designed at two different levels for students from grade three through high school with an optional extension activity for more advanced students. Students will use technology to explore the rate of change of the temperature of hot water over two minutes using different insulation materials. After this exploration, students will use the data they have collected to determine the best materials for designing their own insulator. This insulator will then be judged based on the ability to keep a hot drink hot and on the aesthetic value.

LITERATURE REVIEW

The history of insulated drinkware begins around 1896 when Sir James Dewar invented what would become the first Thermos. When Thermos began making bottles in the U.S. in 1907, it quickly gained popularity with famous users such as President Taft and the Wright Brothers (Insulated Bottle, 2006). It seemed as though everyone was looking for a way to keep their hot drinks hot and cold drinks cold!

DOI: 10.4018/978-1-5225-2334-5.ch012

Copyright ©2017, IGI Global. Copying or distributing in print or electronic forms without written permission of IGI Global is prohibited.

The process for manufacturing commercial insulated drinkware is still very similar to what it was in the early 1900s. Most cups made for hot drinks are a vacuum, sealed between two walls of the same material. Most bottles for cold drinks are lined with a liquid foam that is allowed to harden (Insulated Bottle, 2006). Although this process began in the early 20th century, it hasn't kept people and companies from trying to perfect the insulated bottle. If one were to look up "insulated drinkware patents" on Google Scholar around 16,000 results would be returned. The search is never-ending to make consumers happy with their drink temperature.

When one thinks of insulated drinkware, the disposable coffee cup often comes to mind. These have become ubiquitous with the morning commute of many Americans. The history of the disposable coffee cup is riddled with controversy over the battle between material that is good at insulation (styrofoam) versus material that is better for the environment (paper). This argument was ultimately won by Howard Schultz, owner of Starbucks, in 1987 when he chose the paper cup. In order to insulate more like a foam cup, manufacturers have worked to develop paper cups with walls two or three times as thick as normal paper cups (Park, 2014).

However, the quality of the insulation is not the only consideration for disposable coffee cups anymore. A popular food blog, Eater.com, suggests that starting in about 2014, the visual design of the cups became incredibly important for consumers to feel like they are getting their money's worth for expensive coffee (Baird, 2015). In fact, unique designs of coffee cups have become immortalized on the internet by way of specialized Instagram accounts (Hargreaves, 2016), design blogs (Durso, 2014), and food culture websites (Hohenadel, 2014).

In this activity, students will explore the quality of insulation of different types of materials in order to make educated decisions about how to design an insulator that can compete with the best being manufactured. This is not a solved problem as over 350 of the patents for insulated drinkware on Google Scholar were filed as of 2016. Not only will teams of students have to think about the quality of their insulation, but they will have to consider the aesthetic design qualities as well.

CONTENT INFORMATION

This activity aligns with many national standards across several different subject areas. Most notably, this activity puts to use several standards from the Common Core State Standards for Mathematics (CCSSM). Certain questions in this activity are written at two different levels with an optional extension for students beyond the Algebra 2 level. The questions specifically written for grades 3-5 align with CCSSM standards related to differences and numerical comparisons such as 3.NBT.A.2, 4.NBT.A.2 and 4.NBT.B.4. The questions specifically written for grades 6-12 draw from standards about ratios and rate of change including 6.RP.A.3, 7.RP.A.3, 8.F.B.4, and HSA.CED.A.1. Finally, the optional extension activity relates to standards involving modeling exponential functions found in HSA.CED.A.2 and HSF.LE.A.1 (Common Core Standards Initiative, 2010).

Although this activity takes a mathematically-forward approach to designing insulators, standards from the other STEAM fields are integral to the well-roundedness of the learning experience. From the Next Generation Science Standards, the activity specifically aligns with standards relating to thermal energy transfer and engineering including 3-5-ETS1, MS-PS3-3, MS-ETS1, HS-PS3-3, and HS-ETS1 (NGSS Lead States, 2013). This activity also heavily relies on technology and draws from the International Society for Technology in Education (ISTE) standards 1.c, 4.b, and 4.c (ISTE, 2007). Finally, the arts component of this activity aligns with the creating anchor standards 1, 2, and 3 from the National Core Arts Standards (2014).

This mathematical approach to designing insulators integrates the components of STEAM in many ways. Mathematical modeling is the core of the activity using data collection with a technology component. Science concepts involving energy and heat loss provide the background for why the system is functioning in a specific way and how it can then be improved. Engineering design is incorporated with the arts to design and create an insulator that is effective in retaining heat and is aesthetically pleasing. The activity culminates in students being required to justify their final design through scientific explanation using data they collected.

CASE DESCRIPTION

This lesson is adapted from a PASE (Portable Advanced STEM Exploration) Math lab developed by the South Carolina Governor's School for Science and Mathematics titled *How do we keep it hot?*. (Pedings-Behling, 2015)

Lesson Plan Information

Technology Overview

Throughout this activity, students will perform data collection a minimum of five times, finding the temperature of water at least twelve times during each collection. This data collection can be done one of two ways:

1. **Data Collection Technology:** Companies such as Vernier, Pasco, and Texas Instruments have created hardware specifically for data collection with a temperature probe. If your school has access to this technology, this activity

is the perfect opportunity to make use of these items. The experiment would need to be set up to collect data for 2 minutes, and then an analysis of the data can be done using this technology.

2. **Thermometer:** If your school does not have access to any of this data collection technology, each team of students can use a thermometer and a stopwatch to fill out the table in the lesson. This data would then need to be graphed and analyzed by hand.

Materials per Group:

- 2 disposable plastic cups (e.g., Solo),
- Temperature collection technology (see above),
- Piece of approximately 6" x 6" mesh fabric,
- Piece of approximately 6" x 6" cotton fabric, and
- Piece of approximately 6" x 6" wool fabric.

Materials per Classroom:

- Electric Kettle with constant temperature control,
- Water,
- Collection of mesh, cotton, and wool fabric to use for final design,
- Scissors,
- Glue/tape for final design, and
- Markers.

Safety Concerns

This activity requires students to work with hot water. Teachers may want students, depending on their age, to wear goggles or protective gloves. The teacher may also want to designate an adult to be in charge of the hot water station to supervise. However, only about 1 inch of hot water is needed in the bottom of the cup for the temperature probe to collect correct data. None of this hot water should be used for consumption after using the temperature probe.

Lesson Plan

The following procedures will be carried out five times using the specifications in the following sections.

Procedures

Think about the following scenario: You make hot chocolate in the cup you are testing and accidentally forget it on the kitchen counter for 2 minutes. If you were monitoring the temperature of the hot chocolate the entire time, what would a graph of those temperatures look like? Sketch the prediction of your graph below. Make sure to color in the box next to "Prediction" with the marker you are using.

You will now simulate this scenario using hot water. Gather about 1 inch of hot water in the cup and collect data on the temperature for 2 minutes. It is important to make sure the temperature of the thermometer reaches the maximum prior to starting your data collection (about 20 seconds). Otherwise, you are collecting data about the temperature probe measuring the initial temperature of the water. If using a regular thermometer, use the table below to record the temperature of the hot water every 10 seconds.

Using a different color marker on the graph in Figure 1, graph the actual data on top of the predicted graph. Make sure to color in the box next to "Actual Data" with the marker you are using. Also, make sure to label the markings on the y-axis with the maximum and minimum values from your data collection.

Data Analysis:

1. How well does the graph of your actual data match your predicted graph? Explain any differences between the two graphs.
2. **(Grades 3 - 5):** Find the difference between the highest temperature and the lowest temperature.

Figure 1. Graph of temperature versus time for both predicted data and actual data (Pedings-Behling, 2015)

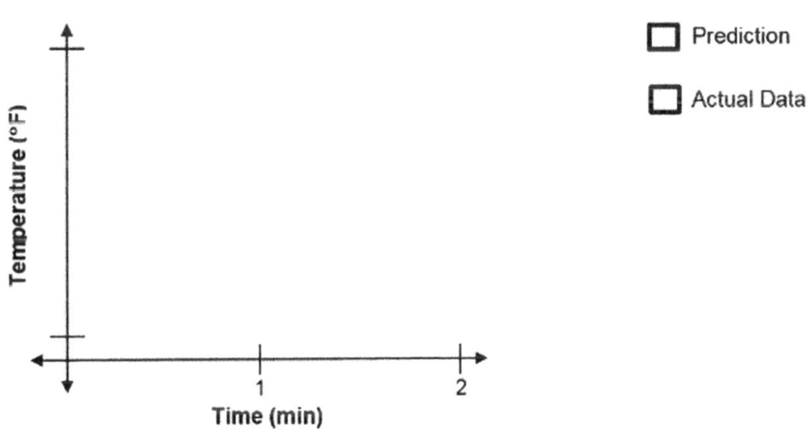

Table 1. Two-minute temperature data collection using a thermometer

Time	Temperature (°F)
0:10	
0:20	
0:30	
0:40	
0:50	
1:00	
1:10	
1:20	
1:30	
1:40	
1:50	
2:00	

3. **(Grades 6 - 12):** Find the average rate of change in the temperature of the water over the two-minute period.
4. What are the units of measurement for your answer? Why? What are some other units that could have been used?
5. What does this value mean in the context of the problem?
6. What can be done to retain more heat in the water?
7. What would happen if you collected data for longer than two minutes?
8. Approximately how long would it take for the water to get back to room temperature? Justify your answer using mathematics.

Advanced Extension Activity

Use Newton's Law of Cooling, either as an exponential function

$$T(t) = T_s + (T_0 - T_s)e^{-kt}$$

$T(t)$ = temperature at time t

T_s = temperature of surroundings

T_0 = initial temperature

k = constant

or as a differential equation

$$\frac{dT}{dt} = -k\left(T - T_s\right)$$

with the same variable definitions to create a more accurate model of the cooling of the water. Use this model to predict how long it would take for the water to reach exactly 0.1°F above the current room temperature.

Part A: Control Data

Follow the procedures using a single plastic cup.

Part B: Testing Different Materials

Repeat the procedures three times, once for each type of material (mesh, cotton, wool). Take two plastic cups and place the material between the two cups to create an insulator. See Figure 2.

Additional Questions:

1. How is each graph different from the graph of the control data?
2. (Grades 6 - 12): Look at the 4 graphs of the data you have collected so far. Is this data linear? What is the difference between this data and linear data?
3. Which fabric retained the most heat? Justify this mathematically and scientifically. Did that surprise you?

Figure 2. Visual process of nesting fabric between two plastic cups

4. (Advanced): How can you tell from the exponential models which fabric retained the most heat?

Part C: Insulator Design

Each group will need to design their final insulator. Insulators must be freestanding, meaning that they do not need to be held together with your hands, and must hold one of the plastic cups which will contain the hot water. These insulators may include a top, but this must be removable to allow for hot water to be added to the cup.

Each group has $20 to buy supplies, and the costs are as follows:

- **Wool:** $1 per square inch.
- **Cotton:** $0.50 per square inch.
- **Mesh:** $0.25 per square inch.

Groups not only need to focus on retaining heat, but also on the aesthetics of their insulator which will also be judged. Groups may design and test as many insulators as their time and money allow. Use Table 2 to keep track of spending.

Part D: Final Insulator Testing

Follow the procedures to test your insulator.

Aesthetic Judging

Option 1: Have the students vote on their favorite insulator design (not their own). Tally the votes to determine the winner.

Option 2: Have either the teacher, the students, or both rate each insulator based on the design qualities below with 1 being the lowest score and 5 being the highest score.

Table 2. Fabric type and cost for insulator design

Fabric Type	Cost Per Square Inch	Number of Square Inches	Total Cost
Mesh	$0.25		
Cotton	$0.50		
Wool	$1.00		
TOTAL			

Table 3. Aesthetic judging for insulators using different design qualities

Design Quality	Score (1 - 5)
Texture: Is the insulator pleasant to hold?	
Superficial Quality: Does the design look like a quality insulator?	
Functionality: Can the insulator be easily held with one hand?	
Transportability: Can the insulator be easily transported in a car cup holder?	
Marketability: Could this insulator be designed with logos or different colors?	
Durability: Does it appear as though the insulator would last for a long time?	
Total Score	

Final Analysis:

1. Justify the design of your insulator citing information gained from your data collection.
2. Who made the best insulator in terms of heat retention? Justify this mathematically. What does it look like and how is it different than yours (if it isn't yours)?
3. Who made the best insulator in terms of aesthetics? What does it look like and how is it different than yours (if it isn't yours)?
4. If you could redesign your insulator, what would you change?

STEAM CONTENT

The overall goal of this activity is to design an insulator in light of the mathematical information gained about the change in the temperature of the hot water using different materials for insulation. Achieving this goal incorporates all aspects of the STEAM model. For the science aspect, students have to question why some materials are better insulators than others. Technology is utilized in this activity through the use of the temperature probes for data collection. The students are asked to design their own insulator which incorporates both the engineering and art components since they have to consider both goals of it retaining as much heat as possible and being as aesthetically pleasing as possible.

The heart of this activity is the mathematical integration. Students in grades 3 - 5 will explore differences in the temperatures of the hot water over the two-minute collection period. Students in grades 6 - 12 will explore the average rates of change in the temperatures of the hot water over the two-minute collection period. If the students have studied exponential models, they can then complete the extension activity about Newton's Law of Cooling. These explorations will lead them to make decisions about what material they will use in making their insulator. The activity culminates in having the students provide a scientific explanation of why they chose the materials they did using their mathematical information as justification.

CURRENT CHALLENGES FACING TEACHERS WITH SOLUTIONS AND RECOMMENDATIONS

My School Does Not Have Any of This Technology

There are many solutions to this challenge. Multiple teachers at a school or in a district can share this technology and create a checkout system for use in their classrooms. The teacher can also choose to collect data as an entire class demonstration or at one station that each group rotates through so that only one set of technology is needed.

I Don't Have Enough Class Time to Complete This Activity

Depending on the length of your class, this project can span over multiple days or class periods. When one considers the large number of standards addressed by this activity, this time in the classroom is justified, but can still be difficult to complete in light of other material needing to be covered in a specific unit. The activity is easily broken into many small portions that can each be completed using much less class time. This will allow the activity to be spread over several classes or weeks. The design portion of the activity can be assigned as out of class work to lessen the amount of class hours spent on the project.

I'm Concerned About the Safety of My Students During This Activity

This is particularly a concern for teachers of younger students, especially in handling the hot water. If the teacher chooses to collect data as an entire class to address the first concern, this challenge is addressed since the teacher can be the only person handling the water. For very young classes, it may be best to bring in additional adults (teachers, administrators, or parent volunteers) to assist in both pouring hot water and handling the hot cups for each of the groups.

I Like Math Because Everything has One Solution. I'm Scared to Do Data Collection Because Everyone's Data Will Look Different and I MIGHT NOT KNOW the "Right" Answer.

The beauty of data collection is that each group of students will have their own unique data with which to work. As one may suspect, the wool *should* be a better insulator than the cotton or the mesh. What if a group's data does not reflect this? This provides a wonderful opportunity to have conversations in your classroom with questions such as: *What could have caused the data to reflect a result that was different from your hypothesized outcome? Was the data collected the same way each time? Do you suspect an issue with your data collection?* Once a teacher embraces the fear that there is no one exact answer with this activity, it may be more freeing to give your students the ability to create truly incredible products.

REFERENCES

Baird, S. (2015). *How Disposable Coffee Cup Design Became High Art*. Retrieved September 30, 2016, from http://www.eater.com/drinks/2015/8/14/9153849/how-disposable-coffee-cup- design-became-high-art

Blinder, S. M. (2007). *The Coffee Cooling Problem*. Retrieved June 07, 2016, from http://demonstrations.wolfram.com/TheCoffeeCoolingProblem/

Bryant, J. (2008). *Newton's Law of Cooling Demonstration*. Retrieved June 07, 2016, from http://demonstrations.wolfram.com/NewtonsLawOfCooling/

Dubson, M., & Loeblein, T. (2010). *Curve Fitting*. Retrieved June 07, 2016, from http://phet.colorado.edu/en/simulation/legacy/curve-fitting

Durso, C. (2014). *Coffee Cups of The World, A Photo Series of Beautiful Takeaway Cups*. Retrieved September 30, 2016, from http://www.foodiggity.com/coffee-cups-of-the-world-a-photo-series- of-beautiful-takeaway-cups/

Hargreaves, H. (2016). *Coffee Cups of the World (@coffeecupsoftheworld) • Instagram photos and videos*. Retrieved September 30, 2016, from https://www.instagram.com/coffeecupsoftheworld/

Hohenadel, K. (2014). *Traveling the World in Search of the Most Unique Disposable Coffee Cup Designs (Photos)*. Retrieved September 30, 2016, from http://www.slate.com/blogs/the_eye/2014/07/03/photographer_henry_hargreaves_coffee_cups_of_the_world_highlights_the_best.html

Howcast. (2013). *How to Demonstrate Insulation Principles*. Retrieved June 07, 2016, from https://www.youtube.com/watch?v=8I96BzSNNqE

Insulated Bottle. (2006). Retrieved June 09, 2016, from http://www.madehow.com/Volume-4/Insulated-Bottle.html

Insulation. (2007). Retrieved June 07, 2016, from http://sciencelearn.org.nz/Contexts/Icy-Ecosystems/Science-Ideas-and-Concepts/Insulation

International Society for Technology in Education. (2007). *National educational technology standards for students*. Retrieved June 7, 2016, from http://www.iste.org/standards/iste-standards/standards-for-students

Khan, S. (2016). *Linear vs. exponential growth*. Retrieved June 07, 2016, from https://www.khanacademy.org/math/algebra/introduction-to-exponential-functions/comparing-exponential-and-polynomial-functions/v/linear-exponential-models

National Coalition for Core Arts. (2014). *National core arts standards*. Retrieved June 7, 2016, from http://www.nationalartsstandards.org/

National Governors Association Center for Best Practices & Council of Chief State School Officers. (2010). *Common Core State Standards for Mathematics*. Washington, DC: Author.

Newton's Law of Cooling. (2015). Retrieved June 07, 2016, from https://www.khanacademy.org/math/differential-equations/first-order-differential-equations/modeling-with-differential-equations/v/newtons-law-of-cooling

NGSS Lead States. (2013). *Next Generation Science Standards: For States, By States*. Washington, DC: The National Academies Press.

Park, M. Y. (2014). *A Brief History of the Disposable Coffee Cup*. Retrieved June 09, 2016, from http://www.bonappetit.com/entertaining-style/trends-news/article/disposable-coffee-cup-history

Pedings-Behling, K. E. (2015). *How Do We Keep It Hot?* Hartsville, SC: South Carolina Governor's School for Science and Mathematics.

The Science of Insulation Explained. (2008). Retrieved June 07, 2016, from http://www.knaufinsulation.com.au/en-au/more/insulation-explained.aspx

KEY TERMS AND DEFINITIONS

Average Rate of Change: The change in the dependent variable divided by the change in the independent variable. For this activity, students will find the rate of change of the temperature of the water with respect to the time of the data collection.

Difference: In this activity, students will find the difference between the highest and lowest temperatures using subtraction.

Exponential Function: A function where the variable is located in the exponent. The rate of change is not constant and is either increasing or decreasing across a graph.

Insulator: A device that does not easily allow for the transfer of heat either from a hot beverage or to a cold beverage.

Linear: A descriptor for a function that indicates that it has a constant rate of change.

Newton's Law of Cooling: A mathematical model that exhibits the physical process that the rate of heat loss of an object is proportional to the difference between the temperature of the object and the temperature of the surroundings.

Temperature: The measure of heat in an object. In this activity, it is specifically the measure of heat in the water.

Thermometer: A tool used to measure temperature.

Chapter 13
Constructing a Marshmallow Catapult

Warren James DiBiase
UNC Charlotte, USA

Judith R. McDonald
Belmont-Abbey College, USA

Kellan Strong
UNC Charlotte, USA

EXECUTIVE SUMMARY

This case will present a project-based scenario where students will take the place of an astronaut stranded on Mars. Like the character in the Disney film The Martian, the astronaut only has a small collection of "spare parts" at his disposal to ensure survival. In this scenario, our astronauts meet predators and in an effort to fend them off, they must design and construct a catapult. During this deep dive process the astronauts working in groups of four, will take an inventory of spare parts available, design and draw a plan for building, build the catapult, test the catapult, and then go through a series of revisions, retesting and sharing their redesigns.

LITERATURE REVIEW

Engineering design activities can be a powerful entry point into science learning. Attempting a design-build project, students use creativity within their project to tackle the problem (Wicklein, 2006). As students develop an understanding of design-build

DOI: 10.4018/978-1-5225-2334-5.ch013

Copyright ©2017, IGI Global. Copying or distributing in print or electronic forms without written permission of IGI Global is prohibited.

projects, teachers can challenge them with an application of a local community based engineering issue to create a relevant issue that students can begin to solve. Engineering design-build projects, "hands-on" or "learning by doing," is grounded in constructivist theory (Fortus, Krajcikb, Dershimerb, Marx, & Mamlok-Naamand, 2005) that is shown to improve student achievement in higher level cognitive tasks, such as scientific processes and mathematical problem solving (Satchwell & Loepp, 2002). Current research indicates that when students are given a project-based task, their interest in STEM can be increased because it requires students to solve authentic problems based on real issues (Fortus et al., 2005). Teachers today are challenged in many ways, from administrative tasks to having students reach goals on the state mandated test scores; this form of instruction provides an engaging methodology that allows the curriculum to be relevant while teaching important skills to your students. Students will be engaged in problem-solving, using critical thinking skills, collaboration, communication, and creativity, all while learning the content you are asked to teach your students. Art and engineering embody creativity by integrating the physics of nature while using the art to creatively design the most effective object to solve a problem or solve an issue. Students will experience, or be introduced to, STEM careers while being challenged to solve a problem or issue.

Figure 1. Engineering design triangle: Next generation science standards, grades 9-12

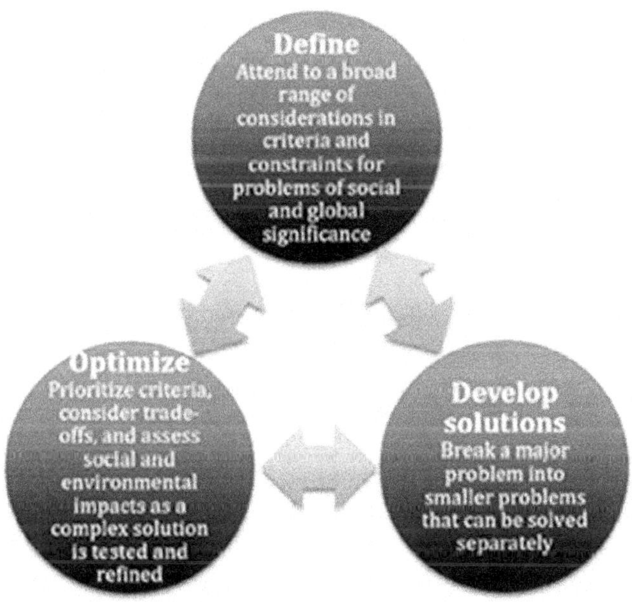

CONTENT INFORMATION

This engineering design-build project will integrate many content areas. Students are first asked to research catapults throughout history and identify effective designs. Students will use good, reliable resources from the internet while identifying quality search terms to help seek the information necessary to design an effective catapult. Students will also identify the time period throughout history when catapults were used and for what purpose. Some students may ponder how this device could be used today. As the students continue designing their catapult, they will need to collaborate and communicate with other group members, sharing their thoughts and ideas.

Mathematics will be an important integration in this activity. Students will investigate many variables, and perhaps identify many more during this investigation. The first variable that students will most likely want to investigate is the angle of launch. If the angle is too great, the projectile will go too high which will result in a shorter distance launched. On the other hand, an angle too low may overshoot the target. It is important to allow students time to test various angles. Figure 2 below explains the angle of launch. Another variable that could be tested is the rubber band resistance. If possible, provide various tensions of rubber bands, which will affect the speed of the projectile. The greater the tension, the greater the speed of the projectile. Another variable that the students can experiment with is the length of the catapult from the target. One final variable, for student exploration, is the length of the catapult launching arm. How will that change the accuracy along with varying the angle of the launch? Encourage your students to explore all these variables and create a data table to collect all these changes in their variables.

Figure 2. Angle of launch using the laws of trigonometry which relate the initial velocity to the maximum height and distance of the projectile. V= velocity and X is the launch angle

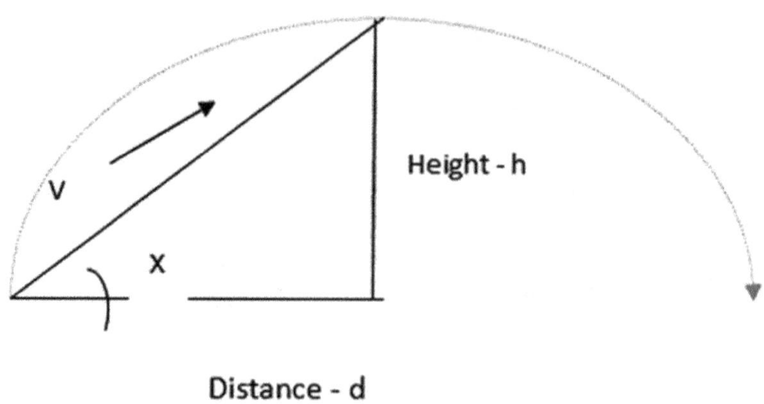

A general summary of the content covered with this activity is:

- **English/LA:**
    - Using good, quality online sources.
    - Identifying search terms.
    - Communicating findings.
- **History:**
    - Researching time period and use of catapult.
- **Mathematics:**
    - Identifying/changing variables.
    - Analyzing angle height vs. distance.
    - Weighing mass of object.
- **Art:**
    - Designing effective catapult.
    - Exploring materials in the design.
- **Physical Science:**
    - **Understanding Force and Motion:** Movement of catapult.

Integration of Standards

Common Core State Standards Connections: Middle Grades

Art is being tapped into with the design of the catapult and creating an object that will launch another object as far as possible or to a target. Leonardo da Vinci saw no boundaries between the arts and science. During this activity many art skills will be acquired: 1. Exploring different ways to create/design a working catapult, 2. Researching designs and materials for maximum effective, efficient use (what makes most sense), and 3. Creating a catapult that is most economic (using the suggestions in the extension part of the lesson plan) (Sousa & Pilecki, 2013).

Engineering Standards that are met by this activity:

- **Students who Demonstrate Understanding Can:**
    - **MS-ETS1-1:** Define the criteria and constraints of a design problem with sufficient precision to ensure a successful solution, taking into account relevant scientific principles.
    - **MS-ETS1-2:** Evaluate competing design solutions using a systematic process to determine how well they meet the criteria and constraints of the problem.

- ◦ **MS-ETS1-3:** Analyze data from tests to determine similarities and differences among several design solutions to identify the best characteristics of each that can be combined into a new solution to better meet the criteria for success.
- ◦ **MS-ETS1-4:** Develop a model to generate data for iterative testing and modification of a proposed object, tool, or process such that an optimal design can be achieved.

English/Language Arts Standards that are met by this activity:

- **Students who Demonstrate Understanding Can:**
  - **ELA/Literacy:** RST.6-8.1 Cite specific textual evidence to support analysis of science and technical texts. (MS-ETS1-1), (MS-ETS1-2), (MS-ETS1-3)
  - **RST.6-8.7:** Integrate quantitative or technical information expressed in words in a text with a version of that information expressed visually (e.g., in a flowchart, diagram, model, graph, or table). (MS-ETS1-3)
  - **RST.6-8.9:** Compare and contrast the information gained from experiments, simulations, video, or multimedia sources with that gained from reading a text on the same topic. (MS-ETS1-2), (MS-ETS1-3)
  - **WHST.6-8.7:** Conduct short research projects to answer a question (including a self-generated question), drawing on several sources and generating additional related, focused questions that allow for multiple avenues of exploration. (MS-ETS1-2)
  - **WHST.6-8.8:** Gather relevant information from multiple print and digital sources, using search terms effectively; assess the credibility and accuracy of each source; and quote or paraphrase the data and conclusions of others while avoiding plagiarism and following a standard format for citation. (MS-ETS1-1)
  - **WHST.6-8.9:** Draw evidence from informational texts to support analysis, reflection, and research. (MS-ETS1-2)
  - **SL.8.5:** Integrate multimedia and visual displays into presentations to clarify information, strengthen claims and evidence, and add interest. (MS-ETS1-4)

Mathematics Standards that are met by this activity:

- **Students who Demonstrate Understanding Can:**
  - **MP.2:** Reason abstractly and quantitatively. (MS-ETS1-1), (MS-ETS1-2), (MS-ETS1-3), (MS-ETS1-4)

- **7.EE.3:** Solve multi-step real-life and mathematical problems posed with positive and negative rational numbers in any form (whole numbers, fractions, and decimals), using tools strategically. Apply properties of operations to calculate with numbers in any form; convert between forms as appropriate; and assess the reasonableness of answers using mental computation and estimation strategies. (MS-ETS1-1), (MS-ETS1-2), (MS-ETS1-3)
- **7.SP:** Develop a probability model and use it to find probabilities of events. Compare probabilities from a model to observed frequencies; if the agreement is not good, explain possible sources of the discrepancy. (MS-ETS1-4)

Social Studies Standards that are met by this activity:

- **Students who Demonstrate Understanding Can:**
  - **D2.His.13.6-8:** Evaluate the relevancy and utility of a historical source based on information such as maker, date, place of origin, intended audience, and purpose.

Integration of Standards

Common Core State Standards Connections: High School

Engineering Standards that are met by this activity:

- **Students who Demonstrate Understanding Can:**
  - **HS-ETS1-2:** Design a solution to a complex real-world problem by breaking it down into smaller, more manageable problems that can be solved through engineering.
  - **HS-ETS1-3:** Evaluate a solution to a complex real-world problem based on prioritized criteria and trade-offs

English/Language Arts Standards that are met by this activity:

- **Students who Demonstrate Understanding Can:**
  - **ELA/Literacy - RST.11-12.7:** Integrate and evaluate multiple sources of information presented in diverse formats and media (e.g., quantitative data, video, multimedia) in order to address a question or solve a problem. (HS-ETS1-1), (HS-ETS1-3)
  - **RST.11-12.8:** Evaluate the hypotheses, data, analysis, and conclusions in a science or technical text, verifying the data when possible and cor-

roborating or challenging conclusions with other sources of information. (HS-ETS1-1), (HS-ETS1-3)
    - **RST.11-12.9:** Synthesize information from a range of sources (e.g., texts, experiments, simulations) into a coherent understanding of a process, phenomenon, or concept, resolving conflicting information when possible. (HS-ETS1-1), (HS-ETS1-3)

Mathematics Standards that are met by this activity:

- **Students who Demonstrate Understanding Can:**
    - **MP.2:** Reason abstractly and quantitatively. (HS-ETS1-1), (HS-ETS1-3), (HS-ETS1-4)
    - **MP.4:** Model with mathematics. (HS-ETS1-1), (HS-ETS1-2), (HS-ETS1-3), (HS-ETS1-4)

Social Studies Standards that are met by this activity:

- **Students who Demonstrate Understanding Can:**
    - **D2.His.5.9-12:** Analyze how historical contexts shaped and continue to shape people's perspectives.

CASE DESCRIPTION

Lesson Plan Information

This case will present a project-based scenario where students will take the place of an astronaut stranded on Mars. Like the character in the Disney film, *The Martian*, the astronauts only have a small collection of "spare parts" at their disposal to ensure survival. In this scenario our astronauts meet predators. In an effort to fend them off, they must design and construct a catapult. The students will be engaged in a deep dive. This deep dive exploration should take a maximum of 11 days but can be adapted to your students' learning needs.

This STEAM inquiry activity is a deep dive exploration. It is a deliberate "dive" into a content area of focus. In this case the area of focus is catapults. During this process the students will research the history, use, and structure of catapults. As a stakeholder in the problem, the students will assess their situation, take an inventory of "spare parts," and brainstorm with the group on how to creatively design a catapult using the best combinations of materials. The deep dive continues as the students test, revise, and retest their designs. Students not only must assess the situation with critical thinking, they also must be able to communicate their ideas to their group and think creatively. Integrating many content areas, this activity will actively engage your students. They will create a catapult using the best materials available and solve design issues as they progress through the activity.

Materials ("spare parts"):

- Cardboard shoe box (one per team),
- Rubber bands (4 for each catapult),
- Popsicle sticks (2 for each catapult),
- Masking tape (one 6-inch piece for each catapult),
- Plastic spoon (1 for each catapult),
- Rulers (1 per team),
- Scissors (1 per team), and
- Marshmallows (1 per catapult).

ENGAGE: Day 1

Begin the lesson by sharing the scenario first.

Scenario

Your team has been selected to explore Mars for the possibility of colonization. You and your team have been up on Mars for approximately 6 months. Regular and routine communication has been occurring between you and NASA. Things have been mostly proceeding according to plan. At 3:00 in the morning your team has been awakened by an alarm, which of course means that something very troubling has been discovered by NASA. Unfortunately, a group of aliens are approaching at a slow pace but they are dangerous and must be stopped. Because NASA didn't foresee this danger you and your fellow astronauts must create a defense. After taking inventory of the materials you have and consulting with the experiences at NASA, it seems that the only extra or spare parts you have are materials to create a catapult. It is impossible to send supplies in the time that the aliens could approach you because you are 49 million miles from Earth and the aliens could approach within the next 30 days. You and your team will need to research catapult design, figure out the best way to use your materials, test, re-design, and prepare for an alien assault.

Next, using some of the resources at the end of this chapter, show pictures and review the history of the catapult contained in this chapter. There are several YouTube videos you could use as well with your introduction to catapults. It would also be a good time to have the students share or research facts they are aware of, their hunches and what questions they have about this engineering task. Also try to elicit the information they might need in order to design, construct, and test a catapult.

EXPLORE: Day 1 - 2

Place the students into groups of three or four. Provide the students with a list of the spare parts available to them. Have the students spend the remainder of day 1 researching about the design of catapults and how to best use their materials. You may want to take a few minutes at the beginning of day 2 to discuss some of the mathematics associated with this activity, such as the angle of catapult, the force to hit a target, and the length of the catapult arm. Allow more research time if necessary.

In a class discussion, have the students share what they learned from their research, perhaps in a jig-saw or a gallery walk. Working in groups of three or four, have the students brainstorm and sketch a design for their catapult. During this step it is important to focus on the details of the material use and predict how it will impact the length and force of the catapult to hit the aliens. Allow plenty of time for this step and demand details on the design sketches. Once their designs are completed and have been approved by the head of NASA design team (teacher), the students will share their ideas at this point. Remind them that scientists often collaborate to gather the best information.

At this point have the students post their groups design and do a graffiti walk. This is a strategy similar to a gallery walk where students post their work and others walk around with post-it notes asking questions about their work or providing compliments. You may need to model appropriate comments for the post-it notes so that students are helping each other in their design portion of this activity. Allow students time to read the comments and perhaps make changes to their design.

Day 3 - 5

Students will begin building their catapults. Tell the students that when engineers, or in this case astronauts, are given a certain amount of materials, they must be used to their best ability. Allow enough time for them to design and create a workable catapult. If any group is struggling allow them to consult with NASA (the teacher) to assist them in their building. Take the students into the hall, gymnasium, or outside and test their catapults to determine the effectiveness of their designs. At this point they are looking for the catapult to launch the marshmallow the farthest, keeping the aliens away. Here is a good time to allow some of the variables, such as angle of launch or length from the target, to be explored. Record all data. Examine the sample data table for students to copy into their lab notebooks.

Have the teams once again brainstorm and sketch a redesign of their catapult on poster paper. These redesigns are informed by their experience testing their catapults. Refer to the data table in the appendix. Review some of the variables prior to the redesigning activity. After the students have completed their redesigns, have them post these designs around the classroom. Have the students do another graffiti walk, placing their comments, written on a post-it note, on each redesign. Review the students' comments with the class.

EXPLAIN: Day 6 – 8

The teacher will assign a number, 1-4 to each member of each group. These teams will be determined from the prior week's observation of your students' strengths. See the table below to see some of the best characteristics for each team member. Create a design team, composed of the #1 member from each group and charge them with sketching a design of a catapult incorporating students' feedback and the best of the entire classes' designs. Create an engineering materials team, composed of the #2 member from each group and charge them with gathering all the materials needed to build the catapult. Group #2 will also collect the data during the testing of the catapult. Create a building team, composed of the #3 member from each group and charge them with building the catapult designed by the design team. Create a testing team, composed of the #4 member from each group and charge them with testing the catapult to see how far the marshmallow is projected.

EXTENSION: Day 9 – 10

Gather the class and challenge all groups to build upon the class design to craft a catapult that will launch the marshmallow even farther. They are to use the current design and add to it to launch the marshmallow further. Allow each group one class period to design and one to build the object. Follow the protocol you used before in testing the projectile.

Table 1.

Design Team Number	Skills Necessary	Assigned Student
1 -Sketching Design	Drawing, collaborative, flexible, creative	
2- Materials Manager and Data Manager	Organized, detailed, good with data tables	
3 – Building Team	Follows plans, asks good questions, creative	
4 – Testing team	Flexible, gets team excited	

EVALUATION: Day 11

Compare to the recorded data of previous designs. Debrief the activity with the students. Have the students compare and contrast their designs with the pictures shared at the beginning of this learning experience. Ask them to write about a scenario in which a catapult could be used today.

STEAM TEACHING

What is STEAM education and why is important? STEM to STEAM is not a large leap at all. After all science, technology, engineering, and mathematics (STEM), done correctly, already involves writing, communication, design, and often creativity that engage the arts. STEAM, includes the arts, which is not limited to the visual arts but inclusive of theatre and music. The National Art Education Association (NAEA) defines arts education as music, drama, dance, media arts, or visual arts, including folk arts. Ruth Catchen (2013), an artist turned STEAM curriculum researcher, is asking the question, what is the learning impact on students? She is currently investigating this question using a new STEAM program targeted in Colorado. She believes that introducing the arts into STEM will benefit the under-represented students as an on-ramp to the STEM curriculum. Traditionally STEM learning focused on qualitative work and memorization and STEAM allows the arts to integrate with STEM which opens more paths for diverse learners without fear (Jolly, 2014). We need students of the 21st century to become global citizens who are problem solvers and critical thinkers, even if they choose a career that does not directly involve STEM; and STEAM education address this for all students.

Art and science have a common theme, discovery! In each of the disciplines we are discovering. The power of the arts aligns with science objective outcomes, discovery and exploration. Both art and science promote curiosity, require observation to ensure accuracy, construct meaning through observation, and allow collaboration. As educators, integration of content areas with the arts makes teaching more interesting, develops cognitive growth, and improves long term memory (Sousa & Pilecki, 2013).

A study conducted at John Hopkins University concluded that arts integration significantly improved retention of learning. They cited seven effects:

1. Practicing the learned information or skills,
2. Expanding meaning to learning,
3. Seeking out more information,
4. Acting out their learning,
5. Talking more about the content,
6. Connecting emotionally to the learned material, and
7. Representing the material through drawing.

All these effects help students learn through multiple instructional strategies using different parts of their brain (Rinne, Gregory, Yarmolinskaya, & Hardiman, 2011).

Examining this activity, *Constructing a Marshmallow Catapult*, we see many skills from the arts that align with the practices of science and engineering. When students are asking questions and defining the problems associated with designing and creating the catapult, they are fully integrating STEAM into this engineering activity. This unit asks the students to design and create a model, which blends art and STEM skills together. Today's NGSS and math common core standards require the quantitative skills to develop a model while using mathematics and computational thinking. These STEM practices dovetail nicely with the art skill of researching the material used for the design and what will work the best for the purpose of the catapult.

The bottom line is we, as educators, need to create individuals who are FULLY literate in the 21st century, using the 4 C's – collaboration, communication, critical thinking, and creativity. And STEAM allows this!

CURRENT CHALLENGES FACING TEACHERS

Along with many obvious challenges, including additional preparation time for teachers, is the need for additional materials and resources, including material storage which on the surface may seem insurmountable. This is a problem for all science classrooms, yet students benefit from the design-build activities. Some of the resources may include computer lab time and perhaps library time, which is often a scheduling nightmare. With a supportive administration and collaborative team approach, these challenges are manageable. This collaborative approach would include not just the science department but perhaps the English, mathematics, technology, and social studies departments as well. Collaboration allows each teacher to share and communicate their content expertise along with the standards that could be highlighted with this activity.

Addressing teacher attitudes towards shifting teaching practices may not seem as obvious, but learning a new pedagogy and how to implement that change can cause a lot of controversy among faculty. Additionally, in integrated programs, teachers from different departments, or areas of expertise, find themselves forced to learn new content, often material that challenges them. Research indicates that implementing these collaborative types of programs found that it had a direct impact on student learning. Again, all the aforementioned challenges can be minimized through a supportive administration, continuous staff development, and perhaps even a contracted specialist or university professor who focuses on the specific needs of teachers transitioning to a new way of teaching (Satchwell & Loepp, 2002). It is very important for the teachers who are implementing the new program to be given the opportunity to practice this instructional strategy prior to teaching it to a group of students, through comprehensive professional development. Leading teachers through the student materials as a teacher development strategy goes a long way toward preparing and motivating teachers to embrace the pedagogical approach (Diaz & King, 2007).

SOLUTIONS AND RECOMMENDATIONS

Research tells us that teachers tend to teach as they were taught and that change is difficult to manage, but once a new innovation is adopted, increased student learning is all that is required for the teacher to sustain the change. This project-based-learning (PBL) deep dive is an inquiry learning experience. The students, in the role of a team of stranded astronauts, work to find a solution to a problem. Furthermore, it is a STEAM activity. While exploring this activity, the students will experience a deep dive in engineering and the science, technology, mathematics, and art of designing and redesigning a catapult. As such, students will not only be exposed to the process of engineering but also to how all the components of STEAM are interrelated. As for teachers, if they decide to take a risk and do an inquiry activity, they will be pleased to see that the students are not only engaged but are also learning the content and concepts.

THE HISTORY OF CATAPULTS

The catapult was one of the most effective and deadly weapons in medieval siege warfare. It has been critical to warfare since ancient times and has been used by Greek, Roman, and Chinese warriors to vanquish their enemies. War was common throughout Europe in the Middle Ages and thick, protective stone walls were often built around cities and castles to prevent attacks from outsiders. This made catapults ideal for launching missiles. Although many different objects were utilized as missiles, the function of the catapult was to launch the object either directly at the wall to cause maximum damage or over the wall to strike the population within the wall's protective barriers.

There are three main types of catapults that were used in medieval warfare. The first is the ballista, which was an attempt at increasing the range and power of another weapon, the crossbow. The word *ballista* comes from the Greek word *ballistes,* which means "to throw." Invented by the Greeks and modified by the Romans, the ballista was created to amplify the range and power of the crossbow. Two wooden arms were attached to a piece of rope, which was made out of human hair or animal sinew. The rope was attached to a wench and pulled, bending the arm back. When released, the ballista would shoot large arrows and darts toward the enemy with deadly precision.

The second type of catapult is the *mangonel* (from the Latin word "manganon," meaning 'engine of war"), which is what many modern people think of when they think of a catapult. The mangonel was invented by the Romans in 400 BCE. The missiles fired with mangonels included anything from stones, human excrement, or rotting animal carcasses. It consisted of a long wooden arm with a bucket (though earlier models used a sling) with a rope attached to the end. The arm is then pulled back from a 90 degree angle, the bucket loaded, and the energy stored in the tension of the rope and arm would be released. Upon release, the mangonel's arm would return to its position, and when it came into contact with the beam, the arm would

Figure 3.

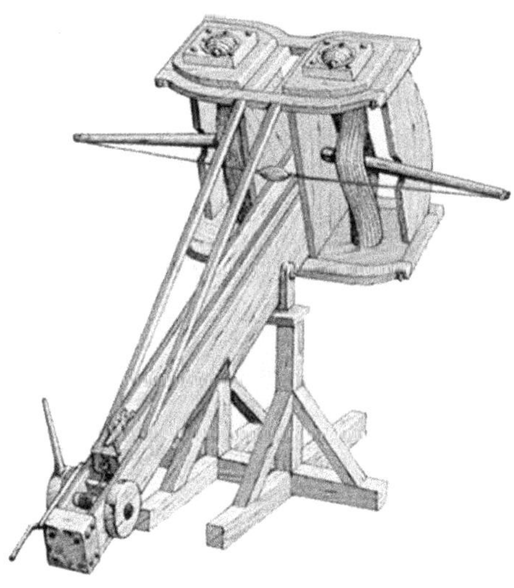

Figure 4.

stop but the missiles stored in the bucket would continue on their path toward the enemy. The mangonel was capable of firing 1,300 feet and was fairly easy to construct. Wheels were sometimes added to the design to increase mobility.

The third type of catapult is the trebuchet, which was designed for maximum force. Created by the Chinese in 300 BCE, the trebuchet was the most powerful of the catapults. It arrived in Europe around 500 CE, and the stones it hurled had a destructive force that could demolish a city and castle walls. The most commonly used ammunition was stones, but darts and sharp wooden poles could be used if necessary. This device consisted of a long arm up to 60 feet long, balanced on a fulcrum that was off center. The short arm was attached to a counterbalance, a heavy

Figure 5.

Table 2. Data table for catapult trials

Variables	Trial 1	Trial 2	Trial 3
Angle of launch 1. angle 2. distance 3. conclusion	 1. 2. 3.	 1. 2. 3.	 1. 2. 3.
Length of spoon launcher 1. length 2. distance 3. conclusion	 1. 2. 3.	 1. 2. 3.	 1. 2. 3.
Rubber band 1. tight - distance 2. loose – distance 3. conclusion	 1. 2. 3.	 1. 2. 3.	 1. 2. 3.
Other variables			

lead weight, or a pivoting ballast box filled with dirt, sand, or stones. A sling was attached to the end of the long arm, and a rope was attached to the long arm and pulled down until the counterbalance was high in the air. The sling was then loaded with projectiles, then the rope released and the energy in the long arm brought to an abrupt stop. The projectile, however, continued its path to its target. The trebuchet was a feared siege weapon and the target of reconnaissance missions sent to burn them to the ground.

Catapults were used extensively throughout Europe (predominantly by the French) until 885-886 CE when new defense systems rendered catapults ineffective.

REFERENCES

Catchen, R. (2013). Reflections: How STEM becomes STEAM. *The STEAM Journal*, *1*(1). doi:10.5642/steam.201301.22

Diaz, D., & King, P. (2007). *Adapting a post-secondary STEM instructional model to K-5 mathematics instruction*. Clemson, SC: Clemson University.

Fortus, D., Krajcikb, J., Dershimerb, R. C., Marx, R. W., & Mamlok-Naamand, R. (2005). Design-based science and real-world problem solving. *International Journal of Science Education*, *27*(7), 855–879. doi:10.1080/09500690500038165

History of Catapults. (n.d.). Retrieved January 17, 2016, from https://sites.google.com/site/physicsofcatapults/home/history-of-catapults

Jolly, A. (2014). STEM vs. STEAM: Do the arts belong? *Education Week Teacher*. Retrieved February 21, 2016, from http://www.edweek.org/tm/articles/2014/11/18/ctq-jolly-stem-vssteam.html

Rinne, L., Gregory, E., Yarmolinskaya, J., & Hardiman, M. (2011). Why arts integration improves long-term retention of content. *Mind, Brain, and Education*, *5*(2), 89–96. doi:10.1111/j.1751-228X.2011.01114.x

Satchwell, R., & Loepp, F. L. (2002). Designing and implementing an integrated mathematics, science, and technology curriculum for the middle school. *Journal of Industrial Teacher Education*. Retrieved February 14, 2014, from http://scholar.lib.vt.edu/ejournals/JITE/v39n3/satchwell.html

Sousa, D. A., & Pilecki, T. (2013). *From STEM to STEAM: Using brain-compatible strategies to integrate the arts*. Thousand Oaks, CA: Corwin Publishing Company.

Wicklein, R. C. (2006). Five good reasons for engineering as the focus for technology education. *Technology Teacher*, *65*(7), 25–29.

KEY TERMS AND DEFINITIONS

Catapult: A device designed to launch a projectile.

Deep Dive: An in-depth examination into a content topic or problem.

Problem-Based Learning: An inquiry instructional strategy where the students work in terms as collaborators on an ill-structured problem.

Science: Both a body of knowledge and a process of studying the natural world.

Technology: Using science to solve problems.

Chapter 14
Engineering and Art:
Putting the EA in STEAM

Sara B. Smith
Goose Creek High School, USA

EXECUTIVE SUMMARY

This chapter per the author discusses challenges faced by first year pre-engineering students. Also discussed are several topics taught within the curriculum including: the engineering design process, sketching, measurement, the elements and principles of design, and three-dimensional modeling. The chapter proposes a design project for engineering students that would tie all of these concepts together to provide an additional learning opportunity for students and more relevant practice of skills like isometric sketching, creating three-dimensional computer aided design models, and measurement. Samples of student work from the project are included.

ENGINEERING AND ART: PUTTING THE EA IN STEAM

I became a teacher after working in industry as an engineer. I knew I needed to learn about teaching and choose to enter a teacher education program and become certified through a traditional route. Once I entered the classroom, aspects about my prior life became more known and I was given the opportunity to teach classes such as engineering and energy in the high school. To me, engineering is the application of math and science concepts to solve problems. When I thought through the difference between teaching engineering and chemistry a few aspects appeared. Such as, engineering solutions to problems could be an invention, an innovation, or

DOI: 10.4018/978-1-5225-2334-5.ch014

Copyright ©2017, IGI Global. Copying or distributing in print or electronic forms without written permission of IGI Global is prohibited.

a more efficient process. Engineers design, build, conduct tests, market, and communicate with others in their pursuit of solutions. Engineering requires creativity and ingenuity. An important part of the creativity that is integral to engineering is art. Art practices lend a hand to an engineer's process of developing solutions. For example, John Maeda is a former president of the Rhode Island School of Design where he was very involved in adding art to the STEM movement. He holds degrees in Electrical Engineering and Computer Science from MIT as well as a Ph.D. in design from Tsukuba University in Japan. In his piece for Scientific American in 2013, Maeda said that "STEAM and arts integration are crucial in K-12 education, engaging students in the STEM subjects and ensuring that creativity doesn't fall by the wayside as we chase innovation..." (Para. 5).

Integration is also important at the college level. So much so, that multiple colleges and universities have designed courses that tie art and engineering together in one. One of these courses is "Enduring Design: The Art of Engineering" being offered at the Rose-Hulman Institute of Technology in Terre Haute, Indiana. The course was described by Mirth and Findley (2015) as being designed by both humanities and engineering faculty. The faculty members saw that the school's engineering students did not take many liberal arts courses and felt that this caused a "potential handicap in their ability to see opportunities from a broader perspective" (p. 1). While the course was open to all students, a majority of those enrolled in the first offering were engineering students. The course is an art course that incorporates examples of engineering. This course is still new and data is still being collected, but initial data show that the art and engineering connections taught throughout the course helped students to become more creative in their ideas for improving both form and function of objects.

At the University of South Florida in Tampa, an engineering professor created a course that "merges his research world with the world of fine art" (National Science Foundation, 2006, Para. 1). The course teaches concepts like electromagnetism and optics while lab activities incorporate lessons from works of fine art to students creating pinhole cameras and learning how to spot a forgery.

As these schools created courses to open the minds and creativity of engineering students by incorporating art, the project described in this chapter is an attempt to do the same for high school students enrolled in pre-engineering programs, most specifically the Project Lead The Way students that I teach.

Project Lead the Way

Project Lead the Way began at Shenendehowa High School in upstate New York in 1987. There was a decrease in enrollment in engineering programs creating a predicted shortfall of over 600,000 engineers in the United States by the year 2000 and a large amount of engineering undergraduates were changing their goals and dropping out of engineering (Kelley & Wyanski, 1987). Richard Blais, the Director of Occupational Education in the school district at the time, set out to answer these problems by instituting a pre-engineering curriculum in high school. The original program consisted of four courses that were designed with input from higher education institutions and industry personnel in the local area. The introductory course included technical drawing both on paper and using computer-aided design (CAD) technology. Students also built scale models from those drawings. Other courses included Digital Electronics, which involves logic and controls, and electrical circuits; and Principles of Engineering in which students built mini manufacturing plants that simulated the automated manufacturing processes used in industry. The courses also taught students about pneumatics, hydraulics, electric motors, automation, and robotics. All of the concepts were delivered through project-based learning. A young man in Kelley and Wyanski's video (1987) said, "It's half teaching ourselves and half of the teacher teaching."

Development of the program continued with input and grant funding from post-secondary institutions like Rensselaer Polytechnic Institute, and from local business and industry partners. Even in 1987 the high schoolers were able to earn college credit for their pre-engineering courses. To help spark interest in STEM at a younger age, the high school students went to local elementary schools and did small projects with the students.

In 1997, Project Lead the Way launched its Pathways to Engineering program in 12 New York State high schools. A partnership with High Schools That Work and the Southern Region Educational Board helped to spread the program to an additional 30 states (Chautauqua Lake School District, n.d.). Today, PLTW reaches over 2 million students in all 50 states and the District of Columbia as well as some U.S. territories. It has trained over 35,000 teachers through its two-week summer training institutes and on-going professional development (PLTW, 2016). There are courses in Aerospace Engineering, Civil and Architectural Engineering, Environmental Sustainability, and Digital Electronics among others for the high school level students. Gateway to Technology is one of the courses available to middle school students. In the past few years, PLTW has extended its mission into the elementary schools. The curriculum provides research-based, problem-based learning experiences for students. I have been teaching the Introduction to Engineering Design course since 2013. This course introduces students to many concepts and skills that will be needed throughout their pre-engineering and college engineering programs.

Course concepts and skills include various types of sketches, the engineering design process, 3D modeling with Autodesk Inventor, and reverse engineering. Of these, many of my students have the most difficulty with sketching. Students are required to be able to generate isometric sketches of objects. Isometric sketches show three faces of a three-dimensional object. The sides extend at 30 degree angles to give the object its three-dimensional appearance. These sketches are completed on isometric graph paper.

Some students can look at that paper and see where the lines go and how to draw the shapes. Others, however, cannot visualize making a three-dimensional shape on this two-dimensional surface. My pondering about this issue raises one question: Why is seeing three dimensions in a two-dimensional space so difficult for high school students? In the Project Lead The Way - Where It Started video (Kelley & Wyanski, 1987), Richard Blais stresses the importance of this skill by saying "Drawing is a language of industry, and before they can really take other courses in technology they need to understand that language."

The Mathematics Link

I taught a homebound student geometry one semester recently and was excited to see that he had to generate isometric sketches from the two-dimensional multi-view sketches as well as the reverse process. This was included in chapter one of the textbook, Prentice Hall's Geometry: Common Core 12th edition (2012). The rest of the curriculum we worked through was traditional geometry: similarity, congruence, triangles, parallel lines and transversals, polygons, etc. While there was overlap between some of this content and the content of the Introduction to Engineering course, none was as helpful to the goal of being able to construct isometric sketches of 3D objects. The high school at which I teach reversed the order of Geometry and Algebra II at the beginning of the 2016-2017 school year. What does this mean

Figure 1. Sample isometric graph paper

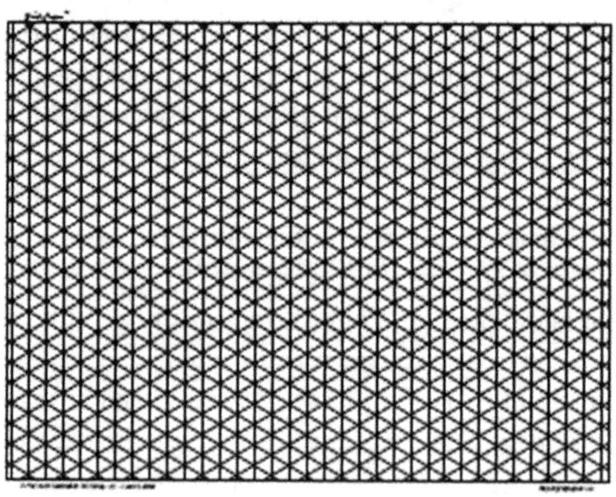

for Introduction to Engineering Design? The course will have both ninth and tenth graders who are enrolled in Algebra I, Algebra II, or Geometry. The combination of all of this information informs me that courses in mathematics will not provide ample support of the skills being taught in this course.

The Fine Arts Link

My high school is the site for the district School of the Arts. We have students from throughout the district that have an aptitude for music, art, dance, creative writing, drama, or a combination of those. Art students tend to excel at the sketching skills required for the engineering course. They can sketch three-dimensionally on isometric paper, graph paper, and, really, any paper you provide. It seems as though they come equipped with an unteachable skill: spatial reasoning. They already have the capacity to see the form on paper. They know how to shade an object to provide a more realistic appearance. These students can take three-dimensional forms and draw them in two-dimensional renderings, cross-sectional views without dissonance.

The Engineering Design Process

Why are the abilities to sketch three-dimensionally and to resolve a three-dimensional object into two-dimensional views so important? The engineering design process requires them. There are many different accepted layouts of this process and the combination of geometry and art dimensional drawing a key content. The process flow taught in the introductory course includes: identification of a problem, the generation of concepts, the development of solutions, the construction and testing of a prototype, the evaluation of solutions, and the presentation of solutions. These steps can and will be iterated as many times as necessary for successful completion of a project. Sketching is incorporated in at least four of the six steps listed. Sketches need to be detailed and accurate for communication between client and designer, designer and manufacturer, and between the multiple designers working on a project team. The art and geometry help with the visualization process and the physical creation of the sketches. Sketches are turned into CAD (computer-aided design) parts and assemblies in Autodesk Inventor. From those parts and assemblies, annotated drawings are generated. These drawings include overall size dimensions, dimensions for the size and location of part features, and can also show how individual parts interact with each other to form the final product.

Reverse Engineering and Elements and Principles of Design

In reverse engineering, the students are given an item such as a kitchen appliance, a toy, or an Automoblox vehicle and are required to fully document the item. Automoblox is a line of wooden toy vehicles that have removable parts and sections that allow children or students to deconstruct them and learn how the parts interact with each other. The vehicles come in a large variety of models from classic cars to modern day SUVs and sedans to emergency response vehicles. I used an Automoblox pickup truck in my summer training institute to learn reverse engineering concepts and the program at my school has several for the students to use in the course. Reverse engineering requires the students to perform structural, functional, and visual analyses on the item. Students documentation of what materials parts are made, how the parts interact with each other, part dimensions, and the inputs and outputs of the product and its individual parts. The students hand-sketch the parts in their notebooks, build and assemble the parts in Inventor, and generate annotated drawings of each part. They also create an explosion drawing with a parts list, see Figure 2, and a video showing how all of the parts go together.

One of the introductory lessons in reverse engineering is the elements and principles of design. These are used in product design, art, architecture, and even landscape design to attract the eye and convey feelings through the work. The elements and principles are used in product design to capture consumer attention and increase the sale of the product. This knowledge is used by the Introduction to Engineering Design students in their visual analysis of an object. Students can conduct an inter-

Figure 2.

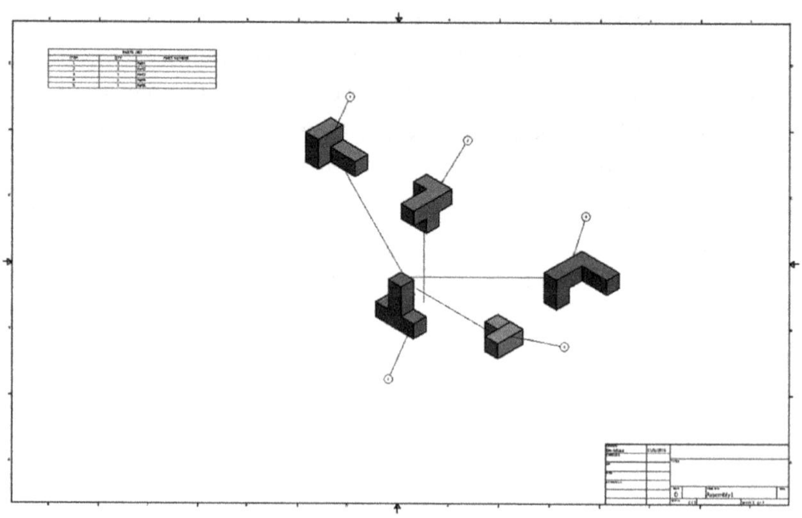

net search to learn what these elements and principles are and what each invokes in a design. The Project Lead the Way curriculum details eight visual design elements: point, line, color, value, shape, form, space, and texture; and seven design principles: balance, emphasis, contrast, rhythm, proportion, unity, and economy. Any combination of these can be found in monuments, buildings, homes, advertisements, gardens, toys, cars, and even tennis shoes! Students in the course are asked to find a variety of objects and analyze them for these elements and principles. They present the information in a poster. My students are required to complete an analysis of six different things and include information on at least three elements and three principles used and how those affect the design of the object they selected. In their reverse engineering project, this would comprise the visual analysis portion. One of the main points that most students struggle with is describing the emotions or feelings invoked by these elements and principles. They can amply identify which are present but not why. The visual arts students, again, seem to excel at these visual analyses; able to verbalize the analysis with more depth and detail.

Computer Aided Design: Autodesk Inventor

Throughout the course, students learn how construct parts in Autodesk Inventor, a computer-aided design program. The parts start out simple and then increase in difficulty. The students create the parts using additive modeling, adding features to a base shape and subtractive modeling, starting with a large generic shape like a rectangular prism and cutting away to achieve the part required; or a combination of both. For many parts, the students are given isometric sketches that they use to guide them in generation of the parts. As they progress through the course, parts are generated by following dimensioned drawings. In reverse engineering, one of the final projects in the course, students generate parts in Inventor based on their own drawings and measurements. As this is a group project, all students are expected to use their measurement and Inventor skills to contribute parts to the final assembly that represents their item. If the parts are not measured and generated correctly, they will not fit together well in the CAD assembly. Sometimes there will be gaps between parts or one part might be wider than its adjacent parts.

Measurement

This brings us to the final concept with which students struggle: measurement. As a high school teacher of students in schools in the United States, I assumed that students have had a lot of exposure to measurement in the U.S. Customary System of Units (inches and feet) and that is why they struggle with the metric system that I teach in Physical Science and Chemistry courses. Engineers have to be able to work and communicate in both systems of units, and, as such, both are taught in the Introduction to Engineering Design course. Students must be able to measure dimensions with the necessary level of precision. On a ruler with sixteenths of an inch increments, students have to be able to measure to the thirty-second of an inch. This takes a bit of practice, but students seem to gain an understanding rather quickly. In addition to a ruler, students learn how to measure with dial calipers. This is a very new instrument for high school students and it introduces an odd concept: inches with decimal places rather than fractions. This is the tool that students use to measure the parts for their reverse engineering project. Precision on a dial caliper extends to the ten-thousandths place. If not calibrated appropriately prior to making a measurement or not read correctly, errors like those described above can find their way into the project. While measurement is crucial in the fields of science and engineering, it is also essential in art. Artists use measurement in the form of scales, ratios, and proportions. While I consider these three forms of measurement very mathematical, they add so much beauty to a work of art whether it be on canvas, a sculpture of a significant person, a monument to celebrate those who have impacted history, or even a beautiful garden.

The engineering design process, sketching, and measurement are all taught in the first half of the course. Elements and principles of design are taught near the end of the course with reverse engineering. Would the students be better served and better practiced in all of the skills described if the latter topic of design were introduced earlier in the curriculum? Not to rewrite well-researched curriculum that has been successfully implemented for almost twenty years, but to provide an additional opportunity for students to master skills and concepts. Why not provide the students with their own design opportunity that will allow them a chance to use the engineering design process, sketching, measurement, and design elements and principles to design their own house or building or monument or product?

Spatial Intelligence

Howard Gardner's theory of multiple intelligences includes spatial intelligence. People who display this intelligence can easily visualize in three-dimensions. They can visualize rotations of objects in space and are able to notice very fine details. Inspiring Breakthrough (2005) lists three skills associated with this intelligence: spatial awareness, working with objects, and artistic design. Navigators, sculptors, and architects are known to have high spatial intelligence. People with high spatial intelligence like puzzles, drawing, and daydreaming. People begin to develop visual-spatial skills as infants and toddlers. Along with verbal and mathematical abilities, visual-spatial intelligence is a strong predictor of future success. Park, Lubinski, and Benbow (2010) describe work done with Jonathan Wai. Together they reviewed data from two longitudinal studies: Project Talent and the Study of Mathematically Precocious Youth (SMPY). Through Project Talent data, Wai and his colleagues found that teens with strong visual-spatial intelligence are very drawn to working with their hands. Those same students went on to pursue education and careers in science and technical fields. They also noted that there are not many programs in high school through which these students can pursue their interests. Similar findings stemmed from the SMPY data.

Engineers employ visual-spatial intelligence when they design a part. The task requires them to understand the part features from multiple perspectives and understand how the part works in conjunction with other parts (Tseng & Yang, 2011). The sketching and CAD modeling of objects in the Introduction to Engineering Design course as described before employ visual-spatial skills. Students must be able to look at an isometric sketch of an object from one angle and understand the features of that object in all three dimensions separately. This understanding is then used to construct multi-view sketches that can then be used to generate a 3-D model in Inventor. The reverse process is also sometimes necessary. Students will receive a multi-view sketch from which they will draw the isometric view. This is particularly difficult as features appear differently in each view. For example, a curved surface on the top of an object may appear as a rectangle in the top view of the object. The only way to know that it is curved is to use the front view of the object at the same time.

Visual-spatial intelligence is also strong for artists, especially sculptors. They are able to visualize their piece in the space occupied by the raw material. They can see in their mind's eye from what areas to remove material. The artist can visualize the form and make mental measurements regarding scale and proportion.

It is in the possession and use of strong visual-spatial intelligence that artists and engineers are similar. While the products of their work serve different purposes, the approach to their design and creation is the same. "Artists and scientists tend to approach problems with a similar open-mindedness and inquisitiveness - they both do not fear the unknown, preferring leaps to incremental steps" (Maeda, 2013, para. 7).

The Standards

There are many sources for engineering education standards. The Next Generation Science Standards include engineering standards as it relates to science education. The International Technology Education Association and its Technology for All Americans Project first published the Standards for Technological Literacy in 2000. The standards were written to "address the mismatch between dependency and understanding" (ITEA, 2007, p. v) of the technology that surrounds us. Project Lead the Way developed standards that guide their curriculum and are only accessible to school systems under contract with Project Lead the Way. Your state may even have their own standards for engineering courses; mine does. Rather than list the many relevant standards here, I have listed several key words and phrases that relate to standards addressed through the included project. These key words include: attributes of design, design process, engineering design, design principles, three-dimensional modeling, isometric sketching, and computer-aided design.

The Project

The first three units in Project Lead the Way's Introduction to Engineering Design Curriculum are the design process, technical sketching and drawing, and measurement and statistics. One key concept to include within these units is the concept of scale. It is important for students to realize that they will be designing a house, building, or monument of some kind but it will have to be sketched and generated in the computer-aided design software on a much smaller scale. By this point in the curriculum, my students have also started using Autodesk Inventor. It breaks up the monotony and the students really want to learn how to use the program. I would incorporate the elements and principles of design, which is included in reverse engineering, immediately after the students learn how to measure in the metric and U.S. Customary systems of units.

Upon completion of the pertinent instruction, students should be assigned the project. It can take as many class periods as you would like to allow. I asked my students to conduct a five-minute brainstorming session to generate concepts. Then we talked about the criteria and constraints of the project. They had to generate an original design. They had to be able to identify the scale that they used in their sketches and CAD renderings. The students also had to be able to explain the incorporation of the elements and principles of design that they used in their design. If they put columns on their monument, for example, they had to be able to explain that the vertical lines of the columns added a sense of formality and distinction to their monument. The students then employed a decision matrix to choose the most viable option. The criteria for their decision matrix included elegance, aesthetics, the time required to complete the requirements for the design, the skill level required to complete the design project, and the ease of including a wide variety of elements and principles of design. As a class, the students determined which of these should be weighted more heavily. They then scored each concept and used the decision matrix results to select a concept.

Once the concept is selected, students should generate an annotated isometric sketch on paper. They should generate sketches for multiple views if they would best show the design details. These sketches need to be drawn to scale; a scale that they have selected.

Figure 3. Student generated decision matrix

	Skyscraper	Total	Statue	Total	House	Total	Fort	Total
Elegance-1	2	2	1	1	2	3	1	1
Aesthetics-2	3	6	3	6	4	8	1	2
Time-4	3	12	3	12	2	8	2	8
Skill-4	3	12	4	12	3	8	2	8
Elements and principles-5	4	20	3	15	3	15	3	15
Total		52		46		41		34

The students will then be able to generate a three-dimensional model in Autodesk Inventor. If Autodesk Inventor is unavailable, students can use Google Sketch Up; a free program with multitudes of tutorials available online.

If the time and supplies are available, students could also build a scale model of their design. This would incorporate more art and make the project more kinesthetic as well as increasing student engagement.

A very important piece of the engineering design process is communication with peers, management, and clients. While creating a PowerPoint presentation does not give today's students much of a challenge, standing in front of the class presenting the information certainly does. The students should include, at a minimum, the results of their brainstorming session, the decision matrix, a picture of their hand-

Figure 4. Student sketch of home design

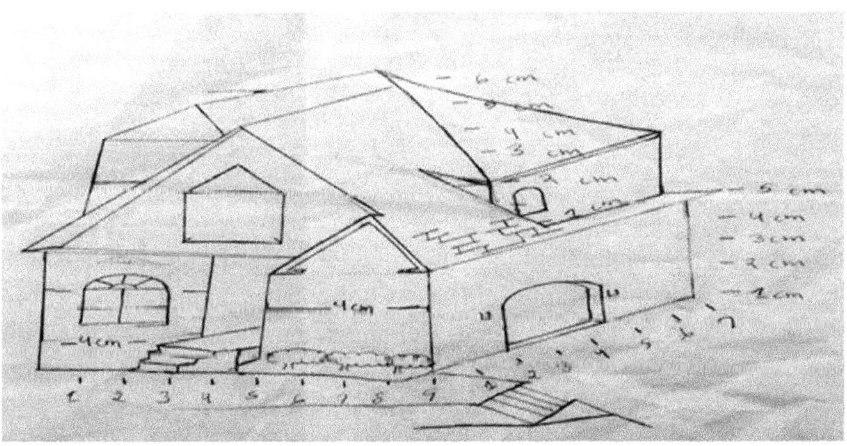

Figure 5. A sketch of the design of a second student group

Figure 6. Google Sketch Up rendering of design sketched in Figure 4

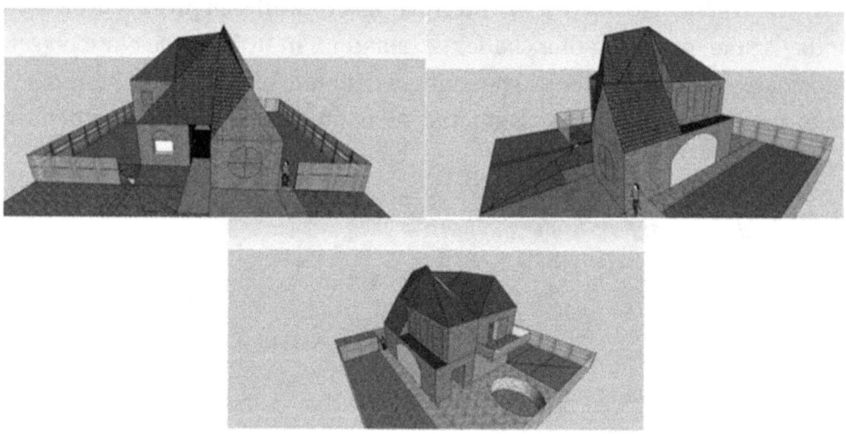

drawn sketches, and a picture of their computer generated design. I like to have my students include a brief biography slide as well. So many students make it through entire semesters and do not even know their classmates' names! Students can use this opportunity to learn how to make a five to ten minute presentation based on slides from which they cannot read.

Inviting guests to participate in this project would be great too. Guests could include other teachers, especially art teachers, administrators, or even representa-

Figure 7. Google Sketch Up rendering of a third student group's design

tives from community organizations and businesses. The guest can be included for as much of the project as you would like. They could be the client who requires a design and provide criteria and constraints of their own. The students could then be required to communicate professionally with the client throughout the process to get design approval or feedback on design concepts. The guests could then attend the presentations. This will provide a more formal atmosphere for the students and a more genuine experience.

REFERENCES

Chautauqua Lake Central School District. (n.d.). *History of project lead the way*. Retrieved from http://www.clake.org/District/Class/28-Engineering-and-Technology-Department/1447-HistoryOfPLTW.html

Inspiring Breakthrough. (2005). *Visual spatial intelligence*. Retrieved from http://www.inspiring-breakthrough.co.uk/learning-styles/visual spatial-learning.html

International Technology Education Association. (2007). *Standards for technological literacy: Content for the study of technology* (3rd ed.). Retrieved from https://www.iteea.org/File.aspx?id=67767&v=b26b7852

Kelley, J. A. (Producer & Director), & Wyanski, J. (Producer & Director). (1987). *Project lead the way - Where it started 1987 Dick Blais* [video]. Retrieved from http://youtu.be/iIdomsEhmYM

Maeda, J. (2013, July 11). *Artists and scientists: More alike than different* [Web log comment]. Retrieved from http://blogs.scientificamerican.com/guest-blog/artists-and-scientists-more-alike-than-different/

Mirth, J., & Findley, A. (2015). Enduring design: Examining the relationship between art, engineering, and creativity. *Proceedings from ASEE '15: 2015 ASEE Zone III Conference.* Retrieved from https://www.asee.org/documents/zones/zone3/2015/Enduring-Design-Examining-the-Relationship-Between-Art-Engineering-and-Creativity.pdf

National Science Foundation. (2006, September 11). *The art of engineering* [press release]. Retrieved from http://www.nsf.gov/news/news_summ.jsp?cntn_id=107990

Park, G., Lubinski, D., & Benbow, C. (2010, November 2). Recognizing spatial intelligence. *Scientific American.* Retrieved from http://www.scientificamerican.com/article/recognizing-spatial-Intel/

Post Illuminance. (n.d.). *Custom graph: Isometric paper*. Retrieved from http://customgraph.com/piart.php?art=590

Project Lead The Way. (2016). *Engaging students from New York to Honolulu.* Retrieved from https://www.pltw.org/about-us

Tseng, T., & Yang, M. (2011). The role of spatial-visual skills in a project-based engineering design course. *Proceedings from ASEE 2011 Annual Conference & Exposition*. Retrieved from http://web.mit.edu/~mcyang/www/papers/2011-tsengYang-ASEE.pdf

KEY TERMS AND DEFINITIONS

Computer-Aided Design: Using a software program to generate a three-dimensional representation of a design that can then be used for documentation or 3D printing a physical model.

Design Principles: How an artist or designer combines the elements of design in their work.

Design Process: Process that guides the development of solutions and designs aimed at solving the world's problems.

Elements of Design: The fundamental aspects of paintings, drawings, and designs.

Engineering Design: The design process followed by engineer's in their pursuit of answers and innovation.

Isometric Sketching: Creating a picture of a design on paper while presenting three faces to show the three dimensions.

Three-Dimensional Modeling: Converting a two-dimensional representation of a design into a three-dimensional representation.

Visual-Spatial Intelligence: The ability to visualize and manipulate an object or design in three-dimensional space.

Chapter 15
Using STEAM in Marine Science:
Incorporating Graphic Design Into an Existing STEM Lesson

Callie (Van Koughnett) Dollahon
Wando High School, USA

EXECUTIVE SUMMARY

This case summarizes two perspectives on inclusion of Arts in STEM/STEAM education, and how they influence the modification of an existing STEM lesson. Teachers are encouraged to use many instruction models throughout their careers, and inclusion of new methods can seem daunting. This case hopes to illustrate how STEAM education can be included in a classroom through intentional use of graphic design in an everyday lesson or a longer unit. Students in the case are asked to design and build a robotic arm that is capable of accomplishing a task such as move or grasp an object. The specific context is Marine Science in nature, but can be adapted to many other content areas.

LITERATURE REVIEW

PBL. PLTW. DBQ. Inquiry. 5E. Backward by Design. STEM. STEAM. Educators are exposed to so many methods, models, and hopefully, best practices, throughout their careers that it's no wonder we sometimes ask why and how we will incorporate them all. One challenge that faces teachers today is how to cover the breadth

DOI: 10.4018/978-1-5225-2334-5.ch015

Copyright ©2017, IGI Global. Copying or distributing in print or electronic forms without written permission of IGI Global is prohibited.

of required content that will be tested by the district or state each year and how to incorporate methods (see list above) that seem to be "bigger" than the time allotted in the course. Specifically, in terms of STEAM (an acronym for Science, Technology, Engineering, Art, and Mathematics) education, we are faced with the thought that in order to incorporate content and lessons from each of the 5 included disciplines, we must create an elaborate project based lesson that might require weeks of instruction.

For example, the Dream Factory at Elizabeth Forward Middle School integrates all of these disciplines by having students set up their very own chocolate business—students work with a local candy factory, print 3-D molds for the chocolate, create artistic labels, analyze cost and sales, and even look at the chemistry of the chocolate itself (Keruskin, 2015). Projects like this are inspirational. They are experiential in nature, provide links between disciplines, allow for connections to be made outside the classroom, and promote rigor in education in an engaging and fun way for students. However, some educators may find challenges such as time, technology, and even support to recreate something similar in their own school.

STEAM education is by nature interdisciplinary, but it doesn't have to be a giant "project" like the Dream Factory. Some teachers are finding ways to inspire the creativity, collaboration, problem solving, and reflective learning that STEAM proponents encourage by simple tweaks to their everyday lessons. Take, for example, Henriksen's (2014) description of a teacher who integrates art into science lessons by having "students create visual advertisements to describe a science idea, or a concept, organism, and so on." The classroom described features daily use of STEAM inspired lesson content, rather than a project or problem based approach, but is still successful in developing the skills sought in quality STEAM education programs. Students are afforded opportunities to use creativity, visual communication, critical thinking (through comparison, explanation, and modeling) and collaboration in this case, just as in the project based approach of the Dream Factory.

Why STEAM? (STEM + A)

My experience as an educator and in education, indicates that it is possible to incorporate quality STEAM learning on both levels—the in depth, time consuming project and the everyday lesson that allows for coverage of the many standards. According to Beers (2011), quality STEM Education "exemplifies the cross-curricular learning that is the foundation of a 21st-Century curriculum," and promotes the 4 C's: Critical thinking and problem solving, Communication, Collaboration, and Creativity (p. 5). STEAM Education, then, should promote these skills, but additionally provide greater opportunity for new ways of seeing, thinking, and learning for each discipline through the additions of arts education.

Take a look at any science textbook, engineer's notebook, NOVA video publication, or architectural installation, and it is clear that art is already a fundamental component of engineering, science, technology and mathematics. The engineering process mimics the artistic process, especially in a commercial setting. The purposeful practice that artists use to improve and enhance their craft is akin to the repeated iterations used by engineers to refine and test solutions. Each reproduction provides new information, from data to muscle memory, whether the learner is an artist, mathematician, scientist, engineer, or otherwise. In a graphic design course, students often keep a portfolio of ideas and variations of a concept, and it is through the process of reflection, evaluation, and sometimes collaboration, that these ideas are narrowed down into one product, whether art or engineering.

My personal experience in graphic design gave me an intuitive window into deeper math concepts in college. After working on a 4-H project involving embroidery design and print graphics in 8th grade, I found myself opening a small business that I operated throughout high school, using graphic digital programs that exposed me to translations, rotations, 3-Dimensional structure, vectors, and color theory that would later be the tangible mental foundation I used to explore and understand the same math, physics, and applied technology concepts in STEM courses. As Vaidyanathan (2012) states, "Digital design is a blend of art and technology, [built] on a foundation of math, science, and engineering" (p. 27). For me, art contributes deeply to my experience as a scientist, teacher, and STEM coordinator. But STEM concepts contribute to my work as a quilter and graphic designer as I determine scale, measurements, working angles, and order of operations (particularly when building a complex quilt design).

Many artists work alone, the creative process and copyright being rather insular in nature. Guyotte, Sochacka, Costantino, Kellam, and Walther (2015) suggest that STEM disciplines can contribute to the arts themselves by allowing for artists to experience collaboration, especially a collaborative approach to creativity. Through the development of a STEAM course, the Transdisciplinary Design Studio, they reinforce the process of collaborative creativity by purposefully crafting an authentic (real-world) and multidisciplinary assignment that forces students from multiple backgrounds to explore and define the problem rather than create a solution. The assignment specifies that students create an art exhibit that illustrates the problem itself, not the solution, which is how problem development rather than solution finding was achieved. Additionally, students in the course kept "visual journals—a hybrid between the writer's reflective journal, artist's sketchbook, or engineer's design notebook" (Guyotte et al., 2015), that offered insight for students and instructors about the learning that took place during the course. Students found that the group exploration and reflection during the course allowed them to develop new understandings of what collaboration, empathy, creativity, and the learning process mean to them as students and as employees in their future careers.

Sketching Holds a Place in STEM

Directions to put together a piece of furniture or use an appliance are made more effective with a visual reference in the form of a picture, sketch, or diagram. Similar to the Transdisciplinary Design Studio students' visual journal, Koester (in press) has developed a professional development program, Project Draw for Science, that applies basic theories of graphic design and sketching to science concepts and inquiry. Teachers and students create explanatory models in the form of drawings called "know"tations that represent the "4 main languages of science, 1) science terminology (the WORDS); 2) explanatory drawings (the IMAGES); 3) the equipment and procedures used during an investigation (the ACTIONS); 4) and mathematical equations, icons, graphs, and tables (the SYMBOLS)" (Koester, 2016). Part of the point of creating a "know"tation is to provide authentic context to the content, which is a recommendation from quality STEM education. Therefore, using visual art forms in this way creates a strong STEM lesson; but by including drawing as a meaningful contributing element, a quality STEAM lesson is created. Ghanbari (2015) suggests that STEM needs arts integration because an integrated arts curriculum promotes collaboration and engagement in learning (fun) for students. Similarly, Koester and Simmons (n.d.) assert, "aesthetic inquiry can move students away from alienation and towards affiliation." Student engagement in any lesson is key, but it is a driving force behind the entire STEAM movement. The hope for educators is that engagement through STEAM lessons will help improve retention in the disciplines during rigorous future coursework and improve achievement gaps by creating entry points to rigorous learning for all learners.

Art in This Case

There are 2 ideas that drive the inclusion of Arts in this case, diversity of student engagement and art as an integral element in STEM at its core. One idea promoting STEAM education is that the Arts can be an entry point for learners, especially reluctant STEM learners (Koester, in press). It is imperative that educators recognize this and use it, especially given the reports on STEM learning and careers (ACT, 2015). Results from the ACT 2015 Report on the Condition of STEM indicate that student (students who took the ACT and participated in the ACT Interest survey) interest in STEM is high, and yet, students expressing an interest in STEM are not prepared for the further rigorous study required for entry into careers in the STEM field. Further, students who expressed an interest in STEM outperformed their peers.

The nature of science and art are often considered quite similar, or even interrelated. Ghanbari compares arts coursework to the process of scientific inquiry stating, "A large facet of arts coursework is inquiry-based, which means it revolves around questioning and understanding concepts versus finding the answer to a given problem" (2015, p 3). Art can be obviously linked to several cross cutting concepts found in the Next Generation Science Standards, including patterns, scale, proportion, quantity, structure and function, and systems models (Koester & Simmons, n.d.). On the other hand, through the inclusion of meaningful aesthetic inquiry, all of the cross cutting concepts can be explored in a deeper way. This is the true nature of a quality STEAM lesson—integrating and transcending multiple disciplines of learning to provide authentic context that is meaningful for all learners.

Creativity is an integral part of art, but it is also a part of great science, technology, engineering, and math. Visual media is by no means the only form of art that can be incorporated into a STEAM lesson, but by giving learners the opportunity to explicitly incorporate elements of graphic design, spatial reasoning, and idea sketching, educators can expand student access to reflective learning, explanations, and communication throughout the learning experience. There are indications that students who develop spatial thinking are more likely to be interested and successful in STEM careers and research (Newcombe, 2010).

For those hoping to incorporate STEAM by modifying existing lessons, STEM or otherwise, the question to ask is: How is this addition meaningful to the learner, with respect to the goals of STEAM education? Specifically, is the modification simply to add an assessment that is artistic in nature, or is it incorporating the arts as an entry point for students to the science, engineering, technology, and math content? Will including basic arithmetic enhance students' spatial reasoning, logic, and ability to think critically about how to solve the problem? By looking at the goals of STEAM education and the objectives selected for the lesson, teachers should be able to avoid the addition of "art for art's sake" or "math for math's sake," respectively.

CASE DESCRIPTION

In the following case, students are asked to plan, design, build, and evaluate a robotic arm that can complete a given task. The lesson is part of a story in which a submarine is lost at sea and must be found in order to rescue the scientists on board. The lab itself is adapted from Lesson 9: Invent a Robot of Volume 2: How Do We Explore of The NOAA Ship Okeanos Explorer Education Materials Collection (National Oceanic and Atmospheric Association, 2014). The context is Marine Science in nature, but can be adapted for virtually any discipline, given the pervasive status of simple machines and robotics in the modern world. Several of the resources included in the case provide examples of the lesson in other content areas. Content applications outside of Marine Science include medicine, factory automation, automotive, construction, and many more.

The examples, perspectives, and results described here are from a high school Marine Science course from a suburban large school setting, 25-30 students per class. Students take Marine Science as an elective, with Honors or College Prep options, and many of them are interested in learning Marine Biology and do not anticipate engineering content. Students in Marine Science have access to 1:1 Chromebook devices during class; and Google Apps for Education is used extensively throughout the entire course, including Google Drive, Classroom, Docs, Slides, Sheets, and Drawings. The lesson falls within an Exploration and Technology Unit that covers content addressing the following essential questions:

1. What's in the deep ocean?
2. How do we use technology to explore the ocean?
3. How do we locate and study phenomena found in the deep ocean?
4. What are challenges to studying the deep sea?
5. How is life supported in the deep Sea?

Standards are adapted from the National Ocean Literacy Principles, but the lesson would address Next Generation Science Standards as well as other Common Core standards. Because the content is adaptable to various disciplines, only a brief list of relevant standards is included here.

- **Next Generation Science Standards:**
    - **MS-ETS1-4 Engineering Design:** Develop a model to generate data for iterative testing and modification of a proposed object, tool, or process such that an optimal design can be achieved. (Grades 6 - 8)
    - **MS-ETS1-2 Engineering Design:** Evaluate competing design solutions using a systematic process to determine how well they meet the criteria and constraints of the problem. (Grades 6 - 8)
    - **HS-ETS1-3 Engineering Design:** Evaluate a solution to a complex real-world problem based on prioritized criteria and trade-offs that account for a range of constraints, including cost, safety, reliability, and aesthetics as well as possible social, cultural, and environmental impacts. Performance Expectation, Grade: High School. (9-12)
- **Ocean Literacy Principle #7:** The ocean is largely unexplored (Ocean Literacy Network, 2013).
    - B. Understanding the ocean is more than a matter of curiosity. Exploration, inquiry and study are required to better understand ocean systems and processes.

- F. New technologies, sensors, and tools are expanding our ability to explore the ocean. Ocean scientists are relying more and more on satellites, drifters, buoys, subsea observatories and unmanned submersibles.

Authentic Context: The Student Challenge

In this unit of study, students progress through a sequence of lessons that are designed to promote understanding of ocean exploration and technology in both 2 and 3 dimensions. The lessons in the unit involve: 1) spatial reasoning topics such as mapping, scale, and elevation above or below sea level, 2) bathymetry topics such as seafloor topography, 2-D and 3-D mapping, sonar, and 3) technology topics such as history, submarine, ROV/AUV, and marine sampling equipment. These lessons emphasize content that help students formulate answers to the Essential Questions from the unit, especially: "How do we use technology to explore the ocean?" and "How do we locate and study phenomena found in the deep ocean?"

Students first learn how to find objects on the surface of Earth using latitude and longitude coordinates, and then learn about technology used to find things beneath the surface, such as sidescan and multibeam sonar. Students are told a story about a submarine, the *Johnson Sea Link*, that has been lost at sea and are asked to "find" it using simulated sonar in the classroom. The story is inspired by actual events involving an incident in 1973 that claimed 2 of the 4 passengers on board the *Sea Link*. In the simulated sonar lab, students search through boxed "models" of the sea floor that represent search quadrants that an underwater mapping device would map, and one of them contains a small replica of the *Sea Link* for students to find. Once the location of the submarine is found, students are asked to design a robotic arm that would be able to assist in the rescue of the trapped *Sea Link* and its passengers. The arm would hypothetically help move the debris or fishing gear that immobilized the submarine. This is only one example of the possible scenarios that could be used to gain buy-in from students. ROV and robotic arm use for the 2010 Deepwater Horizon oil spill solution, general sampling in various marine environments, lost at sea airplanes or ships like the *Titanic* would all provide realistic context for the lesson.

Students are given the following challenge description when they meet in their team groups:

Your group is one of several teams working to design a robotic arm that is able to complete an underwater mission: to rescue the Johnson Sea Link Submarine that is trapped in 200 feet of water with 4 passengers on board. Today, you will test your ideas by building a robotic arm that will perform the task of MOVING an object.

You will submit your final design in the form of a digital model in Google Drawings and an Engineering Report that discusses your design process. Your report should demonstrate why your team's arm is the best choice for the mission.

When students hear that they must build a working robotic arm, the reaction is usually one of two responses: 1) That's really cool (or some variation thereof) OR 2) I can't do that, it's too hard; I'm not an engineer. Too often, they don't realize that the technology used in robotics and what they perceive as "difficult" engineering can be found right in front of them on a daily basis.

To begin the lesson, I ask them to draw a robotic arm. Usually I assign this as a bellwork, exit ticket, or small homework assignment. I don't give them any examples or content, just ask them to draw an arm that can move using parts like nuts, bolts, cardboard, or whatever else they feel might be needed. The purpose for limiting the amount of direction I give here is so that I can really see what they know already about the topic. Based on this pre-assessment, I can make specific student groupings, or determine how difficult I should make the challenge part for the arm. Some students may already have knowledge about hydraulics or joint movements in devices like the robotic arm, so it helps to see where they might be best placed in a group to share that prior knowledge.

The challenge part of the build has two parts: 1) Moving an object and 2) Grasping an object. All groups must pass the first challenge, to move a large piece of coral (weight approximately 4kg) a distance of 10cm. In context, the moving coral rock represents moving the obstacle that entangled the original *Johnson Sea Link* in 1973. Students will need to put effort into creating an arm sturdy enough to hold up the mass of the coral rock when applying force. Further, since the syringe plungers are only 8cm in length, moving the coral 10cm presents an additional challenge, discussed in detail in a later section. For the "Moving an object" challenge, students test their prototypes on a 10cm track that is marked on a central lab table. The rock must move from one taped line and cross the second to pass the "move" test. The second challenge, "Grasping an object," is for groups who finish quickly. When a group finishes well ahead of the others, it is great to have a transition or extension for them to complete while I continue to work with the rest of the class. In this case, I have them build a "grabber" at the end of the arm. This part of the lab can be extra credit, for classroom points, or just part of the overall grade for the lab.

The Lab

According to the NOAA lesson plan used, the materials needed for the device include (NOAA, 2014, p. 133):

- Cardboard or heavy poster board, approximately 12" x 12" (the stiffer the better),
- Duct tape, approximately 2" x 60",
- Machine screw with nut, #8 x 1",
- 2 - Flat washers, #8 hole,
- 2 - Oral syringes,
- Plastic tubing approximately 3/16" inside diameter x 12" (should fit snugly over the end of the syringes),
- Water,
- Heavy scissors to cut cardboard ("bandage scissors" are inexpensive and work well),
- Nail or Phillips screwdriver to punch holes in cardboard, and
- Screwdriver to fit machine screw.

In addition to these materials, I also keep on hand some box cutters and a power drill. Safety lab goggles are obviously a must, and to keep everything cost effective, at the end of the lab, we will break down the devices and reuse the metal parts. Lab materials are kept in a large "sink" tub for each station—a roll of duct tape, scissors, nuts and bolts and washers, ruler, syringes and tubing.

The most common way to build the arm is to join two rectangular sections of cardboard (suggested dimensions: 10in by 3in) at one end by a nut and bolt joint. This joint creates a hinge that will allow one piece of cardboard (the lever arm) to swing out from the other. By attaching the syringe body to one piece of cardboard, and the plunger of the syringe to the lever arm section, the arm is created. The plunger will push out and retract the lever arm away from and to the syringe section of cardboard when the actuator plunger is pressed or pulled on the other end. By altering the angle at which the syringe is attached or the length and shape of the cardboard arm, you can achieve many different design purpose possibilities.

Student Instructions

On the day of the actual lab, each lab group meets and shares their ideas. We discuss the time frame, expectations, and materials they have to use for the lab. One key expectation I give is that all students in the group should be able to answer any question I have for the group during the lab. This means that no one student should be solely responsible for the work and understanding the group is doing. Further, it is my intent to create an inquiry setting, so I do not give them any step-by-step instructions for creating a device. I tell students that the challenges they will face during the lab are purposeful and meant to foster the creativity and collaboration that I want to see from them. We have an established class norm that deep thinking and understanding is hard work, but that I will never give them an assignment they cannot do. There are some scaffolds that I do put in place at various points throughout the lab to help students be successful. This varies by class, and I look for signs of frustration from students that indicate they are giving up before adding additional scaffolds.

Before each group begins, I show the class an "actuator" that will help them build the device. I also show them an example of an existing arm (one that I premake or from another class), but they are not allowed to touch, hold, or look too closely at the example arm. They are allowed to touch, hold, and look closely at the actuator. As a class, we define the term "actuator" as a device that will convert their mechanical energy (pushing a button or controller) into motion on the arm (NOAA, 2014). The actuator consists of two plastic syringes connected by plastic tubing (the plastic tubing is similar to that used in a bubbler for an aquarium and can be purchased in the plumbing section at any hardware store). The example actuator chamber that I show the class is intentionally left empty, without water, so that only air fills the chamber.

I tell students to begin by building an actuator and trying out different motions. From there, it is a matter of observing them work, asking questions, and intervening with additional scaffolds as needed. Some students immediately create an elaborate cardboard extension at the end of the actuator syringe plunger. The key question that I would ask this group is, "Have you tested how far you can move an object with that design?" They will test, discover that the plunger only extends 8cm, and have to regroup to find another solution. At this point I would review the concept of displacement with this group, because they will often attempt to add another "extension" to their initial extension design, which will not change the distance that the arm can actually travel. Depending on the group and their frustration level, or the amount of time left in the class, I might suggest they look for ways to extend the range of motion of the arm by adding a lever or a hinged part. We look at the path a door or their own elbow travels, and they go back to try to incorporate those ideas.

There are some groups who become so frustrated that I feel it is necessary to allow them to look carefully at an existing arm--the original prototype I showed the entire class. I stress to them that this is not the ONLY solution to the problem, but it is one possibility. Once they have an understanding of how the moving joint is created, I take away the arm, and they begin building again. It is important for them to know that not every engineering or technological feat comes from a novel idea; sometimes it is from refining an existing concept. The remaining refinements that they will need to make to this now "known" design, such as adding structural stability to the cardboard, adjusting the angle and distance the lever arm can travel, and creating the joint will still add to their knowledge and understanding of the engineering design process and the science and technology behind lever and robotic arms.

Some groups will design an arm that increases the distance that the arm can move by using 2 actuators in conjunction. This method is successful, but the instructor would have to decide if students would be allowed to use multiple actuators (more materials). Personally, I feel that the purpose of the lab in this case is more about problem solving and engineering design process skills, rather than teaching lever concepts, so I encouraged using more than one solution. These unique designs make for a much more interesting and developed class discussion and reflection at the end of the lab.

Once students have created an arm that "moves" far enough, they begin to tackle the problem of structural stability. Cardboard is not always the most stable building choice, so students can struggle with creating an arm that will hold up to the weight of a large piece of coral "rock." There are not specific scaffolds with which I respond to that type of challenge, but I do send them to check out other groups' solutions around the room. This is sometimes accomplished with a lab meet and greet of sorts, in which one person is elected to stay at their lab station as a spokesperson who explains their design to the other groups' members who rotate through each station. They will usually find many different ways of adding stability to the arm--multiple layers, folded cardboard tubes, and blocks of cardboard attached where the syringe plunger will apply pressure.

The last key scaffold that I implement is encouraging students to consider how well the actuator responds when it is only filled with air. They usually say it feels "weak" or "soft" when they apply pressure to a larger object. Often they will begin thinking about what else could go in the cylinder, and will ask to try water. I make a point to show them how to securely fasten the syringe to the hoses in an effort to avoid water accidents all over the lab when the hoses disconnect under increased, now hydraulic, pressure. I often find students who switch to hydraulic actuators early have discovered, and solved, this problem with additional duct tape at the syringe/tube connection. At this point, the entire class will review prior lessons on gases, density, and compression, and discuss the applications of Pascal's Law, hydraulic, and pneumatic technology in detail.

Students work their way through many problems or challenges with their design throughout the lab. Each time they try their design at the testing station they record areas of the design that need modification. Each change is documented for their reports, and each change puts them closer to creating a working arm. Once the arm is complete, they move on to completing their engineering reports, or with enough time, the second grasping arm challenge.

STEAM Integration Summary

In general, the breakdown of the integrated content from each discipline is as follows.

- **Science:** ROV's, hydraulics, levers, compression, pressure.
- **Technology:** ROV's, syringes, joints, Google Apps, models.
- **Engineering:** Design process, iterations, modeling/build.
- **Art:** Google Drawings or SketchUp, creativity, design, advertisement.
- **Math:** Pressure calculations, measure, angles, displacement.

A more detailed description of how the disciplines are integrated and studied is found in the following sections. It is clear by reading that separation of the disciplines is difficult, which is why STEAM integration is a natural and fitting part of this lesson. The math section discusses science content, the technology section illustrates science applications, the art section highlights the use of technology, the engineering section shows the reflective elements of the scientific and art process, and so on.

Science Content

Students will learn about many kinds of physical concepts during the design and build of a robotic arm, the extent of which can be tailored based on the age, level, time allowed, content area and interest. Generally, students would be expected to learn about hydraulic, or fluid, power and the applications of actuators. The content is extended to touch on the topics of displacement, levers, hydraulic versus pneumatic devices, pressure, compression, and marine technology such as submarines and ROV's.

The use of hydraulic and pneumatic power in technology is quite common. Many public exterior doors operate on a pneumatic cylinder that forces the door to close slowly. Car manufacturers utilize hydraulic cylinders to keep the trunk or hood open without the use of a manual prop. I remember the hydraulic fluid (some form of oil) leaking out of the trunk cylinder of my family's old Ford station wagon as a little girl. I always had to be mindful not to stand under it without careful attention, as, without a functional hydraulic cylinder, the trunk door would come crashing down under the force of its own weight. Robotic arms seen in factories and new medical facilities also use hydraulic and pneumatic devices. Large construction vehicles such as dump trucks, tractors, or front end loaders (think Tonka truck) use hydraulic power to move and lift materials.

For this case, it is essential that students ultimately differentiate between pneumatic and hydraulic, especially in terms of compression and effectiveness in design. This differentiation will help to connect other technology design concepts such as the use of gasoline (liquid is less compressible than air) as a buoyancy device in the 1960 submarine *Trieste*'s voyage to the *Challenger Deep*, 11,000km down in the Pacific Ocean.

To complete the challenge described, moving a coral rock 10cm, students must build a lever arm. This is key because the distance that the syringe plunger moves is limited to 8cm, so, in order to extend that distance it is essential that they use a lever to extend the range of motion of the arm. Thus, it may be useful to review the concept of simple machines, especially the lever, before beginning the lesson. Additionally, depending on how far the teacher would like the mathematics and science content to go, knowledge of Pascal's Law, concerning the pressure exerted in a fluid, would be appropriate.

I purposely limit the background information provided in the lab in order to facilitate inquiry, creativity, and collaboration during the design process. One of the content details that I do not provide students is that hydraulic powered devices are more efficient for the type of work that is need to move the coral rock. Because water is less compressible than air, filling the syringes and tubing with water is an excellent way to achieve better transfer of energy from the actuator to the coral rock. This is where Pascal's Law applies, where as you apply pressure to a liquid such as water, the force is evenly dispersed and without compression, that force is applied directly to the opposite end of the actuator. Students will discover as they work that using the syringe with air allows for some spring or give in the actuator line, and it doesn't seem to move as fast or far as a syringe filled with water. As a facilitator, I ask them to explain WHY the water is a better choice, and be sure to include that in their final report. I also note the responses that are initially made, which often include a focus on the fact that water is denser than air, rather than the concept of compression. This is key to discuss and break down on a group basis and often as an entire class, especially as it relates to other units and lessons such as submarine technology and SCUBA diving.

Technology Content

There are many ways to include technology and improve student understanding of what it means to use technology in addition to electronic devices. This lesson explicitly involves the use of modern electronic technology such as digital design software and Chromebooks, but there are other less explicit technologies that play a role as well. The arm, materials, and process used to build the arm can be considered technology. The actuator is a form of technology used in all the designs. For very sturdy cardboard, it is sometimes helpful to have students use power tools, such as a drill to punch holes for the nut/bolt hinges, as another form of technology in the classroom. The applications that a hydraulic arm can be applied to outside of the context of the student challenge provide numerous other examples of technology. Additionally, students use cell phones for taking pictures of their device, Chromebooks for recording their design process, and SketchUp or Google Drawings to create a diagram of their initial build plan and ideas.

The technical design that combines art (whether media or graphic) with the engineering process is one of the most important parts of this lesson. For many labs, I use a sketch of the lab setup as a pre-assessment for student understanding of the lab before we begin. It is eye-opening sometimes to see how much they understand of the lab directions and can provide me with feedback on how well the instructions are written!

Engineering Content

One of the most important elements of engineering education, particularly for the goals of STEAM education, is the learner's participation in the Engineering Design Process. It is this reflective learning process, geared towards data collection, analysis, and modification that allows for the skills of communication, collaboration, creativity, and problem solving to develop in students. This process can be enhanced through the use of digital technology in the classroom, made more engaging for learners through art, and allow for tangible application of concepts in math and science. There are several "models" of the engineering design process. For this case a combination of a district model and the NOAA lesson's model are summarized below.

1. Identify Problem,
2. Gather Relevant Background Information,
3. Generate Ideas,
4. Build a model,
5. Analyze possible solutions, make modifications as needed, and
6. **Communicate Results:** Document test results and reflect on modifications' effectiveness.

The main difference between the model presented in the NOAA curriculum and here is the addition of Step 4: Build a model and Step 6: Communicate Results. These are included in this lesson to emphasize the build as a hands on, real-world experience and explicit reflection in the student learning process. Students will reflect as they write their final report by assessing their initial design and providing detailed explanations of the changes that were made in order to build a working model by the end of the lab.

The following text is an excerpt from a student reflection on the design process, specifically modifications that were made to the arm in order to make if functional.

Student Reflection

Along the way there were modifications made to the design. One important change was the decision to change from pneumatic air filled actuator to hydraulic water filled actuator at the beginning of the construction process. This change was due to the realization that water is not compressible like air and allows for force to be evenly distributed within the tubes, unlike a pneumatic actuator that would have been less effective given the easily compressible air in the tubes. This change made an improvement in the overall structure of the robotic arm and ensured that the lever would be able to be moved to its fullest potential and distance. Throughout the process adjustments were made as the trials went on. This mostly consisted of strengthening the cardboard by adding additional pieces as reinforcement and ensuring that the board did not bend or cave in with the pressure. The only way to figure out how thick the arm needed to be was to run trial after trial and see how much the arm moved on its own and then the difference when resistance, by way of a heavy rock was applied.

At this point in the lab (using Inquiry practices), this student would learn the term Pascal's Law and make connections between the science content and the intuitive understanding of that content that is described in this reflection from their engineering design process.

Art Content

Vaidyanathan (2012) states that the [creation of digital artwork] is an intuitive way to learn STEM subjects (pg 27). Further, it is a way to increase student participation as most digital editing programs have an "undo" button which allows for freedom to try new ideas without repercussions (Vaidyanathan, 2012). This follows with my own personal experience in graphic design, but is exciting to see in the context of a classroom where students are trying completely new things. In order to promote spatial reasoning skills that can contribute to STEAM learning, Newcombe (2010) recommends that students create sketches to explain their understanding of topics.

Such a sketch must be completed in order to complete this lab in Marine Science. This can be done by hand, or if technology is available (which it is for the examples given for this case) the use of an app or program such as Google Drawings or SketchUp would be a great way to integrate technology into the lesson. (Keep in mind that the design and model itself are a form of technology, and the introduction of electronic devices is just an additional element of technology use in the classroom.) Because I include the pre-design at the beginning, ALL students must use some type of creativity or applied knowledge to create a visual representation of their ideas for a robotic arm. When they first meet in their teams, they will share and discuss each individual design and come up with a team plan and design. Each lab group must present a diagram, completed in either SketchUp or Google Drawings, of their team plan for the arm after collaborating at the beginning of the lab. Finally, they must turn in a final design of what worked. This may be in the form of a photo, improved digital design, or edited (notes on changes) initial team design. This last form is similar to "know"tations (Koester, in press), where arrows and text would explain changes in the diagram/sketch of the idea that were made to create the final working model.

I have included several examples of student work that illustrate their progression of design through a first idea sketch to initial team design to final 3-Dimensional build. Student reflections show some of the thought process that they shared during the build to make modifications.

For students who have no idea how to start in Google Drawings, I share a template with them to help scaffold the digital design process (Figure 1). The template combines elements of digital design (using the program to create a design), science, and engineering (labels, scale, notations).

The context of the challenge is ultimately to design an arm that will be fitted to an ROV that can accomplish a variety of tasks underwater. Students learn about ROV and submarine technology during the unit and can apply that to their arm

Figure 1. Shared template in Google drawings for digital design

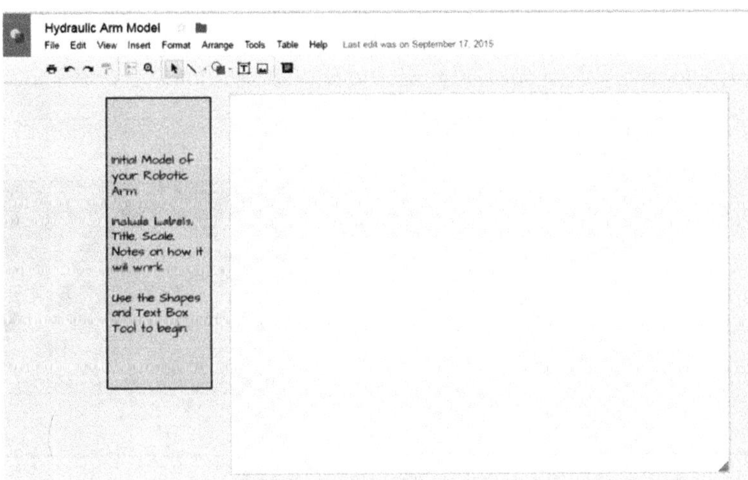

design in context. This student used that context to create a sketch of an ROV that would fit the team's actual model robotic arm (Figure 2). In the student report, they would add reflections about how the design functions and modifications that were made in order to create a functional arm.

The transition from digital diagram to actual working arm is a distinct one for many students. Some students work from the 3-D working model backwards to create a final digital diagram for the team report, and others use the 3-D model to explain changes that were made (Figure 3). As a teacher, I find this illustrative of the diversity of creativity found in each classroom.

Student designs between each team are highly varied, but the way each student represents the design digitally is also distinct (Figure 4). Even in an engineering setting, where the art product would be considered technical and almost scripted, variations in color, theme, symbol, and design are abundant.

Figure 2. ROV design in context with actual working arm. Student sketch provides details of authentic context for the lesson, using symbolics and text

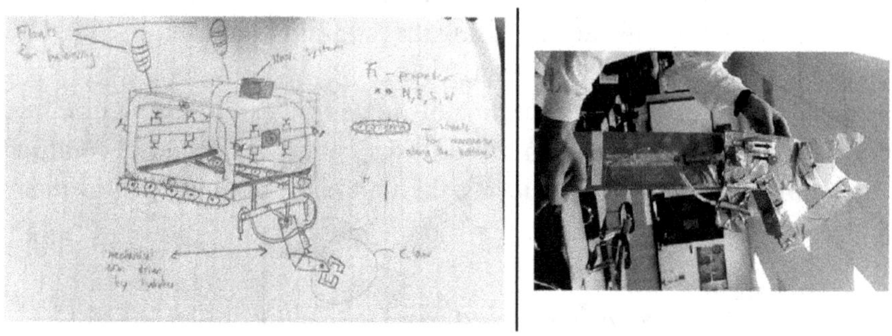

Figure 3. Student progression from digital design to actual 3-D model

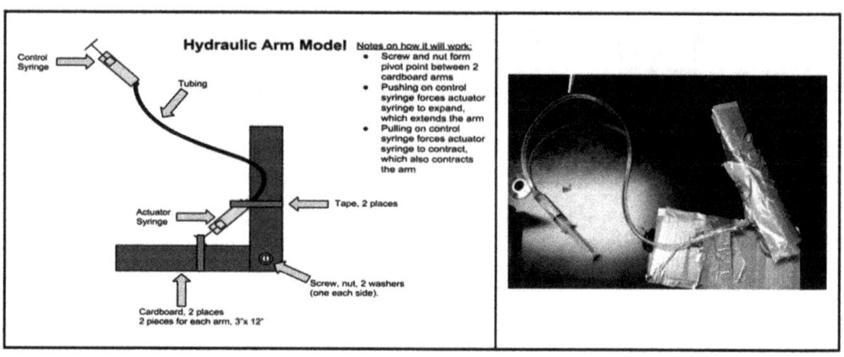

Figure 4. Various student designs highlighting diversity of ideas and methods to illustrate and complete the challenge

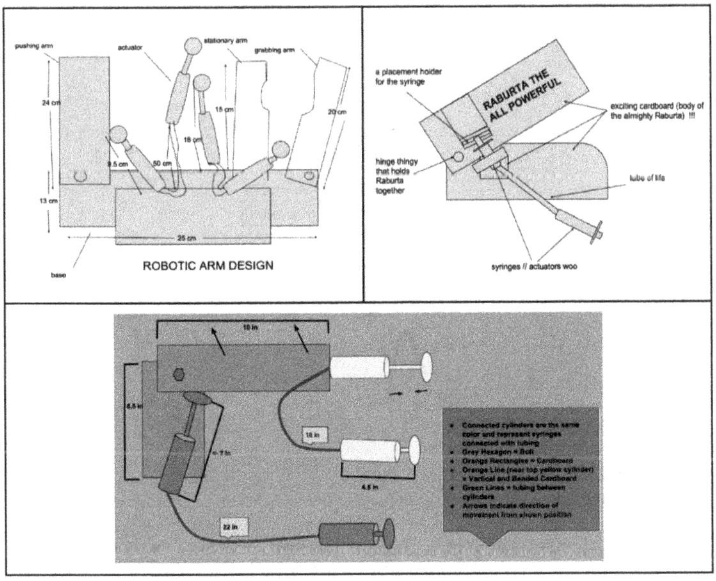

The digital design, coupled with graphic design concepts like labeling, text, arrows, and symbols, is an excellent way for teachers to assess what students learned from a lesson like the hydraulic arm lab. However, the written reflection that accompanies this diagram in the report illustrates key misconceptions that are important to address before concluding the lesson (Figure 5). Art integrated with the other STEM content makes for a powerful assessment tool in this case.

Originally, my thought for this lesson was simply to have students create digital mock-ups of their ideas and creations, but with the use of other integrated technology such as camera phones or digital cameras, they surprised me again with their innovation and creativity. This student felt like it would be more effective to import and edit an actual image of the 3-D model in order to explain the design and modifications for their report (Figure 6).

Even without the digital technology, students will create a variety of products. I encourage them, once they have a working concept, to make the exterior design more complex and to incorporate elements like a control panel. They enjoy decorating the arms and naming them, just like the real-world ROVs we study. Overall, the variety of design and ways of explaining the engineering process, both visually and verbally, provide for an engaging lesson for both student and teacher. Each year we complete the lab, I learn new ways of tackling the design challenge, highlighting the skills of collaboration and innovation that I hope students carry with them to future classes and their careers.

Figure 5. Student design with modifications and reflection. The written report is significant because it illustrates an incomplete understanding of hydraulic versus pneumatic powered devices, that hydraulic is better because water is denser than air, rather than compression. Without a written report, that misconception may have gone unnoticed

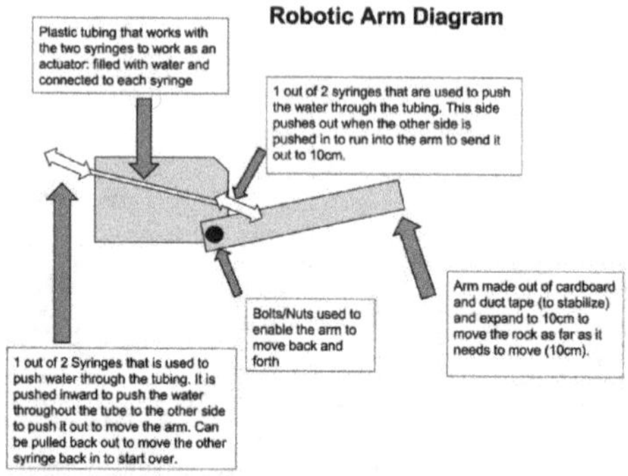

Student Reflection: *Throughout the process of designing and creating the robotic arm, my team and I went through many complications. The Actuator (plastic tubing and syringes) kept popping and leaking water, which caused air pockets to enter that did not allow it to work as well as it could have and was meant to. We had to re-create the actuator many times until we had finally accurately created it...We had wrapped too much duct tape around the part of the arm that held the actuator that pushed the arm out 10 cm (see diagram) and the actuator tube had popped so we had to cut through all the duct tape in order to fix the problem and lighten the weight in order to move the rock easily. My group used water in the actuator to make it a hydraulic rather than air, making it pneumatic, because water is denser than air.*

Figure 6. Final product overlaid with digital notations about the build. Student reflections about design process are included in the written reflection that accompanied the visual diagram

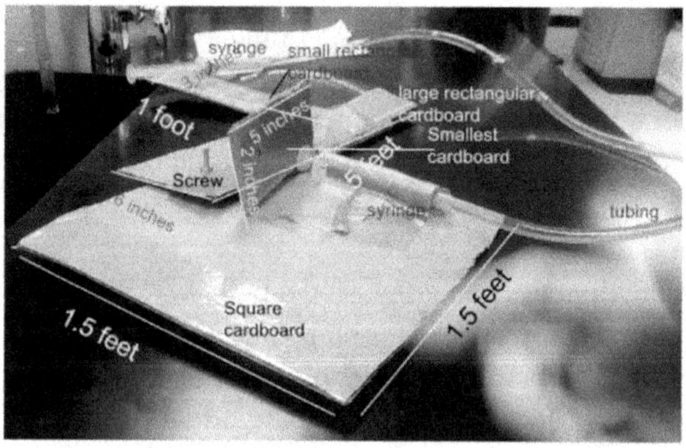

Student Reflection: *We would modify the design to make the arm work better such as; making the (lever) arm smaller to create a shorter distance so no loss of force. Another modification that helped move the rock was adding the screw to create a movement between two cardboard pieces to create a swinging movement. Finally the last major modification was adding water to the actuator, which creating an instant reaction and stronger applied force. We collected data by testing the arm multiple times.*

Math Content

Math is the most implicit content area taught in this particular case. For the purpose of this case example (given the time and content constraints), students are not asked to measure the volume of the cylinders used in the actuator for the purposes of calculating the pressure applied to the arm, although that is a direct application of applied math content that could be used. One explicit use of mathematics that is included is the concept of scale, since student must include scale in their final design proposals. This idea is tied to a previous lesson in the unit sequence in which students create a scale model of the earth by plotting the depth and height (relative to sea level) of features around the earth on the classroom wall.

Additionally, in order to build the arm and create an accurate diagram, they measure the length of each part, and assess the angle at which to place the syringe plunger which will apply force to the moving lever arm. Students test and adjust the angle until they successfully achieve the full range of motion required to move the coral rock 10cm. The concept of displacement is cemented when students measure the distance the syringe plunger moves, a total of 8 cm, and then test the idea of adding an extension to the plunger head to try to increase the distance that the coral will move. Of course, in terms of displacement, adding bulk to the plunger only changes the starting point, and the displacement of the plunger can only be 8cm. This leads them to the "speed bump" in the lesson that forces creativity and reliance on new ideas. Students usually try many angles and types of levers as they work to solve the problem. As each group proceeds, I move around the room asking questions and looking for student understanding of topics.

The Fluid Power Lab (TeacherGeek, 2008) has a great set of diagrams and practice problems for students to use to understand Pascal's Law and the relationship between force and pressure. The lab handouts have many student practice problems for students to use basic math skills of multiplication and division to calculate volume, force, energy, work, and pressure.

A great way to include additional math content that provides real-world skills and context is to have students budget for and "purchase" materials to build the arm. The cost effectiveness of their build would be included in the final report, and groups can compare total cost of each build to how well they worked as a way for the class to reflect on all the solutions to the problem.

CURRENT CHALLENGES FACING TEACHERS

Teacher Concerns About STEAM Education

Does it Have to be a "Big" Lesson?

No. This lesson can be accomplished in 1-2 block periods, depending on the depth of assessment tasks assigned. In the case of the TeachEngineering (Chilton, Carpenito, & Milto, 2013) or Basketbots (Bigelow Middle School, n.d.) resource example provided, you could also create a much longer unit. Just remember that as practitioners we should be reflective and explicit about what and why we are teaching, just as much as it is essential to transdisciplinary learning that students reflect and make explicit the "what" and "whys" of what they did during the lesson.

Isn't it Hard to Grade Something Like This?

I use Google Sheets and an add-on, Doctopus (with Goobric) to grade the reports based on a rubric for student design. There are many ways to assess, but it depends on what your goals for learning are to determine the right assessment. For this case, we were looking for development of a working model, with verbal and visual explanations of how the model was created, and, finally, demonstration of content and technology design skills. These elements were present in the rubric provided to students for their report.

I Don't Know Anything About Building Robots

Neither do your students, in some cases. Go ahead, have fun LEARNING WITH THEM. It will help to create the caring teacher-student relationship that so much literature supports as a best practice!

I Don't Have Access to Technology

A thousand years ago, technology was a formula, or a new way of gardening. Mechanization and computers are NOT the only forms of technology. My students sometimes put forth that written language is the most influential form of technology created. Without getting into a debate about whether language counts, it is true that technology in quality education does not just have to include 3-D printers, expensive 1:1 device programs, robots, or laboratory equipment. Use whiteboards to promote collaboration and sketching or large Post-It posters and markers. Create a gallery walk. Start a BYOD (bring your own device) program. Maybe those little devices in the pockets or purses of your students could be put to use!

It can be helpful to ask students to define the technology they used in a lesson. For this case, the model and diagram itself is a form of technology. The media used count, too, (Guyotte, Sochacka, Costantino, Walther, & Kellam, 2014). Student reflections for this case include that for the real device they would need to use something other than cardboard, or a waterproofing agent. By answering that question, they will begin to reflect on the use of tools and media (in all the disciplines) to define technology, not just the use of a digital device.

It Seems Hard to Include All 5 Disciplines Equally

I have found that not every lesson I teach can possibly encompass all 5 elements. As a curriculum team, we try to diversify the ways we use and connect each discipline. Include art as a performance assessment—have students create a video or advertisement at the end of the lesson that ties together the multiple content areas. Try creating a "know"tation or visual journal on an integrated topic. Start small, think big.

Start with a daily lesson and consider the topic. Guyotte et al. (2015) suggest that using a complex social issue as a topic is a good way to encourage collaboration and deeper understanding and perspective for students, particularly when coupled with artistic reflection. There are so many topics to choose from, especially in the environmental and sustainable design fields. Try finding a problem based lesson and look for ways to incorporate or integrate the arts from there. But, keep in mind the caution of Anne Jolly: "The purpose of STEAM is not to teach art, but to apply it in real situations. Just like the purpose of STEAM is not to just teach math, but to apply math in real situations" (2014).

There are several recommendations for great STEM/STEAM programs in the literature, but as a teacher, my best advice is to just try…take a risk. It might be fun learning something new.

SOLUTIONS AND RECOMMENDATIONS

REFLECT. REFLECT. REFLECT. Be explicit about this. Student outcomes vary over the course of this lesson. One student enjoyed the creative process and challenge so much she changed her major to Engineering. Others determine that engineering isn't for them, but become familiar with the reflective process that can be helpful for any discipline. STEAM is a meta discipline...high quality STEAM/STEM programs do not treat the disciplines separately. It follows that art should not be just a separate discipline in quality STEAM instruction, that all elements should be a relevant part of the learning process. However, the power of the programs described here isn't in the fact that all the disciplines are included equally; it's in the fact that students and teachers took the time to break down the process of what they did and how they did it and what that tells them about the world. That reflection piece can be a group discussion, or a survey, or a written report, or a journal, but it must be a part of the learning process. It's great to include all elements equally, but without explicit breakdown of the learning process itself, the power and meaning of the learning may go unnoticed by the student. This is why it is imperative to include the final steps of the engineering process, communicate results, as a reflective piece. The power of that reflection is evident in both Guyotte et al. (2014) and Koester's (2016) work through visual journals and "know"tations, respectively.

Using the work of Guyotte et al. (2014), perhaps a useful modification to the lesson is to eliminate the restrictions on the challenge and instead, give students an opportunity to define the problem themselves. In their efforts to push students toward collaborative creativity by emphasizing a problem exploration, Guyotte et al. (2014) report that students become more aware of the significance of group dynamics and [possibilities]. Pay careful attention to the "speed bumps" that can be set in place in a carefully constructed lesson to facilitate problem solving rather than solution finding. I find it similar to the twists included on modern cooking competition shows and game shows. Those carefully placed hurdles keep the learner engaged and force (the student) to rely on others, new ideas, or develop emotional intelligence and empathy.

In terms of reaching a diverse group of learners, I see students that struggle in core areas such as language arts, math, and science, have moments to shine during this particular lesson. It is always rewarding as a teacher to watch these interactions. For example, one student who is interested in cars and enrolled in an automotive course created a set of channels in the arm for the syringe plunger stem to rest, adding strength and stability to the moving joint. Other students in the class had previously overlooked his efforts, but in this moment, he had their acknowledgement and praise for expertise. The moments happen in some way in each class, whether a female discovering she enjoys "engineering," a quiet, low performing student being asked to help the "A" students make their Google Drawing better, or a high performing student discovering that their first idea didn't work so they must rely on others to complete the task. The experience creates a change in the classroom dynamic, as more students are willing to listen to others' ideas and hope for creative solutions that stemmed from the mere diversity of experience available in the room.

Through analysis of Texas's TSTEM initiative, Kennedy and Odell (2014) propose that in order to engage more students in STEM, teaching must be altered to incorporate truly interdisciplinary and real-world contextual lessons. Additionally, they state (Kennedy and Odell, 2014):

High quality STEM education programs provide teachers with opportunities to collaborate with one another in unified efforts aimed at integrating the four subjects into one cohesive means of teaching and learning. It is when this objective is achieved that students gain access to meaningful curricular opportunities promoting critical thinking skills that can be applied to their academic as well as everyday lives (p. 256).

It is imperative, then, to achieve the goals of education reform (or improvement) that we provide these opportunities and TIME for teachers to develop, practice, and explore STEAM curriculum and skills at integrating content for curriculum. Hopefully, as teachers gain that experience, daily STEAM lessons will be easier to implement, and, ultimately, the ideal project based lesson like the Dream Factory will seem less daunting and become a reality.

REFERENCES

ACT. (2015). *The condition of STEM*. National ACT, Inc. Retrieved from http://www.act.org/content/dam/act/unsecured/documents/National-STEM-Report-2015.pdf

Beers, S. Z. (2011). 21st century skills: Preparing students for THEIR future. *STEM*. Retrieved from http://cosee.umaine.edu/files/coseeos/21st_century_skills.pdf

Bigelow Middle School. (n.d.). *Engineering technology: Basketbots*. Retrieved June 30, 2016, from http://www.bu.edu/gk12/alex/bioengineering_teacher.pdf

Brain, P. O. P. (2014). *Hydraulics*. Retrieved July 15, 2016, from https://www.brainpop.com/technology/energytechnology/hydraulics/

Chilton, E., Carpenito, K., & Milto, E. (2013). Hands-on activity: Hydraulic arm challenge. *TeachEngineering*. Retrieved June 28, 2016, from https://www.teachengineering.org/activities/view/wpi_hydraulic_arm_challenge

Ghanbari, S. (2015). Learning across disciplines: A collective case study of two university programs that integrate the arts with STEM. *International Journal of Education & the Arts*, *16*(7). Retrieved from http://www.ijea.org/v16n7/

Guyotte, K. W., Sochacka, N. W., Costantino, J. W., Walther, J., & Kellam, N. N. (2014). Steam as social practice: Cultivating Creativity in Transdisciplinary Spaces. *Art Education*, *67*(6), 12–19.

Guyotte, K. W., Sochacka, N. W., Costantino, T. E., Kellam, N. N., & Walther, J. (2015). Collaborative creativity in STEAM: Narratives of art education students' experiences in transdisciplinary spaces. *International Journal of Education & the Arts, 16*(15). Retrieved from http://www.ijea.org/v16n15/

Henriksen, D. (2014). Full STEAM ahead: Creativity in excellent STEM teaching practices. *The STEAM Journal, 1*(2), 15. doi:10.5642/steam.20140102.15

Jolly, A. (2014). STEM vs. STEAM: Do the arts belong? *Education Week Teacher*. Retrieved June 30, 2016, from http://www.edweek.org/tm/articles/2014/11/18/ctq-jolly-stem-vs-steam.html

Kennedy, T. J., & Odell, M. R. L. (2014). Engaging students in STEM education. *Science Education International, 25*(3), 246–258.

Keruskin, T. (2015). Finding the sweet spot: Creativity, candy, and commerce. *Edutopia*. Retrieved June 28, 2016, from http://www.edutopia.org/blog/sweet-creativity-candy-commerce-todd-keruskin

Koester, M. (in press). Getting to "know" STEAM. In M. VanSickle & J. Bazler (Eds.), *Cases in STEAM education in practice*. Hershey, PA: IGI-Global.

Koester, M., & Simmons, S. (n.d.). *The collaborative art of teaching science through drawing* [PowerPoint Slides]. NAEA. Retrieved from http://naeaworkspace.org/naea16/The%20Collaborative%20Art%20of%20Teaching%20Science%20through%20Drawing.pdf

National Oceanic and Atmospheric Association. (2014). Lesson 9: Invent a robot. In *The NOAA ship Okeanos explorer education materials collection, volume 2: How do we explore* (pp. 125-136). Retrieved from http://oceanexplorer.noaa.gov/okeanos/edu/collection/media/vol2_hdwe.pdf

Newcombe, N. (2010). Picture this: Increasing math and science learning by improving spatial thinking. *American Educator*.

NGSS Lead States. (2013). *Next generation science standards: For states, by states*. Washington, DC: The National Academies Press.

Ocean Literacy Network. (2013). *Ocean literacy principle #7*. Retrieved June 30, 2016, from http://oceanliteracy.wp2.coexploration.org/ocean-literacy-framework/principle-7-v2/

TeacherGeek. (2008). *Fluid power lab*. June 30, 2016, from http://teachergeek.org/fluid_power.pdf

Vaidyanathan, S. (2012). Fostering creativity and innovation through technology. *Learning and Leading with Technology*, *39*(6), 24–27.

KEY TERMS AND DEFINITIONS

Actuator: Mechanical device that converts energy into motion, in this case converts muscle energy into motion of the syringe.

Graphic Design: Digital art that is organized using text, shape, and image that communicates a message, idea, or other topic.

Hydraulic: Liquid powered, in this case water.

Lever Arm: Arm that moves around a fulcrum, in this case a hinge or joint.

Pascal's Law: Pressure applied on a fluid is dispersed (evenly distributed) throughout the entire fluid.

Pneumatic: Air powered.

Pressure: Force per unit area.

ROV: An underwater remotely operated vehicle that is tethered to a ship or submarine and operated by a technician, used for various marine expeditions such as sampling and filming.

Section 4
Lessons for Special Needs Students

Chapter 16
Cases on STEAM Education in Practice:
Differentiated Instruction

Kathryn L. Servilio
Monmouth University, USA

EXECUTIVE SUMMARY

This chapter will reflect on the other chapters within the text and have multiple examples of how to differentiate STEAM lessons. Issues that arise such as professional development around differentiated instruction, the time it takes to differentiate (amount of planning), lack of classroom time to complete projects, and lack of support or collaboration with the special education teacher are discussed.

LITERATURE REVIEW

Students differ from one another socially, academically, physically, cognitively, and emotionally. They also learn differently (Gregory & Chapman, 2007). Teachers can no longer just "teach the lesson" and hope that everyone understands. It is important to first take into consideration who each learner is, their strengths and needs, readiness levels, interests, and preferred learning styles (Tomlinson, 2001). This is the first step in differentiated instruction.

Differentiated instruction allows all students to access the same classroom curriculum by providing accommodations tailored to meet their individual needs (Tomlinson, 2001; Watts-Taffe, Laster, Broach, Marinak, Connor & Walker-Dalhouse,

DOI: 10.4018/978-1-5225-2334-5.ch016

Copyright ©2017, IGI Global. Copying or distributing in print or electronic forms without written permission of IGI Global is prohibited.

2012). After teachers know, understand, and assess each learner, it is critical to use that information to design instruction. This can be done through curriculum compacting, grouping, and adjustments to their learning.

Curriculum compacting is a strategy that can sometimes be used if a student has already mastered the content; the teacher then designs the instruction to "dive in deeper" or "compacts it" and moves on to the next skill (Gregory & Chapman, 2007; Logan, 2011). For example, for a first grade student, if the lesson objective is to use observations of the sun, moon, and stars to describe patterns, the student should be able to identify and discuss patterns like how the sun and moon rise in one part of the sky and move across the sky. To compact the curriculum, the teacher would design "deeper instruction" where the student collects data on specific times that the sun rises or data on the location of the sunrise and sunset. The other option for curriculum compacting would be for the teacher to move the student on to the next skill within the curriculum. For example, if they mastered that topic, then the teacher would move the student onto making observations at different times of year to relate the amount of daylight to the time of year (since this is the Next Generation Science Standard) or the next standard that they are going to cover in their curriculum.

An important feature in differentiated instruction is the use of flexible grouping and different grouping strategies. The best instruction is when students have a balance of working alone, with a partner, or in small flexible groups. Flexible grouping is when students are mixed with other students based on their ability and interest level. By incorporating flexible grouping, students "maximize their learning time based on their performance levels" (Gregory & Chapman, 2007). The critical component in flexible grouping is that the students work with a variety of their peers and switch groups often. There are a variety of grouping strategies (see Table 1: Differentiated Grouping Strategies) and, based on the content, the teacher can determine if another grouping strategy would better meet the needs of the students.

The final step in differentiated instruction is making adjustments to students' learning. This can be done when teachers make adjustments to the content, process, and product. When a teacher differentiates the *content*, the teacher identifyies "what we teach or what we want students to learn" (Tomlinson, 2001). There are two ways to differentiate the content: First by adapting "what" we teach. For example, I'm teaching a sixth-grade science class about the solar system. I can differentiate the content because I may have some students first reviewing and relearning the characteristics about the earth while others are working on all of the planets together, in turn, differentiating "what" they learned. The second way to differentiate content is that we can adapt "how we give students access to what we want them to learn" (Tomlinson, 2001). For example, in the same science lesson some students may be reading about the solar system on a sixth grade–level text, some are watching a

Table 1. Differentiated grouping strategies

Grouping Strategy	Brief Overview	Tips
Ability	Students' work with peers that are at the same academic level so material is challenging and stimulating.	Use this strategy less frequently because of the stigma that groups can create as well as the "tracking" of knowledge that can occur when students are only exposed to content at one level.
Cluster	Students are grouped by prior experience or knowledge of a topic.	Make sure to use a pre-assessment to determine current level of skills and experience.
Cooperative Learning	Group work is paired with social skills training. This type of group creates a common goal or an assignment and then assigns specific roles for the task. There is individual as well as group accountability in this type of learning.	Make sure to use heterogeneous groups during this task because the students will typically step up to the challenge.
Interest	Learner identifies they are interested in the topic or subject.	Hold conversations, conduct surveys, and inventories to determine interest level of information to incorporate in learning.
Multiage	Students of differing ages are grouped to learn from each other.	Reach out to a teacher in another grade level and begin discussions on how to create multiaged grouping on a specific project or assignment. The mixed background experiences as well as their current knowledge can really diversify the group to come up with creative ideas!
Peer-to-Peer Tutoring	Based on a specific need, students take responsibility and assist each other in their learning process.	The best tutor is a student the student that just understood the concept (i.e., when the "lightbulb" goes off). This way they are able to practice repeating it and teaching it to someone who may not have understood it yet.
Sharing	Students collaborate about information they will learn or already understand.	Rearrange the room so that the tables or areas are set up so students can talk to each other about specific information.

movie, and others are reading on a fifth-grade level. The access is adapted to how they are acquiring the information.

The second step in making adjustments to students' learning is through differentiating the process. When the *process* is differentiated, the teacher adapts *how* the students are learning the content. For example, I'm teaching a second-grade class about the life cycle of a butterfly. I typically teach it through worksheets where the students read about it, then cut out the different steps, and put them in order. I'm going to adapt the process so that instead of using the worksheet, the students are going to visually see the different models of each step in the process; they are going to act out in small groups each step in the process; then they are going to play a game with a partner to match the different steps in the process; finally they are going to write a story about a butterfly that includes the steps of the life cycle. From this example, a "good" differentiated activity is something students do or make that has a variety of support (or scaffolding), a variety of modes of learning, uses essential skills and essential information, and uses the essential idea to answer an essential question (Tomlinson, 2001).

The final step in making adjustments for students is through differentiating the *product*. This can be done when the teacher makes adaptations to the culminating assignment. By differentiating the product, this should help them rethink, use, and extend what they have already learned. When students differentiate the product, it should be done completed independently and students should take ownership of what they have learned. There are tips in creating a powerful product (see Table 2: Tips to Creating a Powerful Product).

An example of adapting the product is a unit completed with ninth graders where they are designing an iPad case. Typically, at the end of the unit the students make a model in groups and share it with the class. To adapt the product, I instead could give the students a choice to prepare a report of all of the steps, rationale, logistics, and final design of the iPad case; compose a rap song or rhyme that describes their iPad case with specific details; have them describe using a video the sequence or process of how they got to the point and final design of the iPad case; or the original idea of having them create a model and share it. By doing this the students are in charge of their learning because they get to choose how they are going to show what they learned, they will pick based on their learning preference, and the objective of the lesson has not been altered.

To complete this process, you should start small after creating learning profiles and/or gathering information about your students. Then try one component (i.e., content, process, product) with one lesson and build upon your knowledge. It is critical in differentiation to reflect—it should not end with just planning but be a continual process (Parsons, Dodman, & Burrowbridge, 2013). Some ways to do this are to consistently assess student progress, stay up to date on effective pedagogy and how students learn, and finally to reflect "on action" and "in action." When you reflect "on action," you are looking back and critiquing what happened after the instruction has been completed. Reflecting "in action" occurs in the midst of teaching, where you can make immediate changes or interventions based on the students. Upon reflection, the key is to make changes based on what you learned.

Table 2. Tips to creating a powerful product

Step	Be Sure to Include
1. Identify the goals/objectives	• Facts, concepts, and skills. • A list of specific standards in content area • A list of specific standards in other STEAM content areas • Key academic language terminology
2. Brainstorm a list of multiple formats for the product	• Multiple formats based on differing learning styles or interests • Requirements that need to be consistent between all of the products • Ideas that build on students' strengths • A list of exploratory and investigative assignments for critical thinking
3. Identify the specific product you will use & the particular part of the assessment that matches the goals/objectives.	• A chart to identify the specific goal/objective and the exact part of the assessment when that goal/objective is being measured. • Direct connections between the two and double check that in fact that is an accurate way to measure that goal/objective.
4. Determine your expectations	• The content by reflecting on your expectations of the information, materials, ideas, and concepts. • The process by reflecting on your expectations of their participation during the lesson. • The product by reflecting on your expectations of the construction, size, and parts of the assessment.
5. Scaffold the assignment to provide success!	• A criteria for success • Breaking down a large assignment into smaller "steps" or assignments • A timeline • Support to help them in planning • Individual conferencing if appropriate • Peer-to-peer feedback to help support revising and editing
6. Clearly present the assignment	• A reflection of the content to be sure that it is things that you have taught and they have practiced • A review of your steps and process of the assignment • A criteria for how the students will reach the specific level of quality
7. Adapt assignment based on individual students	• Student readiness • Student interest • ELL/ESL learners • Learners with disabilities (including gifted)
8. Facilitate for Achievement!	• The best way to contact you • Specific feedback and growth throughout the process

Through differentiated instruction teachers push students' individual limits to maximize their academic success, and this empowers teachers to be proactive rather than reactive to unique abilities, personalities, and backgrounds in the classroom (Anderson, 2007). This can be completed through knowing the learner and adjusting the curriculum through alterations in the content, process, and product of the lesson.

CONTENT INFORMATION

To gain the most out of this case study, you will first need to think of your students and complete the initial process of differentiation. Make sure you understand what differentiated instruction is and isn't; I would consider using the ASCD Infographic found in the additional resources below (ASCD, 2015). The following section of this case study will have ideas on differentiating the content, process, and product based on the other lesson plans provided in the modules of the book (e.g., Use of Variables, Tower Design, Constructing a Marshmallow Catapult). Integration of science, technology, engineering, art, and mathematics will be demonstrated throughout this case based upon the lessons.

CASE DESCRIPTION

For this case, you will need to refer to other portions of the book: Use of Variables (Part 1), Tower Design, and Constructing a Marshmallow Catapult.

Use of Variables: Part 1

Students are instructed to research the history of flight. One way to differentiate the *content* is to provide a variety of websites, videos, modules, and readings. The key concept here is to check the readability of the content to ensure that students above or below reading level are reading based on their ability. Depending on their grade level and ability, another way to differentiate the *content* is to provide search terms or web links. This way the students are accessing the most appropriate materials and ensuring that it is not too overwhelming based on their individual needs.

Some ways to differentiate the *process* of this lesson are to (1) provide a checklist about how to access relevant information, (2) use videos, (3) review roles and expectations in group work, and (4) use technology. The lesson requires students to research and identify relevant information about airplanes. To differentiate this *process* the teacher could also provide a checklist of how to identify relevant information. This can be in the textbook that they refer to, a chart on the wall, or an individual checklist that is on their desk. This can be reviewed with specific students if necessary. It can be given as an option for students to use if they deem it necessary. Another requirement of the lesson is that they are given an information packet. To differentiate the *process* in this section, the teacher could also provide a video as an option if students wanted to view the content in that format. The third requirement, but not in the exact order, is the use of group work. This is a critical component where differentiating the *process* could really support students. One way to do this is to review the roles of students and the expectations during group work. Another idea would be to have a list of roles and contracts so students would identify their roles and responsibilities with their group and then sign off on the contract, in turn, holding other students accountable. The last option in differentiating the *process* would be the use of technology. This lesson requires the use of a packet; some students may prefer to work using technology to collaborate (e.g., GoogleDocs, GoogleHangout, GoToMeeting), complete their forms (e.g., Weebly, GoogleForms), or simply just to type instead of writing by hand.

The final component in differentiation is the *product*. For this, I would challenge the teacher to extend their learning to have students choose different ways of reflecting on what they have learned. Some students may want to create a rap or song to share their findings, others may want to complete a chart about their results; another option could be to write a few paragraphs on what they learned. These are just a few ways to show what they learned, expanding on the original lesson of a discussion of the evaluation of results and improvements of flight.

Tower Design

In this lesson, students are instructed to research background information about existing towers. The students then design their own tower where they include the height, cost, and time and answer questions that apply to the real world. One way to differentiate the *content* of this lesson is to provide a list of websites and videos for students so that the reading materials are more appropriate for their level.

To differentiate the *process*, there are four ways that would be beneficial: (1) use an online graphing tool, (2) assign groups and roles, (3) create an electronic version of the materials, (4) assign specific questions for students to complete. The first way to differentiate the content is the use of an online graphing tool, so when the students are asked to graph the design, they could be offered the option to use a free online graphing tool (e.g., Tinkercard). This gives the students who may excel in the use of technology differentiated instruction that focuses on their strengths rather than their needs (e.g., poor fine motor skills). Another idea is how the teacher groups the students. It would be important to think about strengths and needs of the students, as well as mixing up the groups. On top of that, to differentiate the *process*, the teacher may want to think about assigning roles to the students while they are in their groups so that the teacher can challenge the students while still holding them accountable. The third way to differentiate the *process* would be providing an electronic version of the *Cost of Materials Data Sheet* as well as the *Questions on Observations*. This can be done through Microsoft Word or an app on their ChromeBook or iPad (e.g., iA Writer, Simplenote, PlainText, Writer). The final way to differentiate the *process* would be to assign specific questions and/or have them create an additional question on their own. This would help the student think about the material in a different way, as well as match them to their strengths (or, in some cases, match them based on their needs on purpose so they can continue working to improve).

In this lesson, the students are asked to write a one-page report on the tower that they designed. To differentiate the *product*, the students could be given the option of creating a brochure, a tourism video, or a website (e.g., Wix, Weebly) that includes all of the information in a report but allows them to choose the format based on their strengths.

Constructing a Marshmallow Catapult

In this case study the students are instructed to research, design, build, and test a catapult. After trial and error the students go through a series of revisions and retesting to build a stronger design. Upon completion they are asked to share their redesigns.

For this case, many differentiated components were already included, or in some cases it would be advised that the teacher complete similar differentiated activities to the *Tower Design* section. To differentiate the *content*, the lesson already has use of videos and a variety of readings, so the only suggestion would be to have the students or teacher write the main ideas on the board. This will help to support the students who are visual learners and may have difficulty with auditory learning.

To differentiate the *process* of this unit, the teacher should allow the use of an online graphing tool to create the design of the catapult (see the Tower Design section). During days 6 to 8, the teacher assigns roles to the group members. One way to support this would be through the use of *Task Bookmarks (*see Table 3: Task Bookmark Example*)*. Task Bookmarks can be given to team members to help them keep focused on their duties for team success. Each section or "role" is given to a specific student to complete.

Since this lesson relies heavily on group work, a way to assess students during that time would be to collect anecdotal observations. One way to do this is by doing a "clipboard cruise" (refer to Table 4: Clipboard Cruise Example). Teachers record their findings on a sheet that's attached to their clipboard. The information that is typically included is the name of the student, the date, time, center or task, and then the behavior that is observed. It is essential to evaluate and provide feedback to students while they are working in groups, and this is one way to have more concrete data to share with students and families.

To differentiate the *process* of the students' final designs, the teacher may choose to have the students use a compare-and-contrast graphic organizer. The graphic organizer will help as a visual aid for students to write out what is similar about their designs as well as what sets each design apart.

Table 3. Task bookmark example

Title	Role
Production Manager	• Oversee the project • Assign final roles after discussion • Ensure that everyone is doing their job • Keep track of the project • Keep the team on track • Give reminders to different students based on their role • Manage deadlines • Communicate with the teacher
Technology Manager	• Provide resources from the Internet • Assist in computer and technology aspects • Design the presentation (if using a technology format)
Information Manager	• Ensure that materials are accurate • Listen and clarify ideas • Follow written directions • Communicate updates to the group based on findings
Resource Manager	• Gather materials • Store all materials for projects • Make sure that materials are available • Communicate needs of materials with production manager

Table 4. Clipboard cruise example

Name	Time	Task	Observed Behavior

The final assessment of the unit is where the students are asked to write a scenario where a catapult could be used today. To differentiate the *product*, the students could create a diorama of the scenario, write a poem about it, or develop a collage. By offering options for students to show what they learn, the students are able to build on their strengths of the unit and have a deeper understanding.

These are just some ideas to differentiate the content, process, and product of different lessons within this book. Continual reflection on research-based practices and pedagogical practices that affect student learning are critical to the success of differentiating instruction.

CURRENT CHALLENGES FACING TEACHERS

Teachers face two main challenges in the differentiated STEAM classroom: time and lack of support. It is important to explore these challenges as well as potential solutions to make this as successful as possible.

Overall, differentiating instruction in the STEAM classroom takes time. The teacher needs to take the time to learn about different ways to differentiate within the STEAM classroom and maybe even attend professional development around differentiation. Another aspect of time is the additional amount of planning that will occur when you begin to differentiate. It takes time to get to know your students, as well as what ways they learn the best, and then takes time to plan. The final aspect of time is the lack of classroom time to complete STEAM projects.

The other challenge that faces teachers is the possible lack of support. This could be from administration and the special education teacher. Administration may not understand the components of a successful differentiated STEAM classroom or ways to support teachers in this setting. Another difficulty is the fact that the special education teacher may not be able to help support as much as needed because of other students on their caseload, common planning time, or teaching time.

It is crucial to keep in mind time and support as you begin to implement the STEAM classroom. Being able to know and recognize these potential challenges are critical so that teachers can plan ahead for what may occur.

SOLUTIONS AND RECOMMENDATIONS

To address the first challenge of the time necessary to learn the ways to differentiate in the STEAM classroom, there are websites that offer professional development online, such as the Association for Supervision and Curriculum Development (ASCD) or the Council for Exceptional Children (CEC). Most of the time these professional development sessions are also recorded so that if you aren't able to attend, you can view it during a time when you can. It's also helpful to search to see what professional development conferences your state offers so that you don't have to travel too far. A lot of times you can also ask your principal for financial support to attend as well.

When thinking about planning time, some ways to help assist would be to force yourself to do it, as well as find an uninterrupted time. This is very hard to do as a teacher—especially during the day—but it does help to stay regimented. I know when I was a teacher I had to close my door during planning to try to minimize the distractions. Another good tip is to have someone hold you accountable or have someone you can share your work with. This will help keep it a priority, and you also may affect their teaching!

There never seems to be enough time in the classroom to cover everything that we want to cover. It's important to think about how you lay out the assignments to make this as smooth and time efficient as possible. Make sure that you create directions and deliver them clearly. Many times multiple directions can be way too much, so it's important to think about how to orally provide them, as well as provide them in a written format, for those to review as you facilitate the students learning. As stated earlier, it is also important to assign students to groups and seating areas; if you list them in the areas on the board, it helps to create smoother transitions that they can refer back to (instead of calling students' names in order and sending them to different areas). Another idea is to have a plan for getting help when you are busy with a group or another student. Some classrooms utilize an "expert for the day" where one or more students who are skilled in that area respond to questions before the teacher. Another idea is to have a question board, where students go up and write their questions on chart paper and then they get answered when the teacher is able to. To try and provide more time, it's also important to think about the students or groups that will get finished first. You want to make sure you have a plan for them so you are promoting on-task behavior and not encouraging rushing through assignments. The final strategy in managing the classroom time would be to break the larger assignments into "chunks"—this means each day you are helping them manage their time because you have only provided a portion of the assignment. This helps students learn time management, and eventually you can scaffold your instruction so that they build up where they have longer tasks and greater amounts of time.

Collaborating with the administration is another critical factor. Invite principals into your classroom to observe (even when it's not a scheduled observation). Stop in their offices and share student work and data that you have collected that show the impacts of the STEAM classroom. You want to make sure that the administration understands the content and depth of your lessons. This way you can approach them to ask for common planning time with other content area teachers, special education teachers, or English as a second language (ESL) teachers. If you aren't able to build this type of relationship prior to meeting, be sure to go with existing data that support why you are requesting what you are requesting. Administrators want to know that what you are doing is supported by data and that it can be successful.

Another challenging factor is the ability to work with the special education or support teachers for students who may be in your STEAM classroom. It's important to ask about their caseload and when the best times to meet are. You don't necessarily need to meet in person, but you need to identify the ways that you both prefer to communicate so you can collaborate to enhance your lessons. Google Docs is a great way to share lessons and materials with special education teachers who may not have the time to physically meet at the same times when you are able to meet. Another option would be to share your curriculum map of upcoming topics so that way the special education teacher could provide potential materials to help you differentiate your lessons. It's not mandatory that the special education teacher be involved when you differentiate, but their expertise is extremely invaluable for students who either are above level or are struggling.

Differentiating in the STEAM classroom is possible! Start small to build confidence, and you will see the positive impacts on the achievement of your students. Just remember to first get to know them and then make adjustments to your curriculum. This way, students are able to access the content, have multiple ways to process the information, and have a variety of assessments to show what they learned. These key ingredients are the recipe for success!

REFERENCES

Anderson, K. (2007). Differentiating instruction to include ALL students. *Preventing School Failure*, *51*(3), 49–54. doi:10.3200/PSFL.51.3.49-54

Autodesk. (2016). *Tinkercad.* Retrieved July 7, 2016, from https://www.tinkercad.com/about/features

Dixon, F. A., Yssel, N., McConnell, J. M., & Hardin, T. (2014). Differentiated instruction, professional development, and teacher efficacy. *Journal for the Education of the Gifted*, *37*(2), 111–127. doi:10.1177/0162353214529042

Gregory, G. H., & Chapman, C. (2007). *Differentiated Instructional Strategies: One Size Doesn't Fit All* (2nd ed.). Thousand Oaks, CA: Corwin Press.

Logan, B. (2011, October). Examining differentiated instruction: Teachers respond. *Research in Higher Education*, 13.

National Center for Educational Statistics. (n.d.). *Create a graph.* Retrieved July 7, 2016, from https://nces.ed.gov/nceskids/createagraph/

Office Applications. (n.d.). *Writer.* Retrieved July 7, 2016, from https://writer.bighugelabs.com/login

Parsons, S. A., Dodman, S. L., & Burrowbridge, S. C. (2013, September). Broadening the view of differentiated instruction. *Phi Delta Kappan*, *95*(1), 38–42. doi:10.1177/003172171309500107

Rock, M. L., Gregg, M. G., Ellis, E., & Gable, R. A. (2008, Winter). REACH: A framework for differentiating classroom instruction. *Preventing School Failure*, *52*(2), 31–47. doi:10.3200/PSFL.52.2.31-47

Tomlinson, C. A. (2001). *How to differentiate instruction in mixed-ability classrooms* (2nd ed.). Upper Saddle River, NJ: Pearson.

Watts-Taffe, S., Laster, B. P., Broach, L., Marinak, B., Connor, C. M., & Walker-Dalhouse, D. (2012). Differentiated instruction: Making informed teacher decisions. *The Reading Teacher*, *66*(4), 303–314. doi:10.1002/TRTR.01126

Zygomatic. (n.d.). *Chart Tool*. Retrieved July 7, 2016, from http://www.onlinechart-tool.com/

ADDITIONAL READING

ASCD. (2008). *Nonnegotiables of differentiated instruction: Respectful tasks*. Retrieved February 23, 2016, from http://www.ascd.org/ascd-express/vol4/425-video.aspx

ASCD. (2015). *What is (and is not) differentiated instruction?* Retrieved February 23, 2016, from http://www.ascd.org/ASCD/pdf/siteASCD/publications/Differentiation_Is-IsNot_infographic.pdf

BER Staff Development. (2011, March 17). *Differentiating instruction, 6-12, Part 1: Student choice and multiple modes of learning*. Retrieved on February 23, 2016, from https://www.youtube.com/watch?v=akvDT9KFZPw

BER Staff Development. (2011, March 17). *Differentiating instruction, 6-12, Part 11: Tiered assignments and grouping*. Retrieved February 23, 2016, from https://www.youtube.com/watch?v=1ob4eGz04G4

Castaneda, R. (2012, July 10). *Getting started on differentiated instruction*. Retrieved February 23, 2016, from https://www.youtube.com/watch?v=LGYa6ZacUTM&NR=1&feature=endscreen

Cox, J. (n.d.). *Differentiated instruction strategies: Learning stations*. Retrieved February 23, 2016, from http://www.teachhub.com/differentiated-instruction-strategies-using-learning-stations

Edutopia. (2012, July 11). *Differentiating instruction through interactive games*. Retrieved February 23, 2016, from https://www.youtube.com/watch?v=XZ0BGXMf83U

Edutopia (2015, September 9). *Differentiated instruction: Resource roundup*. Retrieved February 23, 2016, from http://www.edutopia.org/article/differentiated-instruction-resources

Forest Lake Elementary School. (2010, May 3). *Tech-fueled differentiated instruction engages elementary school students*. Retrieved February 23, 2016, from http://www.edutopia.org/stw-differentiated-instruction-technology-elementary-video

Grant, J. (2014). *Staff development for educators: Differentiated instruction*. Retrieved February 23, 2016, from http://www.sde.com/teacher-resources.asp

Hocket, J. (2015, February 19). *How I learned differentiation*. Retrieved February 23, 2016, from http://www.teachthought.com/learning/personalized-learning/how-i-learned-differentiation/?utm_content=556b54cf04d3014c38000001

Huck, C., & Amaral, B. (2014, October). *Scaffolding for success: Best practices for secondary ELLs*. Retrieved February 23, 2016, from http://newsmanager.commpartners.com/tesolc/issues/2014-10-01/3.html

Kluth, P. (2015). *20 ways to adapt the science lab*. Retrieved February 23, 2016, from http://www.paulakluth.com/readings/differentiating-instruction/20-ways-to-adapt-the-science-lab/

Kluth, P. (2016). *Rewriting history, and nine other ways to adapt textbooks*. Retrieved on February 23, 2016, from http://www.paulakluth.com/readings/differentiating-instruction/rewriting-history-and-nine-other-ways-to-adapt-textbooks/

Roberts, E. J. (2014, October 22). *Differentiating instruction: In the classroom and at home*. Retrieved February 23, 2016, from http://home.edweb.net/differentiated-instruction-classroom-home/

Sousa, D. A. (n.d.). *How can differentiation be achieved-without putting too much burden on teachers?* Retrieved February 23, 2016, from http://www.centerforlearning.org/t-free-resources.aspx

Teaching Channel. (2011). *Content differentiation in 3rd grade science*. Retrieved February 23, 2016, from https://www.teachingchannel.org/videos/content-differentiation-science

Teaching Channel. (2011). *New teacher survival guide: Differentiating instruction*. Retrieved February 23, 2016, from https://www.teachingchannel.org/videos/differentiating-instruction?fd=1

Teaching Channel. (2012). *Differentiating with learning menus*. Retrieved February 23, 2016, from https://www.teachingchannel.org/videos/differentiating-instruction-strategy

The IRIS Center. (2015). *Differentiated instruction: Maximizing the learning of all students*. Retrieved February 23, 2016, from http://iris.peabody.vanderbilt.edu/module/di/

Tomlinson, C. A. (2014, May 15). *Revisiting the differentiated classroom: Looking back and ahead.* Retrieved February 23, 2016, from http://www.ascd.org/professional-development/webinars/revisiting-the-differentiated-classroom-webinar.aspx

Tomlinson, C. A., Brimijoin, K., & Narvaez, L. (2008). *Setting the stage for change toward differentiation*. Retrieved February 23, 2016, from http://www.ascd.org/publications/books/105005/chapters/Setting-the-Stage-for-Change-Toward-Differentiation.aspx

Tomlinson, C. A., & Britt, S. (2012, June 6). *Common Core State Standards: Where does differentiation fit?* Retrieved February 23, 2016, from http://www.ascd.org/professional-development/webinars/tomlinson-and-britt-webinar.aspx

University of Texas. (2013). Differentiating instruction: Key to student success. Retrieved February 23, 2016, from http://buildingrti.utexas.org/online-modules/differentiated-instruction-key-student-success

KEY TERMS AND DEFINITIONS

Adaptation: A change in the environment or curriculum that allows the learner more appropriate access to learning based on his or her abilities and needs. An adaptation can be an accommodation or modification.

Accommodation: A change to the environment or curriculum without changing the objective.

Curriculum Compacting: A strategy that can sometimes be used if a student has already mastered the content, the teacher then designs the instruction to "dive in deeper" or "compacts it" and moves on to the next skill.

Collaboration: When at least two individuals work together towards a common goal.

Differentiated Instruction: A proactive approach to meeting individual's abilities and needs.

Modification: A change to the curriculum that alters the objective so that the student can be in an inclusive setting with more appropriate expectations.

Scaffolding Instruction: A process where teachers model or demonstrate how to solve a problem, then step back, and offer support as needed.

Conclusion

From the beginning of this book, we have conceptualized STEAM education as a form of communal and improvisational *making*—one that unites the head, heart, and hands in teaching and learning performances, whose outcomes cannot be prescribed. Each case presented has the possibility of resulting in a diverse array of learning artifacts as students personally construct knowledge and design solutions to problems. We support the "Yes, and" message of improvisational theatre, and fully recognize that some cases may evoke responses that range from full acceptance to full rejection. It is our hope that your response will be "yes I can do that". May at least a few of these cases of STEAM education in practice inspire you try STEAM in your own classrooms!

Throughout this text, our authors have regarded art practice as an equal partner in STEM education, an essential and necessary component to creative teaching and learning. In all cases, STEAM teaching and learning are regarded as being jointly produced and performed—as a way of bringing something into existence that was not there before. In all cases, STEAM serves as pathway toward an inclusive education because of the practice of arts integration.

Each chapter has a strength and we are choosing one from each author to highlight. In many cases, STEAM education is portrayed as an active metaphor: as musicianship and an ethical act of care (Van Sickle/Koester); as a way of bringing a problem into the light (Dollahan); as a design process (Bazler, Dibiase/MacDonald/Strong, and Behling); as a way of developing mental models in the minds' eye through practice with visualization (Finson, Smith, and Seiverd); as a system of signs and symbols—employing a multimodal form of literacy and semiosis (Koester, and Romagnoli); and as embodied simulation through movement and creative drama (Lindquist/James-Hassan/Lindquist, and Jablon). Groff brings many of these metaphors together in her description of teaching STEAM through the creation of graphic novels. These chapters provide lessons for preK-12 classrooms, and it would be interesting to look at the pollinator project (Rizzuto/Henning/Duckett) because it contains lessons rarely found for hands on science for pre-school students.

Graybill calls for STEAM teaching that helps students practice 21st century skills, urging the creation of partnerships, general educators, and science educators and the purposeful practice of co-teaching. Similarly, Koester advocates for a kind of

STEAM that features research and practice partnerships characterized by a constellation of C's: *Caring Collaboration* among *Content* specialists, *Cultural relevance*, *Cross-training, Curating*, *Creative process*, *Community*, and the achievement of a *Collective* impact. Participants in such STEAM projects would collectively build professional STEAM teaching/learning communities characterized by the sharing of lessons that have "worked" for them, as well as obstacles they may have encountered in the process. Finally, Servillo offers specific examples of how STEAM is a powerful pedagogy for special needs students. Thus, this book is an example of collaboration and promising practice.

We have proposed what we consider many valid reasons for experimenting with STEAM as teaching and learning practice. We propose the letting go of prescribed outcomes and to design lessons that can lead to a construction of deep, personal meaning by students, afforded the freedom associated with making their own ideas, performances, and/or artifacts. We encourage you to adopt artistic habits of mind and emergent mindsets, fully acknowledging that doing so can be scary. Such teaching and learning is not only useful, but empowering.

Diane Cudahy
College of Charleston, USA

Compilation of References

Abell, S. K., & Roth, M. (1992). Coping with constraints of teaching elementary science: A case study of a science enthusiast student teacher. *Science Teacher Education*, *76*(6), 581–595.

ACT. (2015). *The condition of STEM*. National ACT, Inc. Retrieved from http://www.act.org/content/dam/act/unsecured/documents/National-STEM-Report-2015.pdf

Adams, A. E., & Pegg, J. (2012). Teachers enactment of content literacy strategies in secondary science and mathematics classes. *Journal of Adolescent & Adult Literacy*, *2*(56), 151–161. doi:10.1002/JAAL.00116

Adjibolosoo, S. (1995). *The human factor in developing Africa*. Westport, CT: Praeger Publishers.

Ainsworth, S. E. (2006). DeFT: A conceptual framework for learning with multiple representations. *Learning and Instruction*, *16*(3), 183–198. doi:10.1016/j.learninstruc.2006.03.001

Aizen, M. A., Garibaldi, L. A., Cunningham, S. A., & Klein, A. M. (2009). How much does agriculture depend on pollinators? Lessons from long-term trends in crop production. *Annals of Botany*, *103*(9), 1579–1588. doi:10.1093/aob/mcp076 PMID:19339297

American Association for the Advancement of Science. (1993). *Benchmarks for science literacy*. New York, NY: Oxford University Press.

Andersen, C. (2004). Learning in as-if worlds: Cognition in drama in education. *Theory into Practice*, *43*(4), 281–286. doi:10.1207/s15430421tip4304_6

Anderson, K. (2007). Differentiating instruction to include ALL students. *Preventing School Failure*, *51*(3), 49–54. doi:10.3200/PSFL.51.3.49-54

Andrews, E. (2014). 6 Early amusement parks. *History Lists*. Retrieved from http://www.history.com/news/history-lists/6-early-amusement-parks

Archambault, M., Janosz, M., Fallu, J.-S., & Pagani, L. (2009). Student engagement and its relationship with early high school dropout. *Journal of Adolescence*, *32*(3), 651–670. doi:10.1016/j.adolescence.2008.06.007 PMID:18708246

Arieli, B. (2007). *The integration of creative drama into science teaching* (Doctoral dissertation). Curriculum & Instruction College of Education, Kansas State University, Manhattan, KS. Retrieved from https://krex.k-state.edu/dspace/bitstream/handle/2097/531/BrachaArieli2007.pdf?sequence=1

Arts Council. (1992). *Drama in schools: Arts council guidance on drama education.* London: The Arts Council of Great Britain.

Ashley, S. (2001). When the twin towers fell. *Scientific American.* Retrieved from https://www.scientificamerican.com/article/when-the-twin-towers-fell/

Ashton-Hay, S. (2005). Drama: Engaging all learning styles. In *Proceedings 9th International INGED (Turkish English Education Association) Conference.* Economics and Technical University.

Aubusson, P. J., & Fogwill, S. (2006). Role play as analogical modelling in science. In P. J. Aubusson, A. G. Harrison, & S. M. Ritchie (Eds.), *Metaphor and analogy in science education* (pp. 93–104). Dordrecht, The Netherlands: Springer. doi:10.1007/1-4020-3830-5_8

Aubusson, P., Fogwill, S., Barr, R., & Perkovic, L. (1997). What happens when students do simulation-role-play in science? *Research in Science Education, 27*(4), 565–579. doi:10.1007/BF02461481

Aurum, A., & Wohlin, C. (2003). The fundamental nature of requirements engineering activities as a decision-making process. *Information and Software Technology, 45*(14), 945–954. doi:10.1016/S0950-5849(03)00096-X

Ausubel, D. (1968). *Educational psychology: A cognitive view.* New York, NY: Holt, Rinehart and Winston.

Autodesk. (2016). *Tinkercad.* Retrieved July 7, 2016, from https://www.tinkercad.com/about/features

Bachelis, G. F., Maxim, B. R., James, D. A., & Stout, Q. F. (1994). Bringing algorithms to life: Cooperative computing activities using students as processors. *School Science and Mathematics, 94*(4), 176–186. doi:10.1111/j.1949-8594.1994.tb15651.x

Bailey, S. D. (1993). *Wings to fly: Bringing theatre arts to students with special needs.* Rockville, MD: Woodbine House.

Baird, S. (2015). *How Disposable Coffee Cup Design Became High Art.* Retrieved September 30, 2016, from http://www.eater.com/drinks/2015/8/14/9153849/how-disposable-coffee-cup-design-became-high-art

Barton, K., & Smith, L. (2000). Themes or motifs? Aiming for coherence through interdisciplinary outlines. *The Reading Teacher, 54*(1), 54–63.

Bedetti, G. (2015). Collaborative college playwriting and performance: A core course "Trespassing" onto the dramatic arts. *International Journal of Education & the Arts, 16*(19).

Beers, S. Z. (2011). 21st century skills: Preparing students for THEIR future. *STEM*. Retrieved from http://cosee.umaine.edu/files/coseeos/21st_century_skills.pdf

Bennett, J., & Hogarth, S. (2005). *Would YOU want to talk to a scientist at a party? Students' attitudes to school science and science*. University of York, Department of Educational Studies Research Paper 2005/08.

Bianchini, J. A., & Colburn, A. (2000). Teaching the nature of science through inquiry to prospective elementary teachers: A tale of two researchers. *Journal of Research in Science Teaching*, *37*(2), 177–209. doi:10.1002/(SICI)1098-2736(200002)37:2<177::AID-TEA6>3.0.CO;2-Y

Bigelow Middle School. (n.d.). *Engineering technology: Basketbots*. Retrieved June 30, 2016, from http://www.bu.edu/gk12/alex/bioengineering_teacher.pdf

Bigelow, B. (2009). Once upon a genocide: Columbus in children's literature. In W. Au (Ed.), *Rethinking multicultural education: Teaching for racial and cultural justice* (pp. 73–85). Milwaukee, WI: Rethinking Schools.

Blinder, S. M. (2007). *The Coffee Cooling Problem*. Retrieved June 07, 2016, from http://demonstrations.wolfram.com/TheCoffeeCoolingProblem/

Boal, A. (2002). *Games for actors and non-actors*. New York: Routledge.

Bodzin, A. M., & Cirucci, L. (2009). A land-use-planning simulation using Google Earth. *Science Scope*, *32*(7), 30–38.

Bohart, H., Benson-Collick, H., & Charner, K. (Eds.). (2016). *Spotlight on young children: Teaching and learning in the primary grades*. The National Association of Young Children.

Bolton, G., & Heathcote, D. (1995). *Drama for learning: Dorothy Heathcote's mantle of the expert approach to education*. Portsmouth, NH: Heinemann.

Bordieu, P. (2003). The forms of capital. In A. H. Halsey, H. Lauder, P. Brown, & A. S. Wells (Eds.), *Education: Culture, economy, society* (pp. 46–58). Oxford, UK: University Press.

Boyer, E. (1983). *High school: A report on secondary education in America*. New York, NY: Joanna Colter Books.

Brady, J. (2014, September 5). STEM is incredibly valuable, but if we want the best innovators we must teach the arts. *The Washington Post*. Retrieved from https://www.washingtonpost.com/news/innovations/wp/2014/09/05/stem-is-incredibly-valuable-but-if-we-want-the-best-innovators-we-must-teach-the-arts/?utm_term=.7dfe19789cff

Brain, P. O. P. (2014). *Hydraulics*. Retrieved July 15, 2016, from https://www.brainpop.com/technology/energytechnology/hydraulics/

Bransford, J. (2000). *How people learn: Brain, mind, experience, and school*. Washington, DC: National Academies Press.

Brante, E. W., Olander, M. H., & Nystrom, M. (2013). Exploring the impact of contrasting cases in text and picture processing. *Journal of Visual Literacy, 32*(2), 15–38. doi:10.1080/23796529.2013.11674708

Braund, M., Lubben, F., Scholtz, Z., Sadeck, M., & Hodges, M. (2007). Comparing the effect of scientific and socio-scientific argumentation tasks: Lessons from South Africa. *The School Science Review, 88*(324), 67–76.

Bredekamp, S. (2014). *Effective practices in early childhood education: Building a foundation.* Upper Saddle River, NJ: Pearson.

Bresler, L. (2005). What musicianship can teach educational research. *Music Education Research, 7*(2), 169–183. doi:10.1080/14613800500169399

Bresler, L. (2006). Toward connectedness: Aesthetically based research. *Studies in Art Education, 48*(1), 52–69.

Brigante, R. (2010). Want to be a theme park designer? Disney Imagineer and Universal creative experts offer advice. *Inside the Magic.* Retrieved from http://www.insidethemagic.net/2010/04/want-to-be-a-theme-park-designer-disney-imagineer-and-universal-creative-experts-offer-advice/

Brooks, M. G., & Grennon-Brooks, J. (1999). The courage to be constructivist. *Educational Leadership, 57*(3), 18–24.

Brown, J. S., Collins, A., & Duguid, P. (1989). Situated cognition and the culture of learning. *Educational Researcher, 18*(1), 32–42. doi:10.3102/0013189X018001032

Brozo, W. G., Moorman, G., & Meyer, C. K. (2014). *Wham! Teaching with graphic novels across the curriculum.* New York, NY: Teachers College Press.

Brumberger, E. (2010). Visual literacy and the digital native: An examination of the millennial learner. *Journal of Visual Literacy, 39*(1), 19–47.

Bryant, J. (2008). *Newton's Law of Cooling Demonstration.* Retrieved June 07, 2016, from http://demonstrations.wolfram.com/NewtonsLawOfCooling/

BSCS - Biological Science Curriculum Study. (2005). *BSCS Science and technology: Investigating physical systems.* Dubuque, IA: Kendall Hunt.

Buchanan, M. (2016). Why teach drama? *Child Drama.* Retrieved June 3, 2016. Retrieved from http://www.childdrama.com/the-essay.html

Burj Khalifa. (2011). *Civil Engg. Seminar.* Retrieved from http://civilenggseminar.blogspot.com/2011/09/burj-khalifa.html

Burke, R. J., & Mattis, M. C. (Eds.). (2007). *Women and minorities in science, technology, engineering and mathematics: Upping the numbers.* Northampton, MA: Edward Elgar Publishing, Inc. doi:10.4337/9781847206879

Butler, J. (1989). Science learning and drama processes. *Science Education*, *73*(5), 569–579. doi:10.1002/sce.3730730505

Bybee, R. W. (1989). *Science and technology education for the elementary years: Frameworks for curriculum and instruction*. Washington, DC: The National Center for Improving Instruction.

Caldwell, K. (2011). Drama as a teaching tool. *Teaching Through the Arts*. Retrieved June 3, 2016. http://teachingthroughthearts.blogspot.com/2011/07/drama-as-teaching-tool.html

Carlsen, M. (2009). Reasoning with paper and pencil: The role of inscriptions in student learning of geometric series. *Mathematics Education Research Journal*, *21*(1), 54–84. doi:10.1007/BF03217538

Carnine, L., & Carnine, D. (2004). The interaction of reading skills and science content knowledge when teaching struggling secondary students. *Reading & Writing Quarterly*, *20*(2), 203–218. doi:10.1080/10573560490264134

Carter, J. B. (Ed.). (2007). *Building literacy connections with graphic novels: Page by page, panel by panel*. Urbana, IL: National Council of Teachers of English.

Case study: The Leaning Tower of Pisa. (n.d.). *Madrid Engineering*. Retrieved from http://madridengineering.com/case-study-the-leaning-tower-of-pisa/

Catchen, R. (2013). Reflections: How STEM becomes STEAM. *The STEAM Journal*, *1*(1). doi:10.5642/steam.201301.22

Cawthon, S. W., Dawson, K., & Ihorn, S. (2011). Activating student engagement through drama-based instruction. *Journal for Learning through the Arts*, *7*(1). Retrieved from http://www.escholarship.org/uc/item/6qc4b7pt

Cazden, C., Cope, B., Fairclough, N., & Gee, J et al.. (1996). A pedagogy of multiliteracies: Designing social futures. *Harvard Educational Review*, *66*(1), 60–92. doi:10.17763/haer.66.1.17370n67v22j160u

Chaille, C. M., & Britain, L. (2003. The young child as scientist: A constructive approach to early childhood science education (3rd ed.). Boston, MA: Allyn & Bacon.

Chautauqua Lake Central School District. (n.d.). *History of project lead the way*. Retrieved from http://www.clake.org/District/Class/28-Engineering-and-Technology-Department/1447-HistoryOfPLTW.html

Chilton, E., Carpenito, K., & Milto, E. (2013). Hands-on activity: Hydraulic arm challenge. *TeachEngineering*. Retrieved June 28, 2016, from https://www.teachengineering.org/activities/view/wpi_hydraulic_arm_challenge

Christofi, C., & Davies, M. (1991). Science through drama. *Education in Science*, *141*, 28–29.

Clarke, J. (1999). Growing high school reform: Planting the seeds of systemic change. *NASSP Bulletin*, *83*(April), 1–9. doi:10.1177/019263659908360601

Clarke, J. (2001). Teaching each and all: Harnessing dynamic tension in school reform. *NASSP Bulletin, 85*(626), 75–79. doi:10.1177/019263650108562608

Clarke, J. (2003). Personalized learning and personalized teaching. In J. DiMartino, J. Clarke, & D. Wolk (Eds.), *Personalized learning: Preparing high school students to create their futures*. Lantham, MD: Scarecrow Press.

Clarke, J. H. (1990). *Patterns of thinking: Integrating learning skills in content teaching*. Boston, MA: Allyn & Bacon.

Clarke, J., & Agne, R. (1997). *Interdisciplinary high school teaching: Strategies for integrated learning*. Boston: Allyn and Bacon.

Clarke, J., Aiken, J., & Sullivan, M. (1999, September). Interactive leadership in high school innovation. *NASSP Bulletin, 83*(608), 98–111. doi:10.1177/019263659908360811

Clarke, J., Bossange, J., Erb, C., Gibson, D., Nelligan, B., Spencer, C., & Sullivan, M. (2000). *Dynamics of change in high school teaching: A study of innovation in five Vermont professional development schools*. Providence, RI: Northeast and Islands Regional Educational Laboratory at Brown University.

Clarke, J., & Fraser, E. (2003). Making learning personal: Educational practices that work. In J. DiMartino, J. Clarke, & D. Wolk (Eds.), *Personalized learning: Preparing high school students to create their futures*. Lantham, MD: Scarecrow Press.

Clarke, J., Sanborn, S., Aiken, J., Cornell, N., Goodman, J., & Hess, K. (1998). *Real questions/real answers: Focusing teacher leadership on school development*. Alexandria, VA: ASCD Publications.

Clinchy, E. (1996, December). Reforming American education from the bottom to the top: Escaping academic captivity. *Phi Delta Kappan*, 268.

Cobern, W. W. (1993). Contextual constructivism: The impact of culture on the learning and teaching of science. In K. Tobin (Ed.), *The Practice of constructivism in science education* (pp. 51–69). Washington, DC: American Association for the Advancement of Science.

Cochran, K. F. (1991). *Pedagogical content knowledge: A tentative model for teacher preparation*. Paper presented at the annual meeting of the American Educational Research Association, Chicago, IL.

Coffey, H. (n.d.). *Interdisciplinary teaching*. Retrieved on October 3, 2016 from http://www.learnnc.org/lp/pages/5196

Cognition and Development. (n.d.). *Jerome Bruner*. Retrieved from http://psychology4a.com/develop8.htm

Cohen, D. (1995). What is the system in systemic reform? *Educational Researcher, 24*(9), 11-17.

Coleman, J.M., McTigue, E.M., & Smolkin, L.B. (2011). Elementary teachers' use of graphical representations in science teaching. *Journal of Science Teacher Education, 22*(7), 613-644. doi:.10.1007/s10972-010-9204-1

Coleman, J. (2010). Elementary teachers instructional practices involving graphical representations. *Journal of Visual Literacy, 29*(2), 198–222. doi:10.1080/23796529.2010.11674681

Coleman, J. S. (2003). Social capital in the creation of human capital. In A. H. Halsey, H. Lauder, P. Brown, & A. S. Wells (Eds.), *Education: Culture, economy, society* (pp. 80–95). Oxford, UK: University Press.

Collins, J. M. (2013). *The new world champion paper airplane book*. New York, NY: Crown Publishing.

Columbus, C. *(1492). Journal of the first voyage (Diario del primer viaje) 1492* (B. W. Ife, Trans. & Ed.). Warminster, UK: Aris and Phillips Ltd.

Committee on Prospering in the Global Economy of the 21st Century. (2005). *An agenda for American science and technology*. National Academy of Sciences, National Academy of Engineering, and Institute of Medicine.

Conezio, K., & French, L. (2002). Science in the preschool classroom: Capitalizing on children's fascination with the everyday world to foster language and literacy development. *Young Children, 57*(5), 12–18.

Conquergood, D. (1998). Beyond the text: Toward a performative cultural politics. In S. Dailey (Ed.), *The future of performance studies: Visions and revisions* (pp. 25–36). Annandale, VA: National Communication Association.

Cornett, C. E., & Smithrim, K. L. (2001). *Arts as meaning makers: Integrating literature and the arts throughout the curriculum*. Don Mills, Canada: Pearson Education.

Cotton, K. (2001). *New small learning communities: Findings from recent literature*. Seattle, WA: Northwest Regional Educational Laboratory.

Cottrell, J. (1987). Creative drama in the classroom grades 4-6. Lincolnwood, IL: National Textbook Company.

Covell, K., & Howe, R. B. (2001). Moral education through the 3 Rs: Rights, respect and responsibility. *Journal of Moral Education, 30*(1), 29–41. doi:10.1080/03057240120033794

Covell, K., McNeil, J. K., & Howe, R. B. (2009). Reducing teacher burnout by increasing student engagement: A childrens rights approach. *School Psychology International, 30*(3), 282–290. doi:10.1177/0143034309106496

Craven, J. (n.d.). The World Trade Center twin towers, 1973-2001. *About.com*. Retrieved from http://architecture.about.com/od/worldtradecenter/ss/worldtrade.htm#step1

Csikszentmihalyi, M. (1997). *Finding flow: The psychology of engagement with everyday life*. New York, NY: Basic Books.

Cushman, K., Steinberg, A., & Riordan, R. (2001). *Rigor and relevance: Essential ideas about connecting school and work*. Boston, MA: Jobs for the Future Inc.

Darling-Hammond, L., Ancess, J., & Ort, S. (2002). Reinventing high school: Outcomes of the Coalition Campus Schools Project. *American Educational Research Journal, 39*(3), 639–673. doi:10.3102/00028312039003639

Darwin, C. (1909). *The origin of species*. New York, NY: P.F. Collier & Son Company.

De Palma, A., Kuhlmann, M., Roberts, S. P., Potts, S. G., Börger, L., Hudson, L. N., & Purvis, A. et al. (2015). Ecological traits affect the sensitivity of bees to land-use pressures in European agricultural landscapes. *Journal of Applied Ecology, 52*(6), 1567–1577. doi:10.1111/1365-2664.12524 PMID:27546902

Deans, J. (2016). Thinking, feeling and relating: Young children learning through dance. *Australian Journal of Early Childhood, 41*(3), 46–57.

DeCoene, J., & DeCock, R. (1996). The children's rights project in the primary school 'De vrijdagmarkt' in Bruges. In E. Verhellen (Ed.), *Monitoring children's rights* (pp. 627–636). The Hague: Martinus Nijhoff.

DeLoache, J. S. (2000). Dual representation and young children's use of scale models. *Child Development, 71*(2), 329–338. doi:10.1111/1467-8624.00148 PMID:10834468

Dewar, A. (2014). *One minute paper airplanes kit*. Rutland, VT: Tuttle Publishing Company.

Dewey, J. (1938). *Experience and education*. New York, NY: Kappa Delta Pi.

Diaz, D., & King, P. (2007). *Adapting a post-secondary STEM instructional model to K-5 mathematics instruction*. Clemson, SC: Clemson University.

DiMartino, J., & Clarke, J. (2000). *Personal learning plans and high school renewal*. Paper presented at NASSP Annual Conference, San Antonio, TX.

DiMartino, J. (1997). *Study of Dropouts at West Warwick High School*. Providence, RI: Northeast and Islands Regional Educational Laboratory at Brown University.

DiMartino, J., Clarke, J., & Wolk, D. (Eds.). (2003). *Personalized learning: Preparing high school students to create their futures*. New York, NY: Scarecrow Press.

Disney Careers. (2016). *Walt Disney Imagineering*. Retrieved from http://wdi.disneycareers.com/en/default/

Disney Information. (n.d.a). *EPCOT Overview*. Retrieved from http://www.wdwinfo.com/wdwinfo/guides/epcot/ep-overview.htm

Disney Information. (n.d.b). *Mission: Space overview*. Retrieved from http://www.wdwinfo.com/wdwinfo/guides/epcot/epfw-mission.htm

Dixon, F. A., Yssel, N., McConnell, J. M., & Hardin, T. (2014). Differentiated instruction, professional development, and teacher efficacy. *Journal for the Education of the Gifted, 37*(2), 111–127. doi:10.1177/0162353214529042

Dockett, S. (1995). *You make me alive! Developing understanding through play*. Deakin, Australia: Australian Early Childhood Association.

Dorion, K. R. (2009). Science through drama: A multiple case exploration of the characteristics of drama activities used in secondary science lessons. *International Journal of Science Education, 31*(16), 2247–2270. doi:10.1080/09500690802712699

Dow, C. B. (2010). Young children and movement: The power of creative dance. *Young Children, 65*(2), 30–35.

Dubson, M., & Loeblein, T. (2010). *Curve Fitting*. Retrieved June 07, 2016, from http://phet.colorado.edu/en/simulation/legacy/curve-fitting

Duckworth, A. L., Peterson, C., Matthews, M. D., & Kelly, D. R. (2007). Grit: Perseverance and passion for long-term goals. *Journal of Personality and Social Psychology, 92*(6), 1087–1101. doi:10.1037/0022-3514.92.6.1087 PMID:17547490

Duckworth, E. (2006). *The having of wonderful ideas: And other essays on teaching and learning* (3rd ed.). New York, NY: Teachers College Press.

Durso, C. (2014). *Coffee Cups of The World, A Photo Series of Beautiful Takeaway Cups*. Retrieved September 30, 2016, from http://www.foodiggity.com/coffee-cups-of-the-world-a-photo-series- of-beautiful-takeaway-cups/

Duveen, J., & Solomon, J. (1994). The great evolution trial: Use of role-play in the classroom. *Journal of Research in Science Teaching, 31*(5), 575–582. doi:10.1002/tea.3660310510

Dweck, C. (2006). *Mindset: The new psychology of success*. New York, NY: Ballantine Books.

Eckard, S. (Ed.). (2015). *Yin and yang in the english classroom: Teaching with popular culture texts*. Lanham, MD: Rowman & Littlefield.

Ehrenfeld, T. (2011). Reflections on mirror neurons. *Observer, 24*(3). Retrieved from http://www.psychologicalscience.org/index.php/publications/observer/2011/march-11/reflections-on-mirror-neurons.html

Eiffel Tower. (2006). *Engineering.com*. Retrieved from http://www.engineering.com/Library/ArticlesPage/tabId/85/ArticleID/62/Eiffel-Tower.aspx

Eisner, E. (1991). *The enlightened eye: Qualitative inquiry and the enhancement of educational practice*. New York, NY: Macmillan.

Eisner, E. (2002). *The arts and the creation of mind*. New Haven, CT: Yale University Press.

Eisner, W. (1978). *A contract with God.* New York, NY: Baronet Books.

Eisner, W. (1985). *Comics & sequential art: Principles & practice of the world's most popular art form.* Tamarac, FL: Poorhouse Press.

Finson, K., & Pederson, J. (2011). What are visual data and what utility do they have for science education? *Journal of Visual Literacy, 30*(1), 66–85. doi:10.1080/23796529.2011.11674685

Fisher, D., & Frey, N. (2015). Selecting texts and tasks for content area reading and learning. *The Reading Teacher, 68*(7), 524–529. doi:10.1002/trtr.1344

Fitzgerald, A., & Schneider, K. (2013). What teachers want: Supporting primary school teachers in teaching science. *Teaching Science, 59*(2), 7–14.

Fortus, D., Krajcikb, J., Dershimerb, R. C., Marx, R. W., & Mamlok-Naamand, R. (2005). Design-based science and real-world problem solving. *International Journal of Science Education, 27*(7), 855–879. doi:10.1080/09500690500038165

Fosnot, C. T. (1989). *Enquiring teachers, enquiring learners: A constructivist approach for teaching.* New York, NY: Teachers College, Columbia University.

Foucault, M. (1977). *Discipline and punish: The birth of the prison* (A. Sheridan, Trans.). New York, NY: Vintage Books.

Freire, P. (2000). *Pedagogy of the oppressed* (M. B. Ramos, Trans.). New York: Continuum Publishing. (Original work published 1970)

Friedrichs, A. (2000). *Continuous learning dialogue: An ethnography of personal learning plans* (Dissertation). Burlington, VT: University of Vermont.

Frommer's. (2016). *Walt Disney World history.* Retrieved from http://www.frommers.com/destinations/walt-disney-world/683469

Fullan, M. (1993). *Change forces.* London, UK: Falmer Press.

Fullan, M., Bennett, B., & Rolheiser-Bennett, C. (1990). Linking classroom and school improvement. *Educational Leadership, 47*(8), 12.

Gabel, D. (1999). Improving teaching and learning through chemistry education research: A look to the future. *Journal of Chemical Education, 76*(4), 548. doi:10.1021/ed076p548

Gallant, J. (2002). The shape of the Eiffel Tower. *American Journal of Physics, 70*(2), 160–162. doi:10.1119/1.1417530

Gallese, V. (2005). Embodied simulation: From neurons to phenomenal experience. *Phenomenology and the Cognitive Sciences, 4*(1), 23–48. doi:10.1007/s11097-005-4737-z

Gardner, H. (1990). *Frames of mind, the theory of multiple intelligences.* New York, NY: Harper Collins.

Gardner, H. (1991). *The unschooled mind: How children think and how schools should teach.* New York, NY: Basic Books.

Gazzaniga, M. S. (1989). Organization of the human brain. *Science, 245*(4921), 947–952. doi:10.1126/science.2672334 PMID:2672334

Gerde, H., Schachter, R., & Wasik, B. (2013). Using the scientific method to guide learning: An integrated approach to early childhood education. *Early Childhood Education Journal, 41*(5), 315–323. doi:10.1007/s10643-013-0579-4

Ghanbari, S. (2015). Learning across disciplines: A collective case study of two university programs that integrate the arts with STEM. *International Journal of Education & the Arts, 16*(7). Retrieved from http://www.ijea.org/v16n7/

Gibson, D., & Clarke, J. (2000). *Growing toward systemic change: Developing personal learning plans at Montpelier High School.* Providence, RI: Northeast and Islands Regional Educational Laboratory at Brown University.

Gilbert, K. J. (2005). Visualization: A metacognitive skill in science and science education. *Visualization in Science Education, 1*, 9–27. doi:10.1007/1-4020-3613-2_2

Goldberger, S., & Kazis, R. (2000). *Revitalizing high schools: What the school-to-career movement can contribute.* Boston, MA: American Youth Policy Forum, Institute for Educational Leadership, National Association of Secondary School Principals, Jobs for the Future.

Gomes, L. F. A. M. (2016). Teaching decision making analytical skills to engineers: A new paradigm. *Brazilian Institute of Capital Markets and Universidade Federal Fluminense.* Retrieved from http://citeseerx.ist.psu.edu/viewdoc/download?doi=10.1.1.201.3517&rep=rep1&type=pdf

Gonick, L., & Criddle, C. (2005). *The cartoon guide to chemistry.* New York, NY: Collins Reference.

Goodnough, K. (2001). Multiple intelligences theory: A framework for personalizing science curricula. *School Science and Mathematics, 101*(4), 180–193. doi:10.1111/j.1949-8594.2001.tb18021.x

Gopnik, A. (2012). Scientific thinking in young children: Theoretical advances, empirical research, and policy implications. *Science, 337*(28), 1623–1627. doi:10.1126/science.1223416 PMID:23019643

Goulson, D., Nicholls, E., Botías, C., & Rotheray, E. L. (2015, February26). Bee declines driven by combined stress from parasites, pesticides, and lack of flowers. *Science, 347*(6229), 1–16. doi:10.1126/science.1255957 PMID:25721506

Grady, S. (2000). *Drama and diversity: A pluralistic perspective for educational drama.* Portsmouth, NH: Heinemann Drama.

Grant, R. G. (2007). *Flight: The complete history.* New York, NY: Dorling Kindersley.

Greene, M. (2001). *Variations on a blue guitar: The Lincoln Center Institute lectures on aesthetic education*. New York, NY: Teachers College Press, Columbia University.

Gregory, G. H., & Chapman, C. (2007). *Differentiated Instructional Strategies: One Size Doesn't Fit All* (2nd ed.). Thousand Oaks, CA: Corwin Press.

Groff, J. S. (2013). Expanding our frames of mind for education and the arts. *Harvard Educational Review*, *83*(1), 15–39. doi:10.17763/haer.83.1.kk34802147665819

Guinness World Records. (n.d.). Retrieved from http://www.guinnessworldrecords.com

Guyotte, K. W., Sochacka, N. W., Costantino, J. W., Walther, J., & Kellam, N. N. (2014). Steam as social practice: Cultivating Creativity in Transdisciplinary Spaces. *Art Education*, *67*(6), 12–19.

Guyotte, K. W., Sochacka, N. W., Costantino, T. E., Kellam, N. N., & Walther, J. (2015). Collaborative creativity in STEAM: Narratives of art education students' experiences in transdisciplinary spaces. *International Journal of Education & the Arts*, *16*(15). Retrieved from http://www.ijea.org/v16n15/

Halverson, E. (2016). *The art of improvisation in teaching and learning*. Paper presented at the American Educational Research Association Conference, Washington, DC.

Hargreaves, H. (2016). *Coffee Cups of the World (@coffeecupsoftheworld) • Instagram photos and videos*. Retrieved September 30, 2016, from https://www.instagram.com/coffeecupsoftheworld/

Haring, D., & Kelner, T. (2015). Why we got serious about interdisciplinary teaching. *Educational Leadership*, *73*(4), 68.

Hashweh, M. Z. (1987). Effects of subject-matter knowledge in the teaching of biology and physics. *Teaching and Teacher Education*, *3*(2), 109–120. doi:10.1016/0742-051X(87)90012-6

Hassett, D. D., & Scheible, M. B. (2010). Finding space and time for the visual in K-12 literacy instruction. In M. Cappello & B. Moss (Eds.), *Contemporary Readings in Literacy Education* (pp. 325–331). Thousand Oaks, CA: SAGE Publications.

Heathcote, D. (1984). *Collected writings on education and drama*. Evanston, IL: Northwestern University Press.

Heaverlo, C. (2011). *STEM development: A study of 6th-12th grade girls' interest and confidence in mathematics and science* (Doctoral Dissertation). Retrieved from http://lib.dr.iastate.edu/cgi/viewcontent.cgi?article=1035&context=etd

Henriksen, D. (2014). Full STEAM ahead: Creativity in excellent STEM teaching practices. *The STEAM Journal*, *1*(2), 15. doi:10.5642/steam.20140102.15

Hetland, L., Winner, E., Veenema, S., & Sheridan, K. (2013). *Studio thinking 2: The real benefits of visual arts education*. New York, NY: Teachers College Press.

History of Catapults. (n.d.). Retrieved January 17, 2016, from https://sites.google.com/site/physicsofcatapults/home/history-of-catapults

HMI. (1989). Curriculum matters 17: Drama from 5-16. London: HMSO.

Hodson, D., & Reid, D. (1988). Changing priorities in science education: Part I. *The School Science Review*, *70*, 101–108.

Hoffman, J. L., Collins, M. F., & Schickedanz, J. A. (2015). Instructional challenges in developing young childrens science concepts. *The Reading Teacher*, *68*(5), 363–372. doi:10.1002/trtr.1325

Hohenadel, K. (2014). *Traveling the World in Search of the Most Unique Disposable Coffee Cup Designs (Photos)*. Retrieved September 30, 2016, from http://www.slate.com/blogs/the_eye/2014/07/03/photographer_henry_hargreaves_coffee_cups_of _the_world_highlights_the_best.html

Howcast. (2013). *How to Demonstrate Insulation Principles*. Retrieved June 07, 2016, from https://www.youtube.com/watch?v=8I96BzSNNqE

Howe, R. B., & Covell, K. (2005). *Empowering children: Children's rights education as a pathway to citizenship*. Toronto: University of Toronto Press. doi:10.3138/9781442674387

Inspiring Breakthrough. (2005). *Visual spatial intelligence*. Retrieved from http://www.inspiring-breakthrough.co.uk/learning-styles/visual-spatial-learning.html

Insulated Bottle. (2006). Retrieved June 09, 2016, from http://www.madehow.com/Volume-4/Insulated-Bottle.html

Insulation. (2007). Retrieved June 07, 2016, from http://sciencelearn.org.nz/Contexts/Icy-Ecosystems/Science-Ideas-and-Concepts/Insulation

International Society for Technology in Education. (2007). *National educational technology standards for students*. Retrieved June 7, 2016, from http://www.iste.org/standards/iste-standards/standards-for-students

International Technology Education Association. (2007). *Standards for technological literacy: Content for the study of technology* (3rd ed.). Retrieved from https://www.iteea.org/File.aspx?id=67767&v=b26b7852

Jablon, P. (1989). *A Documentation and impact analysis of a project-oriented, interdisciplinary, team-taught, peer-influence program for secondary schools and its effects upon science learning* (Doctoral dissertation). (NYU No. 9016410)

Jablon, P. (2010). *Analogous interrelationship role plays: Engaging the previously disengaged in science*. Paper presented at the Association of Science Teacher Educators International Conference, Sacramento, CA.

Jablon, P. (2013). Get real! Walking the walk to inform talking the talk: Full-time teaching in an urban high school. In M. Dias, C. Eick, & L. Brantley Dias (Eds.), *Science teacher educators as K-12 science teachers: Practicing what we teach* (pp. 39–50). New York, NY: Springer.

Jablon, P. (2014). *The synergy of inquiry: Engaging students in deep learning across the content areas*. Huntington Beach, CA: Shell Education.

Jablon, P., & Born, T. (1988). STS works in the city. *SSTS Reporter.*, *4*(2), 5–8.

Jacobs, H. H. (1989). The growing need for interdisciplinary curriculum content. In H.H. Jacobs (Ed.), Interdisciplinary Curriculum: Design and Implementation (pp. 1-11). Alexandria, VA: ASCD.

Johnson, C. C. (2007). Whole school collaborative sustained professional development and science teacher change: Signs of success. *Journal of Science Teacher Education, 18*(4), 629–662. doi:10.1007/s10972-007-9043-x

Jolly, A. (2014). STEM vs. STEAM: Do the arts belong? *Education Week Teacher*. Retrieved February 21, 2016, from http://www.edweek.org/tm/articles/2014/11/18/ctq-jolly-stem-vssteam.html

Jolly, A. (2014). STEM vs. STEAM: Do the arts belong? *Education Week Teacher*. Retrieved June 30, 2016, from http://www.edweek.org/tm/articles/2014/11/18/ctq-jolly-stem-vs-steam.html

Jolly, A. (2014). *STEM vs. STEAM: Do the arts belong?* Retrieved from http://www.edweek.org/tm/articles/2014/11/18/ctq-jolly-stem-vs-steam.html

Jolly, A. (2014). STEM vs. STEAM: Do The Arts Belong? *Education Week.*

Jones, G. (2004). *Men of tomorrow: Geeks, gangsters and the birth of the comic book*. New York, NY: Basic Books.

Kafai, Y. B. (2006). Constructionism. In K. Sawyer (Ed.), *Cambridge handbook of the learning sciences* (pp. 35–46). Cambridge, MA: Cambridge University Press.

Kakalios, J. (2005). *The physics of superheroes*. New York, NY: Gotham Books.

Kamen, M. (1991, August). *Creative drama and the enhancement of elementary school students' understanding of science concepts* (Doctoral dissertation). The University of Texas at Austin, Austin, TX.

Kamen, M. (1991). Use of creative drama to evaluate elementary school students' understanding of science concepts. In G. Kulm & S. M. Malcom (Eds.), *Science assessment in the service of reform*. Washington, DC: American Association for the Advancement of Science.

Karplus, R. (1964). The science curriculum improvement study. *Journal of Research in Science Teaching*, *2*(4), 293–303. doi:10.1002/tea.3660020406

Kase-Polisini, J., & Spector, B. (1992). Improvised drama: A tool for teaching science. *Youth Theatre Journal*, *7*(1), 15–19.

Keefe, W., & Jenkins, J. (2002). A special section on personalized instruction. *Phi Delta Kappan*, *83*(6), 440–448. doi:10.1177/003172170208300609

Keller, M., & Fuller, N. R. (Illustrator). (2009). Charles Darwin's on the origin of species: A graphic interpretation. Rodale.

Kelley, J. A. (Producer & Director), & Wyanski, J. (Producer & Director). (1987). *Project lead the way - Where it started 1987 Dick Blais* [video]. Retrieved from http://youtu.be/iIdomsEhmYM

Kelly, G. J., Carlsen, W. S., & Cunningham, C. M. (1993). Science education in sociocultural context: Perspectives from the sociology of science. *Science Education*, *77*(2), 207–220. doi:10.1002/sce.3730770208

Kelner, L. B. (1993). *The creative classroom: A guide for using creative drama in the classroom, PreK-6*. Portsmouth, NH: Heinemann.

Kennedy, T. J., & Odell, M. R. L. (2014). Engaging students in STEM education. *Science Education International*, *25*(3), 246–258.

Kentish, B. (1995). Hypotheticals: Deepening the understanding of environmental issues through ownership of learning. *Australian Science Teachers Journal*, *41*(1), 21–25.

Keruskin, T. (2015). Finding the sweet spot: Creativity, candy, and commerce. *Edutopia*. Retrieved June 28, 2016, from http://www.edutopia.org/blog/sweet-creativity-candy-commerce-todd-keruskin

Khan, S. (2016). *Linear vs. exponential growth*. Retrieved June 07, 2016, from https://www.khanacademy.org/math/algebra/introduction-to-exponential-functions/comparing-exponential-and-polynomial-functions/v/linear-exponential-models

Kinsella, K. (1996). Designing group work that supports and enhances diverse classroom work styles. *TESOL Journal*, *6*(1), 24–30.

Klein, A.-M., Vaissiere, B. E., Cane, J. H., Steffan-Dewenter, I., Cunningham, S. A., Kremen, C., & Tscharntke, T. (2007). Importance of pollinators in changing landscapes for world crops. *Proceedings. Biological Sciences*, *274*(1608), 303–313. doi:10.1098/rspb.2006.3721 PMID:17164193

Koester, M. (1989). *Science and art together again* (Unpublished master's thesis). University of Hawaii, Manoa, HI.

Koester, M. (2011). What's your story? How can we bring students to science through reading and the arts? Keynote address delivered at the South Carolina Science Council meeting, Myrtle Beach, SC.

Koester, M. (2014). *Project Draw for Science*. Retrieved from http://merriekoester.wix.com/project-dfs

Koester, M. (in press). Deep learning in science: An argument for an aesthetic paradigm of science education. In F. V. Tochon & D. Busciglio (Eds.), Deep Education Across the Disciplines and Beyond: A 21st Century Transdisciplinary Breakthrough. Blue Mounds, WI: Deep University Press.

Koester, M., & Simmons, S. (n.d.). *The collaborative art of teaching science through drawing* [PowerPoint Slides]. NAEA. Retrieved from http://naeaworkspace.org/naea16/The%20Collaborative%20Art%20of%20Teaching%20Science%20through%20Drawing.pdf

Koester, M. (2015). *Science teachers who draw: The red is always there*. Blue Mounds, WI: Deep University Press.

Koester, M. (2015b). *Agnes Pflumm and the stonecreek science fair* (3rd ed.). Charleston, SC: Read for Science Publishing.

Koester, M. (2015c). *No place like periwinkle* (3rd ed.). Charleston, SC: Read for Science Publishing.

Koester, M. (2015d). *Pond scum and agnes pflumm* (3rd ed.). Charleston, SC: Read for Science Publishing.

Koester, M. (2015e). *Agnes pflumm and the secret of the seven* (2nd ed.). Charleston, SC: Read for Science Publishing.

Koester, M. (in press). Deep learning in science: An argument for an aesthetic paradigm of science education. In F. V. Tochon & D. Busciglio (Eds.), *Deep Education Across the Disciplines and Beyond: A 21st Century Transdisciplinary Breakthrough*. Blue Mounds, WI: Deep University Press.

Koester, M. (in press). Getting to "know" STEAM. In M. VanSickle & J. Bazler (Eds.), *Cases in STEAM education in practice*. Hershey, PA: IGI-Global.

Konie, R. (2011*). A brief overview of laban movement analysis*. Retrieved on April 27, 2016 from http://www.movementhasmeaning.com/wp-content/uploads/2010/09/LMA-Workshop-Sheet.pdf

Kozma, R. B. (2003). Technology and classroom practices: An international study. *Journal of Research on Technology in Education*, *36*(1), 1–5. doi:10.1080/15391523.2003.10782399

Kuenzi, J., Matthews, C., & Mangan, B. (2006). *Science, technology, engineering, and mathematics (STEM) education issues and legislative options*. Washington, DC: Congressional Research Service.

Kyle, R. (1993). *Transforming our schools*. Lessons from the Jefferson County Public Schools.

Lakoff, G., & Johnson, M. (1980). *Metaphors we live by*. Chicago, IL: University of Chicago Press.

Lambert, L. (1998). *Building leadership capacity in schools*. Alexandria, VA: ASCD Publications.

Leaning Tower of Pisa. (n.d.) In *Wikipedia*. Retrieved from https://en.wikipedia.org/wiki/Leaning_Tower_of_Pisa

Lederman, N. (2007). Nature of science: Past, present, and future. In S. Abell & N. Lederman (Eds.), *Handbook of research on science education Abingdon*. Abingdon, VA: Taylor and Francis.

Lederman, N. G. (1992). Students and teachers conceptions of the nature of science: A review of the research. *Journal of Research in Science Teaching*, *29*(4), 331–359. doi:10.1002/tea.3660290404

Lederman, N. G., & Lederman, J. S. (2015). What is a theoretical framework? A practical answer. *Journal of Science Teacher Education*, *26*(7), 593–597. doi:10.1007/s10972-015-9443-2

Lee, V., & Smith, J. (1994a). High school restructuring and student achievement: A new study finds strong links. *Issues in Restructuring Schools, 7*(1-5), 16.

Lee, V., & Smith, J. (1994b). Effects of high school restructuring and size on gains in achievement and engagement for early secondary students. *Issues in Restructuring Schools, 7*(1-5), 16.

Lee, V., Bryk, A., & Smith, J. (1993). The organization of effective secondary schools. Review of Research in Education, 19, 171-267. doi:10.2307/1167343

Lee, O., Fradd, S. H., & Sutman, F. X. (1995). Science knowledge and cognitive strategy use among culturally and linguistically diverse students. *Journal of Research in Science Teaching, 32*(8), 797–816. doi:10.1002/tea.3660320804

Lee, V., & Smith, J. (2001). *Restructuring high schools for equity and excellence.* New York, NY: Teachers College Press.

Lee, V., Smith, J., & Croninger, R. (1995). *Another look at high school restructuring. Issues in Restructuring Schools (Report #9).* Madison, WI: Center on Organization and Restructuring of Schools.

Legters, N., Balfanz, R., & McPartland, J. (2002). *Solutions for failing high schools: Converging visions and promising models. Paper Commissioned by the Office of Vocational and Adult Education.* U.S. Department of Education.

Lcmkc, J. L. (1999). Typological and topological mcaning in diagnostic discoursc. *Discoursc Processes, 27*(2), 173–185. doi:10.1080/01638539909545057

Lemke, J. L. (2002). Enseñar todos los lenguajes de la ciencia: palabras, símbolos, imágenes y acciones. In M. Benlloch (Ed.), *La educación en ciencias: ideas para mejorar su práctica* (pp. 159–186). Barcelona, Spain: Paidos.

Lesh, R. A., & Doerr, H. M. (Eds.). (2003). *Beyond constructivism: Models and modeling perspectives on mathematics problem solving, learning, and teaching.* Mahwah, NJ: Lawrence Erlbaum Associates.

Levine, E. (2002). *One kid at a time: Big lessons from a small school.* New York, NY: Teachers College Press.

Littke, D., & Allen, F. (1999). Whole school personalization: One student at a time. *Educational Leadership, 57*(1), 24–29.

Litts, B. (2015). *Making learning: Makerspaces as learning environments* (Doctoral dissertation). Retrieved from www.informalscience.org/sites/.../Litts_2015_Dissertation_Published.pdf

Llewellyn, D. J. (2002). Inquire within: Implementing inquiry-based science standards in grades 3-8. Thousand Oaks, CA: Corwin Press.

Llewellyn, D. (2004). *Teaching high school science through inquiry: A Case study approach.* Thousand Oaks, CA: Corwin Press.

Logan, B. (2011, October). Examining differentiated instruction: Teachers respond. *Research in Higher Education*, 13.

Long, E. Y., & Krupke, C. H. (2016). Non-cultivated plants present a season-long route of pesticide exposure for honey bees. *Nature Communications Nat Comms*, *7*, 11629. doi:10.1038/ncomms11629 PMID:27240870

Losey, J. E., & Vaughan, M. (2006). The economic value of ecological services provided by insects. *Bioscience*, *56*(4), 311–323. doi:10.1641/0006-3568(2006)56[311:TEVOES]2.0.CO;2

Loucks-Horsley, S. (1996). Principles of effective professional development for mathematics and science education: A synthesis of standards. *NISE Brief, 1*(1). Retrieved from http://eric.ed.gov/?id=ED409201

Maeda, J. (2013, July 11). *Artists and scientists: More alike than different* [Web log comment]. Retrieved from http://blogs.scientificamerican.com/guest-blog/artists-and-scientists-more-alike-than-different/

Maeda, J. (2012). *STEM to STEAM: Art in K-12 Is Key to Building A Strong Economy*. Edutopia.

Maine Commission on Secondary Education. (1998). *Promising Future A call to improved learning for Maine's secondary students*. Augusta, ME: Author.

Mallettt, M. (2010). Choosing and using fiction and non-fiction 3-11: A comprehensive guide for teachers and student teachers. New York, NY: Routledge.

Mandigo, J., Francis, N., Lodewyk, K., & Lopez, R. (2009). *Position paper: Physical literacy for educators.* St. Catherines, ON, Canada: Brock University. Retrieved May 10, 2016 from https://www.phecanada.ca/sites/default/files/pl_position_paper.pdf

Mardirosian, G. H., & Lewis, Y. P. (2009). How to use theater to teach at-risk students. *The Chronicle of Higher Education*, *55*(26), 34–37.

Martin, D. J. (2011). *Elementary science methods: A constructivist approach* (6th ed.). Belmont, CA: Wadsworth Publishing.

Maslow, A. H. (1968). *Toward a psychology of being*. New York, NY: Van Nostrand.

May, R. (1985). *The courage to create*. Toronto, Canada: Bantam.

McAuliffe, C. (n.d.). *BrainyQuote.com*. Retrieved fromhttp://www.brainyquote.com/quotes/quotes/c/christamca134582.html

McCarthy, B. (1990, October). Using the 4MAT system to bring learning styles to schools. *Educational Leadership*, *48*(2), 31–37.

McCaslin, N. (1981). *Children and drama* (2nd ed.). New York: University Press.

McCaslin, N. (1996). *Creative drama in the classroom and beyond*. London: Longman Publishers.

McCloud, S. (1993). *Understanding comics*. Allenspark, CO: Tundra Publications.

McGregor, D., & Precious, W. (2015). Dramatic science: Inspired ideas for teaching science using drama for ages 5–11. New York: Routledge.

McLaughlin, M. (1990). The Rand change agent study revisited: Macro perspectives and micro realities. *Educational Researcher, 19*(9), 11–16. doi:10.3102/0013189X019009011

McLaughlin, M., & Talbert, J. (1990, November). Constructing a personalized school environment. *Phi Delta Kappan*, 230–235.

Meier, D. (1995). *The power of their ideas*. Boston, MA: Beacon Press.

Meier, D. (1996, September). The big benefits of smallness. *Educational Leadership, 54*(1).

Merriam-Webster. (n.d.). *Definition of text*. Retrieved Oct 4, 2016 from http://www.merriam-webster.com/dictionary/text

Metcalfe, R. J. A., Abbott, S., Bray, P., Exley, J., & Wisnia, D. (1984). Teaching science through drama: An empirical investigation. *Research in Science & Technological Education, 2*(1), 77–81. doi:10.1080/0263514840020109

Miller, C., & Saxton, J. (2004). *Into the story: Language in action through drama*. Portsmouth, NH: Heinemann.

Mirth, J., & Findley, A. (2015). Enduring design: Examining the relationship between art, engineering, and creativity. *Proceedings from ASEE '15: 2015 ASEE Zone III Conference*. Retrieved from https://www.asee.org/documents/zones/zone3/2015/Enduring-Design-Examining-the-Relationship-Between-Art-Engineering-and-Creativity.pdf

Morris, D. (1997). Adrift in a sea of innovations: A response to Alexander, Murphy and Woods. *Educational Researcher, 26*(4), 22–26.

Morrison, G., & Quitely, F. (2011). *All star superman*. New York, NY: DC Comics.

Murnane, R., & Levy, F. (1995). *Teaching the new basic skills principles for educating children to thrive in a changing economy*. Cambridge, MA: The Free Press.

NASA. (n.d.). *Fast facts*. Retrieved from http://voyager.jpl.nasa.gov/mission/fastfacts.html

National Aeronautics and Space Administration. (2015). *Welcome to the beginner's guide to aerodynamics*. Retrieved from https://www.grc.nasa.gov/www/k-12/airplane/bga.html

National Association of Secondary School Principals. (1996). *Breaking Ranks: Changing an American institution*. Reston, VA: NASSP.

National Association of State Boards of Education. (2002). *Most likely to succeed Policymaking in support of a restructured high school*. Alexandria, VA: NASBE.

National Center for Educational Statistics. (n.d.). *Create a graph*. Retrieved July 7, 2016, from https://nces.ed.gov/nceskids/createagraph/

National Coalition for Core Arts. (2014). *National core arts standards*. Retrieved June 7, 2016, from http://www.nationalartsstandards.org/

National Commission on Excellence in Education. (1983). *A nation at risk: A report to the nation.* Washington, DC: U.S. Government Printing Office.

National Council for Teachers of English. (1995). *Position statement on interdisciplinary learning, pre-K to grade 4.* Retrieved May 10, 2016 from http://www.ncte.org/positions/statements/interdisclearnprek4

National Education Commission on Time and Learning. (1994). *Prisoners of time: Report of the National Commission on Time and Learning.* Washington, DC: U.S. Government Printing Office.

National Governors Association Center for Best Practices & Council of Chief State School Officers. (2010). *Common Core State Standards for Mathematics.* Washington, DC: Author.

National Oceanic and Atmospheric Association. (2014). Lesson 9: Invent a robot. In *The NOAA ship Okeanos explorer education materials collection, volume 2: How do we explore* (pp. 125-136). Retrieved from http://oceanexplorer.noaa.gov/okeanos/edu/collection/media/vol2_hdwe.pdf

National Research Council. (2012). *A framework for K-12 science education: Practices, crosscutting concepts, and core ideas. Committee on a Conceptual Framework for New K-12 Science Education Standards. Board on Science Education, Division of Behavioral and Social Sciences and Education.* Washington, DC: National Academies Press.

National Research Council. (2012). *A framework for k-12 science education: Practices, crosscutting concepts, and core ideas.* Washington, DC: National Academy of Sciences. Retrieved from http://www.nap.edu/catalog.php?record_id=13165

National Science Foundation. (2006, September 11). *The art of engineering* [press release]. Retrieved from http://www.nsf.gov/news/news_summ.jsp?cntn_id=107990

Newcombe, N. (2010). Picture this: Increasing math and science learning by improving spatial thinking. *American Educator.*

Newmann, F. (1981). Reducing student alienation in high schools: Implications of theory. *Harvard Educational Review, 51*(4), 546–564. doi:10.17763/haer.51.4.xj67887u87l5t66t

Newmann, F., Wehlage, G., & Lamborn, S. (1992). *The significance and sources of student engagement. Student engagement and achievement in American secondary schools* (F. M. Newmann, Ed.). New York, NY: Teachers College Press.

Newton's Law of Cooling. (2015). Retrieved June 07, 2016, from https://www.khanacademy.org/math/differential-equations/first-order-differential-equations/modeling-with-differential-equations/v/newtons-law-of-cooling

NGSS Lead States. (2013). *Next generation science standards: For states, by states.* Washington, DC: The National Academies Press.

NGSS Lead States. (2013). *Next Generation Science Standards: For States, By States.* Washington, DC: The National Academies Press.

NOAA Office of Ocean Exploration. (2016). *Ocean Explorer*. Retrieved from http://oceanexplorer.noaa.gov/

Noddings, N. (1992). *The challenge to care in schools*. New York, NY: Teachers College Press.

Noddings, N. (2003). *Caring: A feminine approach to ethics and moral education*. Berkeley, CA: University of California Press.

Noddings, N. (2005). *The challenge to care in schools: An alternative approach to education* (2nd ed.). New York, NY: Teachers College Press.

O'Neill, C. (1995). *Drama worlds: A Framework for process drama*. Portsmouth, NH: Heinemann.

Oakes, J. (2005). Keeping track: How schools structure inequality. New Haven, CT: Yale University.

Ocean Literacy Framework. (2005-2015). Retrieved from http://oceanliteracy.wp2.coexploration.org

Ocean Literacy Network. (2013). *Ocean literacy principle #7*. Retrieved June 30, 2016, from http://oceanliteracy.wp2.coexploration.org/ocean-literacy-framework/principle-7-v2/

Ødegaard, M. (2003). Dramatic science. A critical review of drama in science education. *Studies in Science Education*, *39*(1), 75–102. doi:10.1080/03057260308560196

Office Applications. (n.d.). *Writer*. Retrieved July 7, 2016, from https://writer.bighugelabs.com/login

Olness, R. (2004). *Using literature to enhance content area instruction: A guide for K-5 teachers*. Newark, DE: International Reading Association.

Olson, J. K. (2013). Education rather than training: Cognitive pluralism and preservice science teacher education. In K. D. Finson & J. E. Pedersen (Eds.), *Visual data and their use in science education* (pp. 95–124). Charlotte, NC: Information Age Publishers.

Onwukwe, E., Akuezuilo, E., & Ijioma, B. (2014). Science drama: Clue to under achievement in secondary school chemistry? *STEMplanet*, 1. Retrieved from http://stemstates.org/journal/stemplanet-2014/2014-vol-11/science-drama-clue-to-under-achievement-in-secondary-school-chemistry.html

Osborne, J., Simon, S., & Collins, S. (2003). Attitudes towards science: A review of the literature and its implications. *International Journal of Science Education*, *25*(9), 1049–1079. doi:10.1080/0950069032000032199

Ottaviani, J., Cannon, Z., & Cannon, K. (2009). *T-Minus: The race to the moon*. New York, NY: Aladdin.

Oxford, R. L., Hollaway, M. E., & Horton-Murillo, D. (1992). Language learning styles: Research and practical considerations for teaching the multicultural tertiary ESL/EFL classroom. *System*, *20*(4), 439–456. doi:10.1016/0346-251X(92)90057-A

Ozcelik, A. T., & McDonald, S. P. (2013). Preservice science teachers uses of inscriptions in science teaching. *Journal of Science Teacher Education, 24*(7), 1103–1132. doi:10.1007/s10972-013-9352-1

Pagnucci, G. S. (2004). *Living the narrative life: Stories as a tool for meaning making*. Portsmouth, NH: Heinemann.

Papert, S., & Harel, I. (1991). Situating constructionism. In I. Harel & S. Papert (Eds.), *Constructionism* (pp. 1–11). Norwood, NJ: Ablex.

Park, G., Lubinski, D., & Benbow, C. (2010, November 2). Recognizing spatial intelligence. *Scientific American.* Retrieved from http://www.scientificamerican.com/article/recognizing-spatial-Intel/

Park, M. Y. (2014). *A Brief History of the Disposable Coffee Cup*. Retrieved June 09, 2016, from http://www.bonappetit.com/entertaining-style/trends-news/article/disposable-coffee-cup-history

Parsons, S. A., Dodman, S. L., & Burrowbridge, S. C. (2013, September). Broadening the view of differentiated instruction. *Phi Delta Kappan, 95*(1), 38–42. doi:10.1177/003172171309500107

Patrick, H. (2009). *Engaging young children with informational books*. Thousand Oaks, CA: Corwin.

Pedings-Behling, K. E. (2015). *How Do We Keep It Hot?* Hartsville, SC: South Carolina Governor's School for Science and Mathematics.

Pestalozzi, J. (1973). *How Gertrude teaches her children*. New York, NY: Gordon Press. (Original work published 1894)

Pestalozzi, J. H. (1894). *How Gertrude teaches her children. Syracuse, NY*. W.: Bardeen.

Piaget, J. (1955). *The construction of the reality in the child.* Basic Books.

Piaget, J. (1964). *Judgment and reasoning in the child.* Patterson, NJ: Littlefield Adams and Company.

Pieczura, P. (2013). Decidedly dramatic! The power of creative drama in social studies. *Social Studies and the Young Learner, 25*(3), 9–12.

Pliskow, T. (2008). *Reaching new heights in middle school science: Straw tower competition builds*. Academic Press.

Polanyi, M. (1966). *The tacit dimension*. New York, NY: Doubleday.

Pomeroy, S. R. (2012). From STEM to STEAM: Science and Art go Hand-to-Hand. *Scientific American Blog.*

Post Illuminance. (n.d.). *Custom graph: Isometric paper*. Retrieved from http://customgraph.com/piart.php?art=590

Potts, S. G., Biesmeijer, J. C., Kremen, C., Neumann, P., Schweiger, O., & Kunin, W. E. (2010, January/February). Global pollinator declines: Trends, impacts and drivers. *Trends in Ecology & Evolution, 25*(6), 345–353. doi:10.1016/j.tree.2010.01.007 PMID:20188434

Pozzer-Ardenghi, L., & Roth, W.-M. (2005). Making sense of photographs. *Science Education, 89*(2), 219–241. doi:10.1002/sce.20045

Prairie, C. (2002). *Draft of a personalization rubric. (Personal Correspondence).* Bloomington, IN: Indiana University.

Pratap, T. (2015, October 29). *What were the engineering challenges in building the Burj Kahalifa?* [Msg 3]. Message posted to https://www.quora.com/What-were-the-engineering-challenges-in-building-the-Burj-Khalifa

Project Lead The Way. (2016). *Engaging students from New York to Honolulu.* Retrieved from https://www.pltw.org/about-us

Pulkkinen, L. (1982). Self-control and continuity from childhood to adolescence. In P. B. Baltes & O. G. Brim Jr., (Eds.), *Life-span development and behavior* (Vol. 4, pp. 64–102). New York, NY: Academic Press.

Raywid, M., & Oshiyama, L. (2000). Musings in the wake of Columbine: What can schools do? *Phi Delta Kappan, 81*(6), 444–449.

Reiff, J. C. (1992). *Learning Styles. What research says to the teacher.* Washington, DC: National Education Association.

Reilly, E. (2015). Superheroes in math class: Using comics to teach diversity awareness. *Works & Days, 32*(1 & 2), 61–74.

Reyes, M.-E. (2012). Increasing diversity in STEM by attracting community college women of color. In B. Bogue & E. Cady (Eds.), *Apply research to practice (APR) resources.* Retrieved from https://www.engr.psu.edu/awe/ARPAbstracts/CommunityCollege/ARP_WomenofColorCommunity%20College_Abstract%20Final.pdf

Rhode, M. (2013). *The Sketchnote handbook: The illustrated guide to visual notetaking.* Berkeley, CA: Peachpit Press.

Rinne, L., Gregory, E., Yarmolinskaya, J., & Hardiman, M. (2011). Why arts integration improves long-term retention of content. *Mind, Brain, and Education, 5*(2), 89–96. doi:10.1111/j.1751-228X.2011.01114.x

Roberts, D. (2011). *Fatal invention: How science, politics, and big business re-create race in the twenty-first century.* New York, NY: The New Press.

Rock, M. L., Gregg, M. G., Ellis, E., & Gable, R. A. (2008, Winter). REACH: A framework for differentiating classroom instruction. *Preventing School Failure, 52*(2), 31–47. doi:10.3200/PSFL.52.2.31-47

Romagnoli, A. (2015). Up, up, and away: Superman in the composition classroom. In S. Eckard (Ed.), *Yin and yang in the english classroom: Teaching with popular culture texts* (pp. 123–135). Lanham, MD: Rowman & Littlefield.

Romagnoli, A., & Pagnucci, G. (2013). *Enter the superheroes: American values, culture, and the canon of superhero literature*. Lanham, MD: Scarecrow Press.

Ross-Nazzal, J. (Interviewer), & Blackburn, G.C. (Interviewee). (2009). *NASA Johnson Space Center oral history project edited oral history transcript.* Retrieved from Johnson Space Center Oral Histories Project Website: http://www.jsc.nasa.gov/history/oral_histories/BlackburnGC/BlackburnGC_5-4-09.htm

Salisbury, B. T. (1986). *Theatre arts in the elementary classroom.* New Orleans, LA: Anchorage Press, Inc.

Sartre, J.-P. (1976). *Anti-Semite and Jew* (G. J. Becker, Trans.). New York, NY: Schocken. (Original work published 1948)

Satchwell, R., & Loepp, F. L. (2002). Designing and implementing an integrated mathematics, science, and technology curriculum for the middle school. *Journal of Industrial Teacher Education.* Retrieved February 14, 2014, from http://scholar.lib.vt.edu/ejournals/JITE/v39n3/satchwell.html

Saunders, A. C. (1994). Graphics and how they communicate. In D. M. Moore & F. M. Dwyer (Eds.), *Visual literacy: A spectrum of visual learning* (pp. 183–192). Englewood Cliffs, NJ: Educational Technology Publications, Inc.

Savage, R. N., Chen, K. C., & Vanasupa, L. (2007). Integrating project-based learning throughout the undergraduate engineering curriculum. *Journal of STEM Education: Innovation and Research, 8*, 1–13.

Scales, P., & Leffert, N. (1999). *Developmental assets: A synthesis of the scientific research on adolescent development*. Minneapolis, MN: Search Institute.

Schwarz, C., & White, B. (2005). Meta-modeling knowledge: Developing students understanding of scientific modeling. *Cognition and Instruction, 23*(2), 165–205. doi:10.1207/s1532690xci2302_1

Scientific American Editors. (2012). *Science education: Back to school.* Washington, DC: National Academy Press.

Senge, P. (1990). *The fifth discipline: The art and practice of the learning organization.* New York, NY: Currency Doubleday.

Settlage, J., & Southerland, S. A. (2007). *Teaching science to every child: Using culture as a starting point.* Abingdon, UK: Taylor & Francis.

Sharapan, H. (2012). From STEM to STEAM: How early childhood educators can apply Fred Rogers' approach. *Young Children, 67*(2), 36–40.

Shaywitz, S. (2003). *Overcoming dyslexia: A new and complete science-based program for reading problems at any level*. New York, NY: Vintage Books.

Shulman, L. S. (1987). Knowledge and teaching: Foundations of the new reform. *Harvard Educational Review*, *57*(1), 1–23. doi:10.17763/haer.57.1.j463w79r56455411

Sibanda, J., & Sibanda, L. (2013). Visual literacy development through the mediation of grade 4 English textbooks. *Journal of Visual Literacy*, *32*(2), 39–66. doi:10.1080/23796529.2013.11674709

Siegesmund, R. (2010). Aesthetics as a curriculum of care and responsible choice. In T. Constantino & B. White (Eds.), *Essays on aesthetic education for the 21st century* (pp. 81–92). Rotterdam: Sense Publishers.

Simply Psychology. (n.d.). *Jerome Bruner – Learning Theory in Education*. Retrieved fromhttp://www.simplypsychology.org/bruner.html

Sizer, T. (1984). *Horace's compromise: The dilemma of the American high school*. Boston, MA: Houghton Mifflin.

Sizer, T. (1992). *Horace's school: Redesigning the American high school*. Boston, MA: Houghton Mifflin.

Sizer, T. (1996). *Horace's hope: What works for the American high school*. Boston, MA: Houghton Mifflin.

Sizer, T. (1999). No two are quite alike. *Educational Leadership*, *57*(1), 12–17.

Smilansky, S. (1968). *The effects of sociodramatic play on disadvantaged preschool children*. New York, NY: John Wiley & Sons.

Smilansky, S. (1990). Sociodramatic play: Its relevance to behavior and achievement in school. In E. Klugman & S. Smilansky (Eds.), *Children's play and learning: Perspectives and policy implications* (pp. 18–42). New York, NY: Teacher College Press.

Snow, C. (2015). Rigor and realism: Doing educational science in the real world. 2014 Wallace Foundation distinguished lecture. *Educational Researcher*, *44*(9), 460–466. doi:10.3102/0013189X15619166

Sousa, D. A., & Pilecki, T. (2013). *From STEM to STEAM: Using brain-compatible strategies to integrate the arts*. Thousand Oaks, CA: Corwin Publishing Company.

Spitzer, H. (2013). Introduction of interdisciplinary teaching: Two case studies: Commentary on Teaching Science, Technology, and Society to Engineering Students: A Sixteen Year Journey. *Science and Engineering Ethics*, *19*(4), 1451–1454. doi:10.1007/s11948-013-9475-y PMID:24085355

Steinberg, A. (2001). *Forty-three valedictorians: Graduates of the Met talk about their learning*. Boston, MA: Jobs for the Future.

Steinert, Y. (1993). Twelve tips for using role-plays in clinical teaching. *Medical Teacher*, *15*(4), 283–291. doi:10.3109/01421599309006651 PMID:8139402

Stencel, J., & Barkoff, A. (1993). Protein synthesis: Role playing in the classroom. *The American Biology Teacher*, *55*(2), 102–103. doi:10.2307/4449595

Stevens, M., & Werner, D. (1996). New elementary teachers' knowledge and beliefs about instructional representations: A longitudinal study. *Journal of Elementary Education*, *22*(3), 327–347.

Stigler, J., & Hiebert, J. (1999). *The teaching gap: Best ideas from the world's teachers for improving education in the classroom*. New York, NY: Simon and Schuster.

Strong, R. W., Silver, H. F., & Perini, M. J. (2001). *Teaching what matters most: Standards and strategies for raising student achievement*. Alexandria, VA: Association for Supervision and Curriculum Development.

Sturm, B. (2009). The drama of science. *Science in School*, *13*, 29-33. Retrieved from http://www.scienceinschool.org/sites/default/files/teaserPdf/issue13_drama.pdf

Tatalovic, M. (2009). Science comics as tools for science education and communication: A brief, exploratory study. *JCOM: Journal of Science Communication*, *8*(4), 1–17.

TeacherGeek. (2008). *Fluid power lab*. June 30, 2016, from http://teachergeek.org/fluid_power.pdf

The Common Core State Standards Initiative. (2016). Retrieved on September 1, 2016 from http://www.corestandards.org/ELA-Literacy/introduction/key-design-consideration/

The National Research Council. (2000). Retrieved from http://www.nationalacademies.org/nrc/

The Science of Insulation Explained. (2008). Retrieved June 07, 2016, from http://www.knaufinsulation.com.au/en-au/more/insulation-explained.aspx

The White House. (2014). *Fact Sheet: The economic challenge posed by declining pollinator populations*. Retrieved from https://www.whitehouse.gov/the-press-office/2014/06/20/fact-sheet-economic-challenge-posed-declining-pollinator-populations

Thompson, J. (2003). *Applied theatre: Bewilderment and beyond*. Oxford, UK: Peter Lang. doi:10.3726/978-3-0353-0309-4

Tochon, F. V. (2013). *Educational semiotics: Signs and symbols in education*. Blue Mounds, WI: Deep University Press.

Tomlinson, C. A. (2001). *How to differentiate instruction in mixed-ability classrooms* (2nd ed.). Upper Saddle River, NJ: Pearson.

Trefil, J. (2008). *Why Science?* New York: Teachers College Press.

Tseng, T., & Yang, M. (2011). The role of spatial-visual skills in a project-based engineering design course. *Proceedings from ASEE 2011 Annual Conference & Exposition*. Retrieved from http://web.mit.edu/~mcyang/www/papers/2011-tsengYang-ASEE.pdf

Tsui, L. (2007). Effective strategies to increase diversity in STEM fields: A review of research literature. *The Journal of Negro Education*, *76*(4), 555–581.

Tu, T. (2006). Preschool science environments: What is available in an early childhood classroom. *Early Childhood Research Journal*, *33*(4), 245–251. doi:10.1007/s10643-005-0049-8

Uhrmacher, P. B., Conrad, B., & Lindquist, C. (2010). The power to transform: Implementation as aesthetic awakening. In T. Constantino & B. White (Eds.), *Essays on aesthetic education for the 21st century* (pp. 183–203). Rotterdam: Sense Publishers.

United Nations Educational, Scientific and Cultural Organization. (2003). *Literacy, a UNESCO perspective*. Retrieved on May 10, 2016 from: http://unesdoc.unesco.org/images/0013/001318/131817eo.pdf

United Nations. (2002). *56/116. United Nations Literacy Decade: education for all. Resolution adopted by the General Assembly, Fifty-Sixth Session*. Retrieved on May 10, 2009 from: http://www.undemocracy.com/A-RES-56-116.pdf

United States Department of Education. (1999a). *New American high schools: In their own words*. Washington, DC: The Kamber Group.

United States Department of Education. (1999b). *Aiming high, strategies to promote high standards in high schools*. Berkeley, CA: MPR Associates.

United States Department of Education. (1999c). *New American high schools at the leading edge of reform*. Berkeley, CA: MPR Associates.

United States Department of Education. (1999d). *Key reform strategies: An overview of research findings*. Berkeley, CA: MPR Associates.

United States General Accounting Office. (2002, February). *School dropouts Education could play a stronger role in identifying and disseminating promising prevention strategies* (GA0-02-240). Author.

USDOE - United States Department of Education. (2006). *The Condition of education*. Washington, DC: National Center for Education Statistics, U.S. Government Printing Office.

Vaidyanathan, S. (2012). Fostering creativity and innovation through technology. *Learning and Leading with Technology*, *39*(6), 24–27.

Valanides, N., Efthymiou, I., & Angeli, C. (2013). Interplay of internal and external representations: Students drawings and textual explanations about shadow phenomena. *Journal of Visual Literacy*, *32*(2), 67–84. doi:10.1080/23796529.2013.11674710

Van Sickle, M., & Spector, B. (1996). Caring relationships in science classrooms: A symbolic interaction study. *Journal of Research in Science Teaching*, *33*(4), 433–453. doi:10.1002/(SICI)1098-2736(199604)33:4<433::AID-TEA5>3.0.CO;2-T

Varelas, M., Pappas, C.C., Tucker-Raymond, E., Kane, J., Hankes, J., Ortiz, I., & Keblawe-Shamah, N. (2010). Drama activities as ideational resources for primary-grade children in urban science classes. *Journal of Research in Science Teaching*, *47*(3), 302–325.

Vasquez, J. A., Comer, M., & Sneider, C. (2013). STEM lesson essential, grades 3-8: Integrating science, technology, engineering, and mathematics. Portsmouth, NH: Heinemann.

Vavra, K. L., Janjic-Watrick, V., Loerke, K., Phillips, L. M., Norris, S. P., & Macnab, J. (2011). Visualization in science education. *Alberta Science International Journal*, *4*(1), 22–30.

Venkateswaran, T. V. (2004). Communicating science through plays, games and simulations. *Indian Journal of Science Communication*, *3*(1).

Vermont Department of Education. (2002). *High schools on the move*. Montpelier, VT: DOE.

Vygotsky, L. (1978). *Mind in society: The development of higher psychological processes*. Boston, MA: Harvard University Press.

Vygotsky, L. (1986). *Thought and language* (A. Kozulin, Trans.). Cambridge, MA: MIT University Press. (Original work published 1926)

Wagner, B. J. (1979). *Dorothy Heathcote: Drama as a learning medium*. London: Hutchinson Education.

Wang, M., Haertel, G., & Walberg, H. (1993). Toward a knowledge base for school learning. *Review of Educational Research*, *63*(3), 249–294. doi:10.3102/00346543063003249

Watters, J., & Diezmann, C. (2007). Multimedia resources to bridge the praxis gap: Modeling practice in elementary science education. *Journal of Science Teacher Education*, *18*(3), 349–376. doi:10.1007/s10972-007-9051-x

Watts-Taffe, S., Laster, B. P., Broach, L., Marinak, B., Connor, C. M., & Walker-Dalhouse, D. (2012). Differentiated instruction: Making informed teacher decisions. *The Reading Teacher*, *66*(4), 303–314. doi:10.1002/TRTR.01126

Webb, N. (1980). An analysis of group interaction and mathematical error in heterogeneous abilities groups. *The British Journal of Educational Psychology*, *50*(3), 266–276. doi:10.1111/j.2044-8279.1980.tb00810.x

Weld, J. D. (2000). Fly Guys. *Teacher Magazine*, *11*(8), 56–58.

Westby, C. (2010). Multiliteracies: The changing world of communication. *Topics in Language Disorders*, *30*(1), 64–71. doi:10.1097/TLD.0b013e3181d0a0ab

White, H. (2011). Our education system is not so much "broken" - as it is totally outdated! In *STEAM*. Retrieved September 16, 2016, from http://steam-not-stem.com/articles/our-education-system-is- not-so-much-broken-as-it-is-totally-outdated/

Wicklein, R. C. (2006). Five good reasons for engineering as the focus for technology education. *Technology Teacher*, *65*(7), 25–29.

Willis Tower. (n.d.). Retrieved from http://www.willistower.com

Willis, S. (1993). Educators seek 'developmental appropriateness'. *Teaching Young Children.*, *45*, 35–42.

Windschitl, M. (2002). Framing constructivism in practice as the negotiation of dilemmas. An analysis of the conceptual, pedagogical, cultural and political challenges facing teachers. *Review of Educational Research*, *72*(2), 131–175. doi:10.3102/00346543072002131

Woodrow Wilson Foundation National Commission on the Senior Year. (2001). Raising our sights: No high school senior left behind. Princeton, NJ: Author.

Wright brothers. (n.d.). In *Wikipedia.* Retrieved August 2016, from https://en.wikipedia.org/wiki/Wright_brothers

Wright, P. (2016). *The unmixed question: the spatial music of Charles Ives*. American Mavericks. Retrieved from http://musicmavericks.publicradio.org/features/feature_unmixed_question.html

Wu, H.-K., & Krajcik, J. S. (2006). Exploring middle school students use of inscriptions in project-based science classrooms. *Science Education*, *90*(5), 852–873. doi:10.1002/sce.20154

Xerces Society. (2011). *Attracting Native Pollinators*. Portland, OR: Timber Press.

Xerces Society. (2016). *Gardening for Butterflies*. Portland, OR: Timber Press.

Yager, R. E. (1993). The constructivist learning model. *Science Teacher (Normal, Ill.)*, *60*, 27–31.

Yeh, H.-T., & Lohr, L. (2010). Towards evidence of visual literacy: Assessing pre-service teachers perceptions of instructional visuals. *Journal of Visual Literacy*, *29*(2), 183–197. doi:10.1080/23796529.2010.11674680

Yopp, R. H., & Yopp, H. K. (2012). Young childrens limited and narrow exposure to informational text. *The Reading Teacher*, *65*(7), 480–490. doi:10.1002/TRTR.01072

Zygomatic. (n.d.). *Chart Tool.* Retrieved July 7, 2016, from http://www.onlinecharttool.com/

About the Contributors

Judith Bazler earned her Ed.D with a focus on Science Education and Curriculum from University of Montana. Having taught in secondary, and university settings in Missoula, Montana and at Lehigh University in Bethlehem, Pennsylvania, where she founded a hands-on science museum, she is now a full professor at Monmouth University in New Jersey, mainly in charge of the elementary and secondary science education programs. Her major research interest includes equity in the science classrooms, informal science, technology in the classroom and global education. Her work has been published in national and international journals such as Journal of Research in Science Teaching, Journal of Science for Persons with Disabilities, Journal of Science Teacher Education, Proteus, Journal of Career Development, Journal of General Education, The Clearing House, The American Biology Teacher, The Science Teacher, Technology and Teacher Education Annual, School Science and Mathematics, GASAT (Women and Science and Technology), Science Scope, Journal of Career Assessment, The Informal Sciences Review and Worldspeaker. She also has written three books for Greenwood Press and authored and edited numerous reference books for Facts On File, Inc and co-authored two chapters in Cases on Instructional Technology in Gifted and Talented Education.

Meta Van Sickle, Ph.D., is a Full Professor at the College of Charleston. A teacher with over 20 years of experience, Meta earned her Ph.D. in Science Education at the University of South Florida. Her two most recent IGI Global publications are about teacher education candidates who are gifted and talented. The first book chapter reviews the history of gifted and talented teaching through the science and mathematics education policies and practices. The second book chapter is a cross analysis of gifted and talented teacher education candidates.

* * *

Warren DiBiase is the middle/secondary science educator at UNC Charlotte, a position he has held for 19 years. In addition to his work in higher education, he has 20 years experience as a secondary science teacher.

Callie (Van Koughnett) Dollahon received her BS in Marine Science from the University of South Carolina Honors College in 2007 and a MEd in Science and Math for Teachers from the College of Charleston in 2013. In 2010 she moved to Copenhagen, Denmark, as a Fulbright Scholar, studying the integration of national science standards in Danish teaching. She has taught high school courses in Anatomy, Biology, Biomedical Science, and Marine Science, and was the Marine Science curriculum leader at Wando High School in Mount Pleasant, SC, for 3 years until 2016. She is currently the STEM Coordinator for East Cooper Montessori Charter School in Mount Pleasant.

Catherine Duckett is a professor at Monmouth University.

Kevin D. Finson, Ph.D., is a professor of science education and co-director of the Bradley University Center for STEM Education. He received his degrees from Kansas State University, including a B.S. in Science Education (1975), M.S. in Curriculum and Instruction with a focus on science (1980), and Ph.D. in Science Education (1985). He began his career teaching earth and physical sciences at the high school level, then moved into teaching earth, life, and physical sciences at middle school. His college teaching has included earth science and geology, physical science, energy curriculum, teaching pedagogy and methods, and instructional theory. Among the professional organizations of which he is a member, Finson is very active in the Association for Science Teacher Education (ASTE). Other organizations in which he is active include the National Science Teachers Association (NSTA) and the National Earth Science Teachers Association (NESTA). He has authored over seventy journal articles, one book, several book chapters, and co-edited several books. A major focus of his work has been on students' perceptions of scientists (such as Draw-A-Scientist-Test research), and more recently on models of conceptions of scientists in learners' schema. He has also been active in publishing and work related to making science more accessible to students having special needs. Extramurally funded projects for which he has been a co-director include those on science and special needs, and development of science potential in middle school students.

Letitia Graybill received an AB degree from the City University of New York in Chemistry and Zoology, an M.Ed. Degree in science education from the Graduate School of Education at Harvard University and Doctor of Education in science education and cognitive psychology from Rutgers University. She joined the faculty of Monmouth University as a Lecturer after a career in teaching in public schools in New York and New Jersey. Dr. Graybill's responsibilities in the public schools included supervising science and mathematics classes in grade seven through twelve and teaching several science content areas including developing, teaching and scoring advanced placement courses in chemistry. While still teaching in in the public schools Dr. Graybill was responsible for developing a program whereby gifted secondary science students were placed with practicing scientists in the pharmaceutical laboratories in New Jersey. Here they received invaluable experience in working in real world science research environments. As a result of this work Dr. Graybill name was added to the list of distinguished national science educators in Washington DC. All of the students in the program went on to have significant careers in various fields of science including engineering, medicine and sociology. At Monmouth University, she has taught a variety of courses with her main responsibility being in the teaching of science methods courses at the graduate and undergraduate levels. Her major interests at Monmouth have been and are the integration of technology within lesson planning and the emphasis on science process skills at the elementary level. This is Dr. Graybill's second experience in writing for IGI publications. She, along with her colleagues Professors Meta Van Sickle and Judith Bazler coauthored two chapters of a book on teaching gifted educators. The first chapter reviews the history of gifted and talented education through science and mathematics policies and practice. The second chapter is an analysis of gifted and talented teacher education candidates.

Carolyn Groff is an associate professor of K-8 Literacy Instruction and a department chair at the Monmouth University School of Education.

John Henning is Dean of the School of Education at Monmouth University in West Long Branch, New Jersey. Previously, he served as the Associate Dean in The Patton College and the Chair of the Teacher Education Department at Ohio University. Dr. Henning has more than 50 publications; including three books, over thirty refereed journal articles, and seven book chapters. His current research interests include teacher development, clinically-based teacher education, and classroom discourse. Dr. Henning has more than 50 publications; including three books, over thirty refereed journal articles, and seven book chapters.

Paul Jablon, Ph.D., taught in the New York City public schools for 19 years where he and his colleagues created classrooms, programs, and schools that utilized inquiry teaching and interdisciplinary, project-based learning to engage those students previously disengaged from school. As a faculty member at Brooklyn College, the University of Massachusetts, and Lesley University, he had the opportunity to work for over two decades with both preservice and practicing teachers. Leading systemic-change projects in urban and suburban school districts, he has collaborated with thousands of teachers to understand how inquiry teaching is best implemented with a diverse set of students from kindergarten through college.

Martha James-Hassan, EdD earned her undergraduate degree from the University of Wisconsin-Madison in Kinesiology, her Master's from Hamline University in St. Paul, Minnesota, and her Doctorate in Critical Pedagogy from the University of St. Thomas in Minneapolis, Minnesota. She is a graduate of the AASA/Howard University Urban Superintendents Academy and the Saint Paul Public Schools Leadership Academy. She is a researcher, teacher, and leader with nearly 25 years of experience in urban public education. Dr. Hassan's is currently an Assistant Professor at Morgan State University in Baltimore, Maryland. Her research agenda involves investigating in-service teacher professional development, cultural fluency, and interdisciplinary education. Dr. Hassan serves her community as a Commissioner on the Baltimore City Board of School Commissioners.

Merrie Koester, Ph.D., author of the nationally implemented Agnes Pflumm science education novels, is the director of Project Draw for Science action research initiative at the University of South Carolina Center for Science Education. Her book, Science Teachers Who Draw: The Red Is Always There, is the narrative of teacher researchers who learned to teach science through drawing and visualization in order to engage previously failing students—primarily struggling readers and/or special needs learners. As a science and arts integration specialist, Merrie serves on the advisory council of the Innovation Collaborative.

Nathan Lindquist. PhD, received his M.S. in Physics and his Ph.D. in Electrical Engineering from the University of Minnesota. In addition to teaching courses on computer programming, classical mechanics, electricity and magnetism, and materials physics, his research interests include interdisciplinary nanotechnology and super-resolution imaging. In his labs at Bethel University, he tinkers with lasers, microscopes and molecules. His research with undergraduate students has received major grant support from the National Science Foundation including a Faculty Early Career Development (CAREER) award.

William Lindquist, PhD, received his MA in Curriculum and Instruction: Learning Technology from the University of St. Thomas in Saint Paul, MN and PhD in Curriculum and Instruction: Elementary Science Education from the University of Minnesota. Dr. Lindquist can to Hamline University where he serves as an Assistant Professor after a 25 year career in public elementary settings. He teaches "Teaching Science in the Elementary School" and "Educational Psychology". His research interests include inquiry-based science, science and literacy, and science notebooks.

Kathryn Pedings-Behling earned her BS in Mathematics with a minor in Secondary Education in 2008 from the College of Charleston. She then earned her MS in Mathematics from the College of Charleston in 2012. Kathryn was a high school mathematics educator from 2008 - 2013. She currently works for the Outreach Department at the SC Governor's School for Science and Mathematics as a curriculum developer. In 2011, she was named the SC Public Charter School Educator of the Year, and in 2013, she received her National Board Certification.

Kerry Carley Rizzuto has worked as a classroom teacher in general and special education, a Reading Specialist for grades K-12, a Reading Recovery teacher, a staff developer, and school administrator in both New York and New Jersey. In addition to her teaching position at Monmouth University, Dr. Rizzuto has completed the Learning Disabilities Teacher-Consultant program at Monmouth University and earned her doctorate from Rowan University in educational leadership, with a focus on early childhood. Dr. Rizzuto is the Program Director for the early childhood program at Monmouth University. In addition to developing science education in early childhood classrooms, her research interests include developing social justice dispositions in pre-service teachers and developing culturally responsive pedagogical practices.

Alex Romagnoli is an Assistant Professor of English Education and the Master of Arts in Teaching Program Director at Monmouth University. He received his Ph.D. in Composition at TESOL (Teaching English to Speakers of Other Languages) at Indiana University of Pennsylvania where he also taught Composition. Previously, Alex taught high school English at East Stroudsburg High School South. He is the co-author of Enter the Superheroes: American Values, Culture, and the Canon of Superhero Literature. Additionally, he received the IUP Composition and TESOL program's "Exemplary Teaching of Literacy and Language Award" in 2013. His research interests include multimodality, multiliteracies, graphic novels in academic contexts, and popular culture.

Kyle Seiverd is a science educator currently living in New Jersey. Since his childhood he has been passionate about wildlife, mainly elephants. He has traveled to Kenya multiple times to observe African elephants and plans to see all the great apes in the wild. For the past 6 years he has taught High School, but has also been a guest speaker at elementary schools, senior centers, county libraries, and presented at academic conferences. His heroes include Jane Goodall, Daphne Sheldrick, Neil DeGrasse Tyson, and Walt Disney. He shares the belief of Jane Goodall that "Every individual matters. Every individual has a role to play. Every individual makes a difference".

Kathryn L. Servilio earned a doctorate from West Virginia University in special education. Her B.S. in multi-disciplinary studies and M.A. in elementary education were also from West Virginia University. She acquired additional certification in multi-categorical special education and early childhood education. She has taught inclusion for reading and mathematics in grades K, 1, and 2, and has taught in a self-contained setting for grades 9-12. Dr. Servilio worked as an Assistant/Associate Professor at Elmhurst College for 6 years and is currently an Assistant Professor at Monmouth University. Dr. Servilio is a member of the Council for Exceptional Children. She served as an executive board member of Esperanza Community Services Center and also as an executive board of the Teacher Education Division for the Council for Exceptional Children. Her current scholarship focuses on differentiating instruction, The Collaborative Student Teacher Training Model and the impacts of technology on student engagement.

Sara Smith is a teacher of Project Lead the Way Introduction to Engineering and Advanced Career Clean Energy Technology courses. She has been teaching for ten years of which the last six have been at Goose Creek High School. Ms. Smith holds a B.S. in Chemical Engineering from the University of Pittsburgh, an M.Ed. in Science and Math for Teachers from The College of Charleston, and an M.Ed. in Curriculum and Instruction from Concordia University. Her favorite part of teaching is providing her students with an engaging and rigorous education in engineering and clean energy technologies.

Index

M

N

O

P

R

S

Purchase Print + Free E-Book or E-Book Only*

Purchase a print book through the IGI Global Online Bookstore and receive the e-book for free or purchase the e-book only! Shipping fees apply.

www.igi-global.com

Recommended Reference Books

ISBN: 978-1-4666-9752-2
© 2016; 324 pp.
List Price: $148

ISBN: 978-1-4666-9749-2
© 2016; 318 pp.
List Price: $148

ISBN: 978-1-4666-9746-1
© 2016; 354 pp.
List Price: $148

ISBN: 978-1-4666-5162-3
© 2014; 480 pp.
List Price: $252

ISBN: 978-1-4666-6615-3
© 2015; 458 pp.
List Price: $260

ISBN: 978-1-4666-6154-7
© 2014; 349 pp.
List Price: $156

*IGI Global now offers the exclusive opportunity to receive a free e-book with the purchase of the publication in print, or purchase any e-book publication only. You choose the format that best suits your needs. This offer is only valid on purchases made directly through IGI Global's Online Bookstore and not intended for use by book distributors or wholesalers. Shipping fees will be applied for hardcover purchases during checkout if this option is selected.

Should a new edition of any given publication become available, access will not be extended on the new edition and will only be available for the purchased publication. If a new edition becomes available, you will not lose access, but you would no longer receive new content for that publication (i.e. updates). The free e-book is only available to single institutions that purchase printed publications through IGI Global. Sharing the free e-book is prohibited and will result in the termination of e-access.

Publishing Information Science and Technology Research Since 1988

IGI GLOBAL
DISSEMINATOR OF KNOWLEDGE

www.igi-global.com | Sign up at www.igi-global.com/newsletters | facebook.com/igiglobal | twitter.com/igiglobal

Stay Current on the Latest Emerging Research Developments

Become an IGI Global Reviewer for Authored Book Projects

The overall success of an authored book project is dependent on quality and timely reviews.

In this competitive age of scholarly publishing, constructive and timely feedback significantly decreases the turnaround time of manuscripts from submission to acceptance, allowing the publication and discovery of progressive research at a much more expeditious rate. Several IGI Global authored book projects are currently seeking highly qualified experts in the field to fill vacancies on their respective editorial review boards:

Applications may be sent to:
development@igi-global.com

Applicants must have a doctorate (or an equivalent degree) as well as publishing and reviewing experience. Reviewers are asked to write reviews in a timely, collegial, and constructive manner. All reviewers will begin their role on an ad-hoc basis for a period of one year, and upon successful completion of this term can be considered for full editorial review board status, with the potential for a subsequent promotion to Associate Editor.

If you have a colleague that may be interested in this opportunity, we encourage you to share this information with them.

www.igi-global.com

InfoSci®-Books

A Database for Progressive Information Science and Technology Research

Maximize Your Library's Book Collection!

Invest in IGI Global's InfoSci®-Books database and gain access to hundreds of reference books at a fraction of their individual list price.

The InfoSci®-Books database offers unlimited simultaneous users the ability to precisely return search results through more than 75,000 full-text chapters from nearly 3,400 reference books in the following academic research areas:

Business & Management Information Science & Technology • Computer Science & Information Technology
Educational Science & Technology • Engineering Science & Technology • Environmental Science & Technology
Government Science & Technology • Library Information Science & Technology • Media & Communication Science & Technology
Medical, Healthcare & Life Science & Technology • Security & Forensic Science & Technology • Social Sciences & Online Behavior

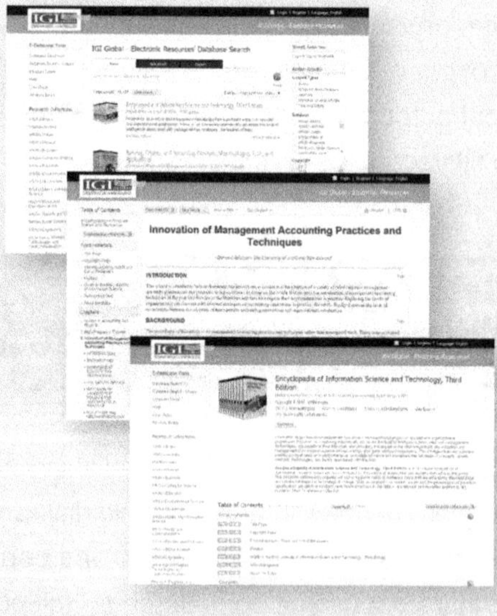

Peer-Reviewed Content:
- Cutting-edge research
- No embargoes
- Scholarly and professional
- Interdisciplinary

Award-Winning Platform:
- Unlimited simultaneous users
- Full-text in XML and PDF
- Advanced search engine
- No DRM

Librarian-Friendly:
- Free MARC records
- Discovery services
- COUNTER4/SUSHI compliant
- Training available

To find out more or request a free trial, visit: www.igi-global.com/eresources

IGI GLOBAL DISSEMINATOR OF KNOWLEDGE
www.igi-global.com

IGI Global
Proudly Partners with

eContent Pro

eContent Pro specializes in the following areas:

Academic Copy Editing
Our expert copy editors will conduct a full copy editing procedure on your manuscript and will also address your preferred reference style to make sure your paper meets the standards of the style of your choice.

Expert Translation
Our expert translators will work to ensure a clear cut and accurate translation of your document, ensuring that your research is flawlessly communicated to your audience.

Professional Proofreading
Our editors will conduct a comprehensive assessment of your content and address all shortcomings of the paper in terms of grammar, language structures, spelling, and formatting.

IGI Global Authors, Save 10% on eContent Pro's Services!

Scan the QR Code to Receive Your 10% Discount

The 10% discount is applied directly to your eContent Pro shopping cart when placing an order through IGI Global's referral link. Use the QR code to access this referral link. eContent Pro has the right to end or modify any promotion at any time.

Email: customerservice@econtentpro.com

econtentpro.com

www.ingramcontent.com/pod-product-compliance
Lightning Source LLC
LaVergne TN
LVHW080607300926
849162LV00005B/1065